METHODS IN MOLECULAR BIOLOGY™

Series Editor
John M. Walker
School of Life Sciences
University of Hertfordshire
Hatfield, Hertfordshire, AL10 9AB, UK

For other titles published in this series, go to
www.springer.com/series/7651

Ligand-Macromolecular Interactions in Drug Discovery

Methods and Protocols

Edited by

Ana Cecília A. Roque

REQUIMTE, Department of Chemistry, Faculdade de Ciências e Tecnologia,
Universidade Nova de Lisboa, 2829-516 Caparica, Portugal

✳ Humana Press

Editor
Ana Cecília A. Roque
REQUIMTE
Department of Chemistry
Faculdade de Ciências e Tecnologia
Universidade Nova de Lisboa
Caparica
Portugal

ISSN 1064-3745   e-ISSN 1940-6029
ISBN 978-1-60761-243-8   e-ISBN 978-1-60761-244-5
DOI 10.1007/978-1-60761-244-5
Springer New York Dordrecht Heidelberg London

Library of Congress Control Number: 2009933594

Printed on acid-free paper

Springer is part of Springer Science+Business Media (www.springer.com)

Preface

Drug research, guided by pharmacology, chemistry, and clinical sciences, has contributed enormously to the progress of medicine over the past century. Drug discovery is an exciting field lately transformed by the "omics revolution" and the advances in computational tools, combinatorial chemistry, and high throughput screening techniques (HTS). The process of drug discovery involves the identification of target molecules (mostly proteins) that serve as points of attack for future medicines, the design and synthesis of potential lead compounds, and further characterization, screening, and assays for therapeutic efficacy and toxicity. Drug development is not only of scientific interest; it entails a fascinating interplay among a variety of economic, social, and political institutions.

The aim of this edition of *Methods in Molecular Biology* is to present the techniques for the study of ligand–macromolecule interactions as tools in the drug discovery process. *Ligand–Macromolecular Interactions in Drug Discovery: Methods and Protocols* is divided in two sections. The first section comprises reviews and starts with an historical perspective of drug research focusing on the contribution of genomics, proteomics, high-throughput methods, and computational developments in drug discovery. The principles of molecular recognition are introduced by Nolan, Singh, and McCurdy to facilitate comprehension and set the scene for the next chapters. The contributions from Romão and Ramos laboratories (Carvalho et al. and Cerqueira et al.) cover two important and complementary fields in the drug discovery process, namely the macromolecular X-ray crystallography and the virtual screening of compound libraries. The protocols section begins with a report from Breinbauer and Mentel about solid phase organic synthesis (SPOS) and how it can be used to generate diversity in compound libraries. Viegas, Macedo, and Cabrita introduce NMR spectroscopy as a tool to screen for novel compounds and to characterize ligand binding. The importance of isothermal titration calorimetry (ITC) and differential scanning calorimetry (DSC) in drug discovery, particularly in the understanding of the thermodynamics of molecular recognition processes, is further explored by Holdgate. The remaining protocols chapters are more detailed, and intended to present several high-throughput methodologies, utilized alone or in combination, either for the selection of lead drug candidates for specific molecules or for the discovery of new target molecules for drug design. Schneider and Schüller start by presenting the "adaptive" compound library design for the optimization of a focused Ugi-type product library toward trypsin inhibitors. Liang et al. introduce the basic techniques and applications of chemical microarrays for the identification of inhibitors for different enzymes, namely kinases, proteases, histone deacetylases, and phosphatases. Horicuchi and Ma show how fluorescence polarization (FP) and time-resolved fluorescence resonance energy transfer (TR-FRET) techniques were utilized in the screening of 400 compounds from a natural product library against an important drug target, the kinase PI3K. Following on the microarrays technology, Meng et al. used a known kinase inhibitor to profile a protein microarray containing over 8,000 functional human proteins. Other methods for the selection of bioactive compounds against biological targets are presented in the remaining chapters. Quaglia and De Lorenzi discuss the screening of 208 compounds against target proteins by affinity capillary electrophoresis (ACE). Minuni and Bilia exemplify the use of surface plasmon resonance (SPR) for the search of natural products (from plant extracts) interacting with DNA. Slon-Usakiewicz and Redden showed how FAC-MS (frontal affinity chromatography with mass spectrometry) can be useful for the screening of intact racemic compounds from diverse libraries against protein targets. Siegel elaborates on the secondary screening of small compound libraries by GPC spin column HPLC-ESI-MS methods, whereas Wang and Jones present a novel scintillation proximity assay (SPA) to identify compounds inhibitors of a human protein. Luesch and Abreu explain how genomic tools can be incorporated in natural product approach to drug discovery, exemplifying drug susceptibility screens employing cDNA libraries. Finally, Pramanik further explores cell assays in drug discovery by the use of fluorescence correlation spectroscopy (FCS) to monitor ligand–macromolecule interactions in live cells.

This volume on *Ligand–Macromolecular Interactions in Drug Discovery: Methods and Protocols* was only possible due to the great support, participation, and contributions from all authors. It attempts to cover the main principles and methodologies currently utilized in the study of molecular interactions between compounds, either natural or synthetic, and complementary biological targets, within the scope of drug discovery. As the interaction "ligand–macromolecule" sustains the drug discovery practice, it is our hope that the current volume shall be of use to scientists in academia and in industry, working in areas as diverse as medicine, biology, biotechnology, pharmacology, medicinal chemistry, or nanotechnology.

Caparica, Portugal   *Ana Cecília A. Roque*

Contents

Contributors

PEDRO ABREU • *REQUIMTE, Department of Chemistry, Faculdade de Ciências e Tecnologia, Universidade Nova de Lisboa, Caparica, Portugal*

ANNA RITA BILIA • *Dipartimento di Scienze Farmaceutiche, Facoltà di Farmacia, Università degli Studi di Firenze – Polo Scientifico, Sesto Fiorentino, Italy*

ROLF BREINBAUER • *Institute of Organic Chemistry, Graz University of Technology, Graz, Austria*

EURICO CABRITA • *REQUIMTE, Department of Chemistry, Faculdade de Ciências e Tecnologia, Universidade Nova de Lisboa, Caparica, Portugal*

ANA LUÍSA CARVALHO • *REQUIMTE, Department of Chemistry, Faculdade de Ciências e Tecnologia, Universidade Nova de Lisboa, Caparica, Portugal*

NUNO M.F.S.A. CERQUEIRA • *Theoretical and Computational Chemistry Research Group, REQUIMTE, Departamento de Química, Faculdade de Ciências, Universidade do Porto, Porto, Portugal*

PEDRO A. FERNANDES • *Theoretical and Computational Chemistry Research Group, REQUIMTE, Departamento de Química, Faculdade de Ciências, Universidade do Porto, Porto, Portugal*

GEOFF HOLDGATE • *Lead Generation, Discovery Capabilities & Sciences, AstraZeneca Pharmaceuticals, Mereside, Alderley Park, Macclesfield, Cheshire, UK*

KURUMI HORICUCHI • *Reaction Biology Corporation, Malvern, PA, USA*

ABID HUSSAIN • *REQUIMTE, Department of Chemistry, Faculdade de Ciências e Tecnologia, Universidade Nova de Lisboa, Caparica, Portugal*

PHILIP JONES • *Neuroscience Discovery Research, Wyeth Research, Princeton, NJ, USA*

SHUGUANG LIANG • *Reaction Biology Corporation, Malvern, PA, USA*

ERSILIA DE LORENZI • *Department of Pharmaceutical Chemistry, School of Pharmacy, University of Pavia, Pavia, Italy*

HENDRIK LUESCH • *Department of Medicinal Chemistry, University of Florida, Gainesville, FL, USA*

HAICHING MA • *Reaction Biology Corporation, Malvern, PA, USA*

ANJOS L. MACEDO • *REQUIMTE, Department of Chemistry, Faculdade de Ciências e Tecnologia, Universidade Nova de Lisboa, Caparica, Portugal*

DAWN MATTOON • *Life Technologies Corporation, Eugene, OR, USA*

CHRISTOPHER R. MCCURDY • *Department of Medicinal Chemistry and Department of Pharmacology, School of Pharmacy, University of Mississippi, University, MS, USA*

LIHAO MENG • *Life Technologies Corporation, Carlsbad, CA, USA*

MATTHIAS MENTEL • *EMBL Heidelberg, Heidelberg, Germany*

MARIA MINUNNI • *Dipartimento di Chimica Facoltà di Scienze Matematiche Fisiche e Naturali, Università degli Studi di Firenze – Polo Scientifico, Sesto Fiorentino, Italy*

TAMMY NOLAN • *Department of Medicinal Chemistry, University of Mississippi, University, MS, USA*

ANA SOFIA PINA • *REQUIMTE, Department of Chemistry, Faculdade de Ciências e Tecnologia, Universidade Nova de Lisboa, Caparica, Portugal*

ALADDIN PRAMANIK • *Department of Medical Biochemistry and Biophysics, Karolinska Institute, Stockholm, Sweden*

PAUL PREDKI • *Life Technologies Corporation, Carlsbad, CA, USA*

MILENA QUAGLIA • *LGC Ltd, Teddington, UK*

MARIA JOÃO RAMOS • *Theoretical and Computational Chemistry Research Group, REQUIMTE, Departamento de Química, Faculdade de Ciências, Universidade do Porto, Porto, Portugal*

PETER REDDEN • *Transition Therapeutics Inc. Toronto, ON, Canada*

MARIA JOÃO ROMÃO • *REQUIMTE, Department of Chemistry, Faculdade de Ciências e Tecnologia, Universidade Nova de Lisboa, Caparica, Portugal*

ANA CECÍLIA A. ROQUE • *REQUIMTE, Department of Chemistry, Faculdade de Ciências e Tecnologia, Universidade Nova de Lisboa, Caparica, Portugal*

GISBERT SCHNEIDER • *Institute of Organic Chemistry and Chemical Biology, Goethe-University, Frankfurt am Main, Germany*

ANDREAS SCHÜLLER • *Duke-NUS Graduate Medical School, National University of Singapore, Singapore*

MARSHALL M. SIEGEL • *Chemical and Screening Sciences Division, Wyeth Research, Pearl River, NY, USA*

NIDHI SINGH • *Department of Medicinal Chemistry, University of Mississippi, University, MS, USA*

JACEK SLON-USAKIEWICZ • *1124 Barr Crescent, Milton, ON, L9T 6Y3, Canada, jusakiewicz@cogeco.ca*

SÉRGIO F. SOUSA • *Theoretical and Computational Chemistry Research Group, REQUIMTE, Departamento de Química, Faculdade de Ciências, Universidade do Porto, Porto, Portugal*

JOSÉ TRINCÃO • *REQUIMTE, Department of Chemistry, Faculdade de Ciências e Tecnologia, Universidade Nova de Lisboa, Caparica, Portugal*

ALDINO VIEGAS • *REQUIMTE, Department of Chemistry, Faculdade de Ciências e Tecnologia, Universidade Nova de Lisboa, Caparica, Portugal*

YUAN WANG • *Reaction Biology Corporation, Malvern, PA, USA*

YUREN WANG • *Neuroscience Discovery Research, Wyeth Research, Princeton, NJ, USA*

WEI XU • *Reaction Biology Corporation, Malvern, PA, USA*

Part I

Reviews

Chapter 1

An Historical Overview of Drug Discovery

Ana Sofia Pina, Abid Hussain, and Ana Cecília A. Roque

Summary

Drug Discovery in modern times straddles three main periods. The first notable period can be traced to the nineteenth century where the basis of drug discovery relied on the serendipity of the medicinal chemists. The second period commenced around the early twentieth century when new drug structures were found, which contributed for a new era of antibiotics discovery. Based on these known structures, and with the development of powerful new techniques such as molecular modelling, combinatorial chemistry, and automated high-throughput screening, rapid advances occurred in drug discovery towards the end of the century. The period also was revolutionized by the emergence of recombinant DNA technology, where it became possible to develop potential drugs target candidates. With all the expansion of new technologies and the onset of the "Omics" revolution in the twenty-first century, the third period has kick-started with an increase in biopharmaceutical drugs approved by FDA/EMEA for therapeutic use.

Key words: Drug discovery, Proteomics, Genomics, High-throughput screening, Drug target, Recombinant proteins

1. Introduction

"Drug research, as we know it today, began its career when chemistry had reached a degree of maturity that allowed its principles and methods to be applied to problems outside of chemistry itself and when pharmacology had become a well-defined scientific discipline in its own right".

Jürgen Drews

The alliance between Chemistry, Biology, and Pharmacology has enabled great improvements in Medicine over the last century *(1)*, facilitating the design and discovery of new compounds which has always been the main goal in Medicinal Chemistry *(2)*. These

Ana Cecília A. Roque (ed.), *Ligand-Macromolecular Interactions in Drug Discovery: Methods and Protocols,*
Methods in Molecular Biology, vol. 572,
DOI 10.1007/978-1-60761-244-5_1, © Humana Press, a part of Springer Science+Business Media, LLC 2010

compounds are designated as drugs because of their controlled use in the cure or prevention of disease. Natural compounds, isolated from natural sources such as plants, micro-organisms, vertebrates, and invertebrates *(2, 3)*, represent the major class of molecular drugs and these are involved in the treatment of 87% of all categorized human diseases *(3, 4)*. They were also the starting point for the discovery of important anticancer agents (e.g. paclitaxel and camptothecin), immunosuppressive agents (e.g. cyclosporins and rapamycin), and cholesterol-lowering agents (e.g. lovastatin and mevastatin) *(3–5)*. In the past, serendipity played an important role in drug development, and the creativity and intuition of the medicinal chemist was the basis for the drugs' success *(2, 5)*. Since the drug discovery challenge is related to the identification and development of molecules that elicit a certain desired effect in a living organism, proteins involved in key biological pathways represent potential drug targets *(6–10)*. Therefore, a study of the set of proteins expressed by a genome, together with the development of high-throughput methods has had a major impact in drug discovery *(6, 7, 9–13)*.

Nowadays, drug research comprises several stages relying on expertise from a wide range of disciplines such as biology, biochemistry, pharmacology, mathematics, computing, and molecular modelling. The first stage includes combinatorial chemistry and high-throughput screening (in silico or in vitro) for the selection of potential "lead" compounds *(14)*. When a chemical structure shows activity and selectivity in a pharmacological or biochemically relevant screening protocol, it can be considered as a potential "hit" compound *(5)*. Ensuing steps are concerned with lead optimization and development selection. At each stage, an evaluation of the structure-activity relationships (SARs) must be addressed *(15, 16)*. "Drug-like" properties need to be studied in vivo through pharmacokinetics studies to evaluate the absorption, distribution, metabolism, excretion (ADME), and interactions of a drug. These studies ascertain physicochemical properties (solubility, permeability, lipophilicity, and stability in vitro) and molecular properties (molecular weight, hydrogen bonding, and polarity studied in silico or in vitro) *(5, 15, 17)*. Beyond this strategy-defined pharmaceutical profiling, a separate set of criteria, correlating physical properties with oral bioavailability has been formulated by Lipinski et al. and designated as the "rule-of-five" *(4, 15, 18)*. This rule is associated with the solubility and permeability of a compound. Specifically it states that when a compound possesses poor absorption or permeability, there are >5 hydrogen-bond donors, the molecular mass is >500, calculated log P is >5 (P-partition coefficient – the ratio of concentrations of a compound in the two phases of a mixture of two immiscible solvents at equilibrium), and the sum of nitrogen and oxygen atoms in a molecule >10 *(4, 5, 18)*. Once the drug has shown to be a good candidate, clinical trials are required for its approved use by the Food and Drug Administration (FDA) in North America or the European Agency for the Evaluation of Medicinal Products (EMEA) in Europe *(5)*. In these clinical trials, the candidate drug is administrated to normal human volunteers for toleration (Phase I), then to patients having the condition that the drug has been designed to treat (Phase II), and finally to a large number of patients (Phase III) *(2, 5)*.

Genomics and Proteomics have both contributed towards developments in the drug discovery process, by facilitating the cloning, expression, identification, and study of target proteins for specific diseases *(19)*. Typical proteomics technologies as 2-D electrophoresis combined with mass spectroscopy (MS) and high performance liquid chromatography (HPLC) coupled with MS (HPLC/MS) provide good separation of proteins *(6, 9, 20)*. Increasingly, affinity chromatography and micro-array technologies have emerged as powerful techniques for probing drug-protein interaction. In the former case, the elucidation of drug–protein interaction involves direct interrogation of the substrates through covalent binding between protein and drug candidates immobilized into a suitable solid-support *(11)*. In the latter, the monitoring of the binding of a small molecule to a wide range of proteins can be performed rapidly and simultaneously *(7)*. X-ray diffraction and nuclear magnetic resonance (NMR) are also important techniques for the prediction of the 3D structure of an expressed and purified target protein, either alone or when complexed with a specific drug *(21, 22)*. NMR spectroscopy has also proved to be an effective technique for generation of multivariate metabolites in order to improve the lead compound selection in drug toxicity screening *(23)*. The so-called Omics revolution has also contributed towards an increase in biopharmaceutical drugs, with currently over 165 biopharmaceuticals products (recombinant proteins, monoclonal antibodies, and nucleic acid-based drugs) meeting FDA and/or EMEA approval. Several thousand more lie await in the discovery and pre-clinical development stages *(13)*.

2. Historical Perspective

Drug discovery started in ancient times, when natural products, mainly extracted from plants, were used for medicinal purposes in traditional medicine *(24)* (**Fig. 1**). In the following centuries, these compounds from plant sources led to the development of substances. This development and refinement strategy had a great impact in medicinal chemistry. One such example is the highly potent analgesic morphine, isolated from opium extracts

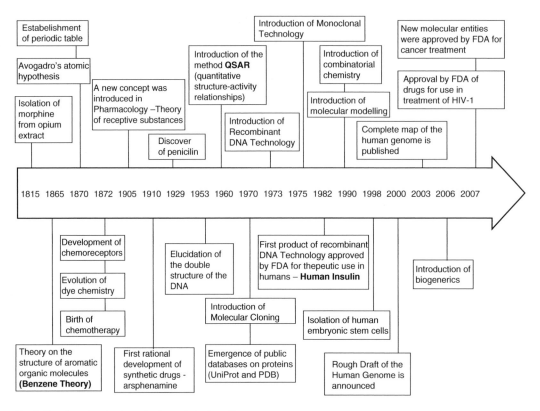

Fig. 1. Timeline showing important events in drug discovery.

by Sertüner in 1815 *(1)*. The emergence of analytical chemistry in the late nineteenth century made possible the isolation of individual bioactive ingredients from plant materials (although the purity levels were not of pharmaceutical grade by modern standards). Nevertheless, this approach enormously advanced nineteenth-century medicine *(1, 2)*. As the 20th century dawned, new advances in chemical knowledge (such as the concept of aromaticity in benzene, proposed by August Kekulé in 1865) profoundly influenced the fledgling theory of "chemoreceptors", originally postulated by Paul Ehrlich in the period 1872–1874 *(1)*. In 1905, J. N. Langley formulated a more functional concept of receptive substances, where the receptor could accept signals and then generate responses that resulted in a biological or pharmacological effect *(1, 2)*. This line of research stimulated subsequent investigations in medicinal chemistry, and led to the development of novel concepts such as ligand–receptor interaction; in other words, the very idea of the existence of a pharmacophore as an interactive entity *(2)*.

In 1908, Ehrlich's investigations paved the way for the first rational synthetic drugs such as arsphenamine. Additionally, Ehrlich's research team laid the foundations for reliable biological screening and evaluation procedures *(2)*.

At this stage, in the pre-NMR, pre-computational chemistry era, the chemists' tools were somewhat limited. Hence, Institutional support and instrument development were fundamental in making the transition from chemical creativity to drug discovery. As such, one can say that "chance" was the basis for the discovery of penicillin by Alexander Fleming in 1928 *(1, 2, 24)*. Penicillin consists of a mixture of related beta-lactams sharing the same core structure. The strong antibiotic characteristic of penicillin was realized immediately. Mass production of the antibiotic started around 1942 and had a significant impact on controlling sepsis during World War II. Penicillin was carried by Allied Medial Personnel during the D-Day landings. It is still used today in combating Gram positive micro-organism infections *(24)*.

The elucidation of the structure of the penicillin molecule led to a new successful era of antibiotics discovery, greatly improving healthcare in the treatment of bacterial infections *(1, 24)*. Contemporaneously, Chain, Florey, and their collaborators selected a metabolite from penicillium mould that could lyse *Staphylococci (1)*. However, some of the antibiotics based on penicillin were degraded by protective enzymes produced by bacterial cells. The discovery of new antibiotics through modification of the penicillin basic structure was intensified, in order to prevent bacterial degradation. In 1948, Brotzu reported a new molecule, referred as cephalosporin, that was used to treat infections resis-tant to penicillium *(24)*. By then, many companies such as Merck, Sandoz, and Taked increased their microbiological facilities in order to find other drugs with different pharmacological and chemotherapeutic properties *(1)*.

From 1950, the role of Medicinal Chemistry changed and can be divided into two important periods: From 1950 to 1980, Medicinal Chemistry was distinguished by *in vivo* tests; and the second era dating from 1980 to the present is distinguished by the emergence of new design and screening technologies *(5)*.

In the late 1960s and early 1970s, Beecham and Pfizer found new molecules with similar pharmacokinetic properties to penicillin – Beecham discovered clavulanates and Pfizer produced semi-synthetic molecules known as sulfactams, in which the thiazoles sulphur was oxidized to the sulfone *(24)*. Starting from these compounds and following different pathways, it was possible to develop various drugs having different therapeutic effects such as hypoglycaemic agents, diuretics, and antihypertensive drugs *(1)*. One priority during this period was the evaluation of drug behaviour in the human body. Unfortunately, even when a lead molecule showing *in vivo* activity was identified in appropriate animal model systems, problems could still arise in pharmacokinetics studies (absorption, distribution, metabolism, excretion, and interaction of a drug in the clinic) *(5)*.

From the late 1980s onwards, the development of new technologies together with the advent of computational chemistry, meant that some of the aforementioned problems could be circumvented with relative ease *(1, 5, 25, 26)*. Indeed, since the early 1990s, rapid advances in molecular modelling tools, as well as the application of combinatorial chemistry and automated high-throughput screening, has brought immediate benefits in going from "lead discovery" to "lead optimization". Rationally designed libraries of compounds, based on known drug scaffolds can be generated in a relatively short space of time *(5, 12)*. Rapid automated screening results in much shorter delay between delivering compounds and ascertaining the results of the screening. Screening data is then fed back into the SAR design process, leading to an iterative cycle of refinement, synthesis, and screening until the desired properties are achieved **(Fig. 2)**.

By the close of the twentieth century, advances in molecular and cell biology such as recombinant DNA techniques helped to revolutionize the pharmaceutical industry *(12)*. With the completion of the human genome project it became possible to gain a better understanding of the various molecular pathways responsible for certain diseases, and to use this information as a starting point in the search of new drugs based on proteins *(12, 26)*. Another technological breakthrough was the advent of hybridoma technology in 1975, which boosted the use of monoclonal antibodies as therapeutics *(12)*. As such, a transition

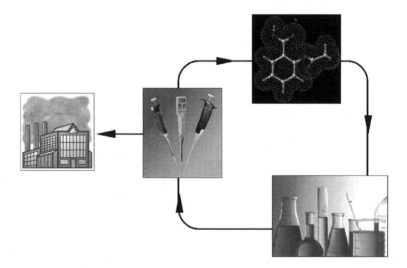

Fig. 2. The process of modern drug discovery. Commencing at the top with a protein of known structure and function, a virtual drug is designed in silico. The next step is the actual synthesis of the drug (in reality, hundreds of related compounds). The compounds are then screened for efficacy. If the desired criteria are met, the drug can be trialled in animal models and eventually in humans before obtaining FDA/EMEA approval and large-scale production can begin. But if the lead candidates requires further refinement, then the entire process is repeated.

period occurred between the twentieth and twenty-first centuries when the paradigms of molecular biology were associated with drug discovery *(26)*. Genetic engineering led to the replacement of natural proteins by recombinant versions. This afforded several advantages in therapeutics. One such advantage was the greater safety of using recombinant-derived biopharmaceuticals, as opposed to proteins derived from human or animal sources which had the potential to cause transmission of infective agents. Additionally, it was possible to produce quantities of recombinant proteins in large scale; to develop new drugs directed to a target and the creation of engineered proteins with improved therapeutic properties *(27, 28)*. Systems utilized to produce the majority of approved recombinant proteins were *E. coli, S. cerevisiae,* and animal cell lines such as Chinese hamster ovary (CHO) or baby hamster kidney (BHK) *(28)*. The first product resulting from DNA recombinant technology with approval for medical application in humans was human insulin produced in 1982, also known as "Humulin" *(27)*. A wide range of proteins for therapeutic purposes has been produced since then and by mid-2002, 120 biopharmaceuticals had obtained approval in the USA and/ or the EU. These included a range of hormones (e.g. growth hormone), blood factors (e.g. factors VIII and IX), and thrombolytic agents (e.g. tissue plasminogen activator, tPA) *(28)*. By 2000, nearly 25% of all new drugs in development were monoclonal antibodies *(12)*; in 2006, one of the top 200 prescribed drugs was a recombinant protein used to treat diabetes, as well as rheumatoid arthritis, Gaucher's disease, and multiple sclerosis *(29)*. New types of oral anti-HIV drugs – CCR5 antagonist Selzentry (maraviroc; Pfizer) and the HIV-1 integrase inhibitor Isentress (raltegravir; Merck) – were approved in 2007 by FDA for the treatment of HIV-1 *(30)*. Also new molecular entities as Torisel (temsirolimus; Wyeth) were approved in 2007 for the treatment of advanced renal cell carcinoma *(30)*.

Presently, the biopharmaceutical industry has introduced Biogenerics, i.e. products having similarity to an approved biopharmaceutical *(31)*. Thus, in 2006, biogenerics received approval in Europe for the first time, with products such as two recombinant human growth hormones, Holzkirchen, Germany-based Sandoz's Omnitrope and Valtropin, offered by Biopartners of Baar, Switzerland *(32)*. Sandoz's Omnitrope is a biogeneric follow-on to Pfizer's recombinant growth hormone product Genotropin (somatropin), also approved in the USA *(31, 32)*. Nowadays, there is an alliance between world-wide generics industry and several companies in order to follow up types of insulin, growth hormone, interferon, and erythropoietin, and colony-stimulating factor. In this way, in 2007, three new drugs similar to erythropoietin were approved (Binocrit, Abseamed, and Epoetin alfa Hexal), and are being used in the treatment of anaemia, secondary to chronic renal failure or chemotherapy *(29, 31)*. Embryonic stem cells had equally contributed to drug development due to their unique abilities of self-renewal and indefinite division in in vitro cultures, and the potential to generate every cell in the body *(33–36)*. The isolation of the first human embryonic stem cells derived from human blastocysts occurred in 1998 *(34, 35)*, and since then several opportunities related with drug responses and/or toxicities have emerged *(35)*. In 2005 only 20 new drugs were approved by the US FDA when compared to 2004 where 40 new drugs were approved *(33)*. This difference is related to the failure of clinical trials but this can be overcome with the use of human stem cell technologies. With these new tools, the entire drug discovery process could become more efficient, with a decrease in need of *in vivo* experimentation *(33)* and could reveal the toxicity of certain drugs at an earlier stage, thus avoiding conventional tests in animal models *(35)*. Due to their characteristics, stem cells could be an alternative in regenerative medicine as well *(33)*.

In general, the entire drug discovery process results from a strong alliance between several disciplines. Each discipline offers key knowledge and expertise, which when brought together, can be directed towards a common goal. To obtain a successful drug, three important issues are addressed: the target, the molecule, and the delivery of that molecule into the target. Associated with these issues are specific considerations, such as the interactions between the drug and its molecular target, the pathways that determine the access of the drug to that target, and the interactions that represent interindividual variability *(26)*.

3. Conclusions

In times gone by, drug discovery was dependent on the serendipity of the medicinal chemist. Nowadays, with the knowledge-base obtained from several disciplines, drug discovery has a new route, where the identification of a disease and its related biochemical pathway can become the starting point for the development of a successful drug.

In the first era of drug discovery, drugs were mainly derived from natural sources without prior knowledge of compounds' toxicity. Thus, many drugs failed to serve as useful therapeutics. To improve the success rate of finding suitable drug candidates, techniques upstream of clinical trials have been developed in order to support the clinical safety of the drug. Cell models have contributed towards an understanding of the interaction between the drug candidate and drug target. The developments at proteomics and genomics have also improved disease target identification. In this century so far, with all the advances in science and with the closer collaboration between the different scientific disciplines and the pharmaceutical industry, a multifaceted approach is being applied to drug discovery, thus improving the chances of finding new drugs.

References

1. Drews, J. (2000). Drug discovery: a historical perspective. *Science* **287**, 1960–1964

2. Thomas, G. (2000). *Medicinal Chemistry – An Introduction*. Wiley, England

3. Chin, Y., Balunas, M.J., Chai, H.B., and Kinghorn, A.D. (2006). Drug discovery from natural sources. *AAPS J* **2**, E239–E253

4. Zhang, M. and Wilkinson, B. (2007). Drug discovery beyond the "rule-of-five". *Curr Opin Biotechnol* **18**, 478–488

5. Lombardino, J.G. and Lowe, J.A. III. (2004). The role of the medicinal chemist in drug discovery – then and now. *Nat Rev Drug Discov* **3**, 853–862

6. Niwayama, S. (2006). Proteomics in medicinal chemistry. *Mini Rev Med Chem* **6**, 241–246

7. Sleno, L. and Emili, A. (2008). Proteomic methods for drug target discovery. *Curr Opin Chem Biol* **12**, 1–9

8. Ahn, N.G. and Wang, A.H.-J. (2008). Proteomics and genomics: perspectives on drug and target discovery. *Curr Opin Chem Biol* **12**, 1–3

9. Veenstra, T.D. (2006). Proteomic approaches in the drug discovery. *Drug Discov Today Tech* **3**, 193–195

10. Kopec, K.K., Bozyczko-Coyne, D., and Williams, M. (2005). Target identification and validation in drug discovery: the role of proteomics. *Biochem Pharmacol* **69**, 1133–1139

11. Katayama, H. and Oda, Y. (2007). Chemical proteomics for drug discovery based on compound-immobilized affinity chromatography. *J Chromatogr B* **855**, 21–27

12. Gershell, L.J. and Atkins, J.H. (2003). A brief history of novel drug discovery technologies. *Nat Rev Drug Discov* **2**, 321–327

13. Gupta, P. and Lee, K.H. (2007). Genomics and proteomics in process development: opportunities and challenges. *Trends Biotechnol* **25**, 324–330

14. Bleicher, K.H., Böhm, H., Müller, K., and Alanine, A.I. (2003). Hit and lead generation: beyond high-throughput screening. *Nat Rev Drug Discov* **2**, 369–378

15. Kerns, E.H. and Di, L. (2003). Pharmaceutical profiling in drug discovery. *Drug Discov Today* **8**, 316–323

16. Di, L. and Kerns, E.H. (2003). Profiling drug-like properties in discovery research. *Curr Opin Chem Biol* **7**, 402–408

17. Yengi, L.G., Leung, L., and Kao, J. (2007). The evolving role of drug metabolism in drug discovery and development. *Pharm Res* **24**, 842–858

18. Lipinski, C.A., Lombardo, F., Dominy, B.W., and Feeney, P.J. (2001). Experimental and computational approaches to estimate solubility and permeability in drug discovery and development settings. *Adv Drug Deliv Rev* **46**, 3–26

19. Han, S. and Kim, S.H. (2007). Introduction to chemical proteomics for drug discovery and development. *Arch Pharm* **340**, 169–177

20. Ryan, T.E. and Patterson, S.D. (2002). Proteomics in drug target discovery – high-throughput meets high-efficiency. *Drug Discov World* **3**, 43–52

21. Edwards, A.M., Arrowsmith, C.H., Christendat, D., Dharams, A., Friesen, J.D., Greenblatt, J.F., and Vedadi, M. (2000). Protein production: feeding the crystallographers and NMR spectroscopists. *Nat Struct Biol* **7**, 970–972

22. Schmid, M.B. (2002). Structural proteomics: the potential of high-throughput structure determination. *Trends Microbiol* **10**, S27–S31

23. Jain, K. (2002). New routes for drug discovery. *Drug Discov Today* **7**, 900–902

24. Newman, D.J., Cragg, G.M., and Snader, K.M. (2000). The influence of natural products upon drug discovery. *Nat Prod Rep* **17**, 215–234

25. Balaram, P. (2004). Drug discovery: myth and reality. *Curr Sci* **87**, 847–848

26. Triggle, D.J. (2007). Drug discovery and delivery in the 21st century. *Med Princ Pract* **16**, 1–14

27. Kayser, O. and Müller, R.H. (2004). *Pharmaceutical Biotechnology, Drug Discovery and Clinical Applications*. Wiley, England

28. Walsh, G. (2003). *Biopharmaceuticals: Biochemistry and Biotechnology.* Wiley, England

29. Dudzinski, D.M. and Kesselheim, S. (2008). Scientific and legal viability of follow-on protein drugs. *N Engl J Med* **358**, 843–848

30. Hughes, B. (2008). 2007 FDA drug approvals: a year of flux. *Nat Rev Drug Discov* **7**, 107–109

31. Tucker, J., Yakatan, S., and Yakatan, S. (2008). Biogenerics 2007: how far have we come? *J Commer Biotechnol* **14**, 56–64

32. Lawrence, S. (2007). Billion dollar babies - biotech drugs as blockbusters. *Nat Biotechnol* **25**, 380–382

33. Sartipy, P., Björquist, P., Strehl, R., and Hyllner, J. (2007). The application of human embryonic stem cell technologies to drug discovery. *Drug Discov Today* **12**, 688–699

34. Friel, R., van der Sar, S., and Mee, P.J. (2005). Embryonic stem cells: understanding their history, cell biology and signalling. *Adv Drug Deliv Rev* **57**, 1894–1903

35. Keller, G. (2005). Embryonic stem cell differentiation: emergence of a new era in biology and medicine. *Genes Dev* **19**, 1129–1155

36. McNeish, J.D. (2007). Stem cells as screening tools in drug discovery. *Curr Opin Pharmacol* **7**, 515–520

Chapter 2

Ligand Macromolecule Interactions: Theoretical Principles of Molecular Recognition

Tammy Nolan, Nidhi Singh, and Christopher R. McCurdy

Summary

Molecular recognition is mediated by three main factors: surface complementarity, thermodynamics, and associated physicochemical properties. These principles are responsible for ligand–target binding and therefore serve as the foundation for the design of new biologically relevant chemical entities. As these principles are involved in nearly all biological processes, a firm understanding of the details involved in binding is necessary for drug design. The consideration of these factors individually has proven useful; however, the combined effect of these governing principles is most important. And despite extensive studies, there are still many gaps in our understanding of this recognition process. The aim of this chapter is to introduce the basic concepts of ligand binding to set the stage for the following chapters, while briefly discussing fundamental techniques of drug design, including the indispensable tools of molecular modeling.

Key words: Protein–ligand interactions, Molecular recognition, Drug design, Ligand binding, Noncovalent interactions

1. Introduction

Molecular recognition is central to the interactions between two molecular entities and is highly dependent upon surface complementarity, hydrophobicity, and electrostatic forces. These interactions were first illustrated in the late nineteenth century by Emil Fischer's famous lock-and-key model which proposed that a substrate acts as a key that is specific for a certain lock or macromolecule *(1)*. The observation that macromolecules are not static entities led Daniel Koshland to propose the currently accepted theory of induced fit. Therein both ligand and macromolecule

Ana Cecília A. Roque (ed.), *Ligand-Macromolecular Interactions in Drug Discovery: Methods and Protocols,*
Methods in Molecular Biology, vol. 572,
DOI 10.1007/978-1-60761-244-5_2, © Humana Press, a part of Springer Science+Business Media, LLC 2010

are described as dynamic components that can readily undergo shape adjustments in order to optimize their fit *(2)*. This theory, while proposed for enzyme complexes, is applicable to most ligand-macromolecular complexes in that both large and small molecules rarely have a fixed three-dimensional (3D) nature. For the purpose of this chapter, biological targets (receptors, enzymes, nucleic acids, etc.) will simply be referred to as receptors or macromolecules.

The noncovalent interactions of molecular recognition are fundamental for nearly all biological processes *(3, 4)*. The interactions between ligands and macromolecules are therefore of pivotal importance for the design and discovery of new drugs. The stability of the ligand-macromolecule complex is directly related to the observed biological activity. As a result, molecular modeling studies such as docking and 3D Quantitative Structure Activity Relationship (QSAR) methods are based on our current knowledge of molecular recognition, as discussed in subsequent chapters.

Supramolecular chemistry and host–guest chemistry are synthetic studies focused on providing insight into the specific noncovalent interactions of molecular recognition *(4)*. Synthetic host macromolecules, such as crown ethers, calixarenes, porphyrins, etc., are used to direct studies aimed at filling the gaps in our current understanding of molecular interactions in addition to shedding light on important but currently overlooked principles *(5, 6)*. A recent review by Oshovsky et al. provides the particulars of these studies *(7)*. It is worth mentioning that the 1987 Nobel Prize in Chemistry was awarded to the three founders of this field, Pedersen, Lehn, and Cram *(8)*. The wealth of knowledge gained from these studies, in terms of energetic and structural analyses, can be directly applied to understand molecular recognition in biological systems.

2. Principles of Molecular Recognition

Experimental data clearly indicate the important role that surface complementarity plays in the tight and selective binding of a ligand to a target macromolecule. Steric, electrostatic, and lipophilic properties of both the substrate and the target-binding site determine the stability of the subsequent ligand–macromolecule complex. Enthalpic and entropic factors also contribute to the process of complex formation. To date, the solvation and desolvation energetic contributions to the system are unclear and of debate among scientists of different fields. The energy contribution resulting from the ligand binding conformation is also an important factor considered in molecular recognition.

2.1. Thermodynamic Contributions

Thermodynamic properties, including free energy of binding, enthalpy (ΔH), and entropy (ΔS), are directly involved in the formation of the ligand-macromolecule complex. The stability of a biological complex is determined by the binding free energy (ΔG). Binding enthalpy represents the loss of noncovalent associations, such as hydrogen bonds with water molecules, and the subsequent formation of interactions between the ligand and macromolecule, along with the reorganization of displaced solvent molecules *(9, 10)*. The binding entropy term considers the displacement of solvent from the binding surfaces as well as the loss of conformational freedom *(11)*. In general, higher binding affinities resulting from increased intermolecular interactions correspond to increased order and thus more negative values for enthalpic and entropic properties *(12)*. Unfavorable contributions include desolvation, strain, and restricted rotation. There are several methods by which these properties can be quantified, some of which will be discussed in the following chapters.

2.1.1. Molecular Flexibility

As described by the induced fit theory, molecular flexibility is indispensable for complex formation and therefore, biological function. Conformational changes in both ligand and macromolecule contribute to the overall energy of the system. Complexation is dependent on these dynamic changes as well as the solvation properties, **Fig. 1**. In terms of energy, complex formation is characterized by the loss of translational and rotational freedom with the subsequent gain of vibrational freedom *(13, 14)*. These changes result in a decrease in entropy, with an effect on free energy ranging from 9 to 45 kJ/mol for a typical ligand-protein complex *(15)*. It is widely known that the biologically relevant conformation is often not the lowest energy conformation. Conformational strain associated with high energy conformations may be surmounted by the favorable interactions between ligand and receptor upon binding. However, these additional unfavorable contributions to the binding energy must be taken into

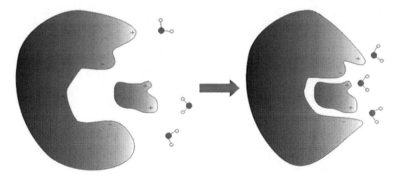

Fig. 1. Cartoon schematic of complex formation, in terms of solvation and changes in conformation.

account. To this end, the design of conformationally restrained ligands is an important consideration for the development of new ligands. Because rigid molecules have restricted conformational movement, their unfavorable contribution to free energy is less. Moreover, biologically active conformations can be readily inferred from rigid ligands.

2.1.2. Solvation in Complex Formation

Molecular recognition is highly dependent on the aqueous environment in which it takes place. The displacement and subsequent reorganization of water upon complex formation contributes to the energy of the system; however, it is important to note that energetic contributions of solvent are not fully understood or agreed upon. Due to the complexity of solvent contributions in molecular recognition, much of the current knowledge is based on molecular studies in the vacuum state. However, the critical nature of an aqueous environment has encouraged the study of solvation contributions.

The classical view suggests that the entropic factors serve as the driving force for hydrophobic interactions. The extensive hydrogen-bonding network of bulk water is disrupted by the inability of hydrophobic ligands to form hydrogen bonds (16). This results in an unfavorable decrease in entropy due to the reorganization of water around the compound. The phenomenon is known as the "iceberg effect" which suggests that the hydrogen bonds of water molecules surrounding the surface of hydrophobic ligands are stronger than in the bulk phase, with respective free enthalpies for dissociation of 2.6 and 2.0 kJ/mol, thus balancing the contributions from enthalpy (5, 17). However, an alternative view suggests that hydrophobic interactions may be driven by the enthalpy contribution resulting from the dissociation of hydrogen bonds of the bulk phase and the reorganization of water upon ligand binding. Solvent reorganization was reported to contribute between 25 and 100% of the enthalpy of complex formation (9). The resolution of water's role and significance in ligand binding is the focus of many current studies (18).

In addition to its role in hydrophobic interactions, solvation is also involved in the modulation of electrostatic interactions. The dielectric constant of the aqueous environment, around 80, effectively reduces these interactions; whereas the dielectric constant ranging from 1 to 20 of the binding cavity allows for strong interactions. The strength of hydrogen bonds is also sensitive to the dielectric constant. Prior to binding, water serves as a shield for electrostatic regions affecting the conformational range. Furthermore, water can play a structural role in certain enzymes and has been found to make up greater than 70% of the atoms of some protein crystal structures (19). Water can be covalently bound to certain enzymes and can serve a number of roles, including acting as a nucleophile or hydrogen-bonding partner.

Displacement of these important structural water molecules has proven to be an effective approach in drug design, affording high affinity ligands. Recent studies focus on developing methods for defining "relevant" water molecules within protein active sites, as well as their function and significance (20).

2.2. Interactions in Molecular Recognition

Surface complementarity, as mentioned, is vital for the interactions between ligand and macromolecule. Noncovalent intermolecular forces are largely responsible for molecular recognition and complex formation; while covalent bonds are seldom involved. The relevant interactions are illustrated in **Table 1**. Due to the weak nature of noncovalent forces, multiple interactions are usually necessary for the formation of a stable complex. The short range of most noncovalent forces makes it necessary for interacting groups to be in close proximity, illustrating the importance for complementarity, **Fig. 2**. It is generally believed that electrostatic forces are responsible for the initial recognition of ligand and receptor.

Table 1
Interactions involved in molecular recognition

Interaction	Strength (kcal/mol)	Example
Hydrogen bonding	3–5	δ^- δ^+ =O......H-O-
Ionic	5–10	-NH$_3^+$......$^-$O
Van der Waals	0.5–1	δ^+ δ^- -H$_2$C-H......CH$_3$-
CH–π		......H-CH$_2$-
Cation–π		......+H$_3$N-
Hydrophobic		CH$_3$ H$_3$C / CH$_3$ H$_3$C
Covalent	40–120	-H$_2$C-S-S-CH$_2$-

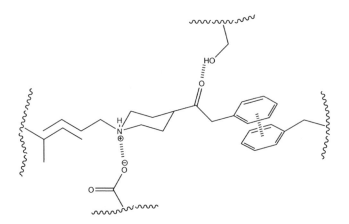

Fig. 2. Noncovalent interactions important in ligand binding.

2.2.1. Hydrogen Bonding

Hydrogen bonds are electrostatic attractions between two dipoles and are often referred to as dipole-dipole interactions. The proton, covalently bonded to the electronegative donor group, carries a δ^+ charge and thus is capable of forming a dipole–dipole interaction with the nonbonded pair of electrons of an electronegative acceptor *(21)*. While nitrogen and oxygen are most commonly involved in hydrogen bonding, fluorine and sulfur have also been shown to be capable of participating in these interactions. Hydrogen bonds involving sulfur bases differ from ordinary hydrogen bonds in that they tend to be weaker and less frequent, they originate from charge stabilization via the dipole or quadrupole of the sulfur base, and they have been shown to prefer a perpendicular directionality *(22)*. Additionally, carbon and silicon, though nonelectronegative, as well as several transition metals and the π-orbitals of aromatic systems can also function as hydrogen-bonding participants *(16)*.

Despite the weak nature of these attractions, contributing only 3–5 kcal/mol/bond in most cases, hydrogen bonding is crucial for molecular recognition. As is the case with electrostatic interactions, bond strength is dependent on the local medium, in particular the dielectric constant of the surrounding environment, the distance between donor and acceptor, and the direction of the interaction *(23)*. The directionality of hydrogen bonds is based on steric effects of the interacting species and neighboring atoms and the total electrostatic fields or mulitpoles of the system. In general, the most energetically favorable hydrogen bonds result from the colinearity of the two interacting dipole moments, with a 12° standard deviation. However, hydrogen bonding involving a carbonyl oxygen acceptor, which preferentially participates in two hydrogen bonds, tends to be in the direction of the lone pair electrons on the oxygen *(24)*.

The 3D structures, and therefore function, of many biological molecules are highly dependent on intramolecular hydrogen bonding. These interactions are responsible for the DNA double helix structure which is held together entirely by hydrogen bonds between base pairs on opposing strands. They are also indispensable in the formation of both secondary and tertiary structures of proteins, giving rise to α-helices and β-sheets. Atoms capable of hydrogen bonding but left unpaired are unfavorable situations. This concept is reinforced by a study in which less than 2% of buried amide backbone units did not participate in hydrogen bonding *(25)*. The importance of hydrogen bonding in the process of ligand–macromolecule complex is to balance the adverse energetic effects caused by the displacement of water. With this, hydrogen bonding is the largest contributor to the stability of the complex.

2.2.2. Ionic Interactions

Ionic bonds result from the attraction of two oppositely charged atoms and are considered long-range interactions. It has been suggested that this type of interaction is likely responsible for the initial recognition of ligand and receptor. The bond strength, varying between 5 and 10 kcal/mol, is inversely proportional to the distance between the two charged atoms and the dielectric constant of the surrounding medium. The characteristic long-range attraction likely plays a role in initial recognition between a ligand and its target receptor. These interactions are particularly important due to the ionization of both acids and bases at physiological pH. The amino acids that are predominantly ionized: arginine and lysine anions, and aspartate and glutamate cations, are often essential for molecular recognition within the active site of a protein. Many endogenous ligands, including neurotransmitters and hormones, rely heavily on ionic interactions for specific binding. An example of this is the ionic interaction of the protonated amine function of classic opiates with an active site aspartate residue. Mutagenesis studies involving this critical residue clearly indicate the importance of this interaction with the complete loss of biological activity in most cases.

2.2.3. Van der Waals Interactions

Optimal van der Waals (VDW) forces are dependent on shape complementarity of the protein active site and the interacting ligand. Gaps in the active site of the bound complex are unfavorable and in several cases, both the protein and ligand can undergo dramatic changes in order to minimize this space. This adversity is also true for groups that are too close, resulting in repulsion. VDW or London dispersion forces are often considered hydrophobic interactions. However, these interactions result from the attraction between temporary dipoles produced by the uneven distribution of electron density. These short-range interactions are weaker than both ionic and hydrogen bonds with an energy of 0.5–1 kcal/mol. If surface complementarity is high, these interactions can be numerous and thus contribute significantly to complex stability. The importance of matched surfaces is illustrated in **Fig. 3** which shows (a) a mutant HIV-1 protease (Protein Data Bank (PDB): 1MRW)

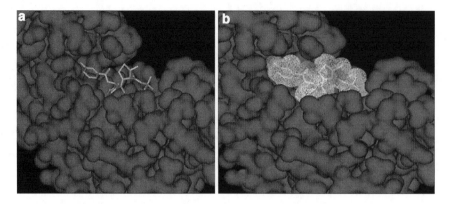

Fig. 3. HIV-1 protease with bound ligand (PDB: 1MRW) **(a)** showing the protein's VDW surface, and **(b)** the VDW surface of the ligand.

with a bound ligand *(26)* displaying the protein's VDW surface and (b) the paired VDW radii of ligand and protein (a portion of the protease has been removed for better viewing of the active site and ligand).

2.2.3.1. CH–π Interactions

To date, "VDW forces" has been used as a general term encompassing a variety of nonspecific interactions such as dipole/dipole interactions, multipole/multipole, induced dipole/induced dipole, and even cation/dipole interactions. But, current studies are revealing the details behind these distinguishable interactions. Specifically, the interactions involving π electron systems have been shown to include cations and CH groups. Groups containing π electrons have also been shown to engage in interactions with aliphatic functionalities. Unlike the cation/π interactions, these CH/π interactions are characterized by significant contributions from dispersion and delocalization effects with negligible electrostatic contributions. The CH/π interaction is described as a hydrogen bond between a soft acid (aliphatic portion) and soft base (π system) *(27)*. Both T- and L-shaped geometries are favorable for these interactions. Interestingly, the aforementioned acetylcholinesterase (AChE) crystal structure also provided evidence for the existence of these interactions. A network of CH/π interactions within the active site was determined to be important for stabilization, a mechanism referred to as "aromatic guidance" by Sussman et al. *(28)* and the "CH/π-piloted pathway" by Nishio and Hirota *(27)*.

2.2.3.2. Cation–π Interactions

Cation/π interactions occur between a positively charged molecule and a π system, such as aromatic rings, alkenes, alkynes. Though there are several factors giving rise to this interaction, it is largely due to electrostatic forces *(29)*. Experimental studies with these interactions have demonstrated that the amino acids phenylalanine, tyrosine, tryptophan, and occasionally histidine can act not only as hydrophobic residues but also as polar quadrupoles in molecular recognition, with over one-fourth of tryptophan residues in the PDB involved in cation/π interactions *(30)*. The importance of these interactions in biological systems was revealed with the determination of the AChE crystal structure which illustrates the interaction of the quaternary ammonium charge of acetylcholine with a conserved tryptophan residue *(28)*. Additional information for these interactions is contained in the review of Ma and Dougherty *(31)*.

2.2.4. Hydrophobic Interactions

Hydrophobic interactions occur between two nonpolar regions, namely alkyl chains and aromatic rings. They are critical for ligand-macromolecule complexation driven by the increase in entropy. The unfavorable organization of water around exposed hydrophobic regions is reduced by the association of these nonpolar regions with one another *(32)*. The conformational changes that occur upon complex formation are driven by these hydrophobic interactions, with the macromolecule enclosing the ligand in order to reduce the exposure of hydrophobic surfaces to the aqueous environment. A dramatic example of this structural change is seen with aldose reductase which can undergo significant conformational adjustments in order to accommodate a variety of substrates *(33)*.

2.2.5. Covalent Bonds

Covalent bonds are minor contributors to molecular recognition. The formation of a covalent bond between ligand and macromolecule most often results in irreversible binding due to the bond strength, ranging from 40 to 120 kcal/mol. Consequently, these stable complexes must be destroyed and new macromolecules produced. However, irreversible binding is sometimes desirable, as seen with DNA alkylating agents and some enzyme inhibitors. Active site residues such as cysteine, serine, lysine, aspartate, and glutamate can be involved in covalent bonding with a ligand. Antimetabolite agents, who resemble natural nucleotides, act by forming covalent bonds with terminal nucleotides, causing truncation of the chain. Intentional design of covalent ligands has been successful in the area of affinity labels. These affinity labels resemble natural substrates for a specific receptor but also contain reactive groups through which it covalently binds to the macromolecule, usually via a nucleophilic residue. Due to this reactivity and corresponding toxicity as well as the inherent nonselectivity of these agents, they are not therapeutically relevant. However, these ligands have been useful in determining the binding sites and conformations of active and inactive proteins, as seen by the opioid affinity labels which afforded useful information into active site residues *(34)*. Affinity labels have been useful for a number of receptors.

2.2.6. Metal Chelation Interactions

Another important protein-ligand interaction is complexation with metal ions. Two types of electrostatic bonds are involved in metal ion reactions with substrates or proteins. The first of these, an interionic attraction forming an ionic bond between two opposite charges, may be important in the structural organization of the system, the physical order of which is dependent on the presence of certain ions. The second type of bond formed is the ion–dipole bond which results from electrostatic attraction between the positively charged metal ion and a dipolar molecule, either permanent or induced.

The role of metal ion binding in the folding/stability and function of a variety of metalloenzymes has long been recognized (35). These ions may serve as the catalytic center of the enzyme or may be required, as a binding group, to allow interaction of the enzyme and substrate. They may also be required to maintain physiological control by antagonizing the activating effect of some other metal on an enzyme system. The metal ions may also be responsible for promoting a variety of reactions, including bond formation and scission, electron transfer, and atom transfer in these enzymes. The reactions catalyzed are essential for several biological processes like DNA repair and synthesis, cell replication, energy production in cells, generation of oxygen through photosynthesis, sensing, and transporting, as well as oxygen activation for reaction with numerous molecules.

All biologically important metal ions can form complexes by simultaneously coordinating with a large number of different chemical species, **Fig. 4**. However, these interactions are limited to only a few elements from Group V and VI in the periodic table, namely oxygen, nitrogen, and sulfur. For the interactions between

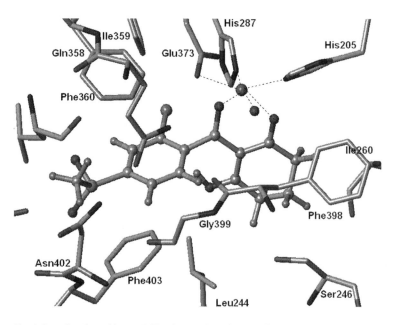

Fig. 4. Complexation with a metal ion, interacting with six different entities in the protein active site.

metal ions and ligands, a convenient classification has been proposed by Ralph Pearson who introduced the "hard" and "soft" acids and bases (HSAB) concept *(36)*. Accordingly, metal ions which are hard acids are usually small in size, highly charged, and have low polarizability. Bonds formed between "hard" metal ions and ligands are mainly ionic in nature. "Soft" metal ions are usually large in size with a low charged state and they form bonds that are considerably covalent in nature. "Hard" ligands are generally small in size and have low polarizability. "Hard" metal ions form their strongest complexes with "hard" ligands (e.g., magnesium ions and the oxygen atoms in pyrophosphate). "Soft" metal ions form their strongest complexes with "soft" ligands (e.g., mercury ions and thiols). Interestingly, metal ions such as iron and zinc are capable of forming complexes with both "hard" and "soft" ligands and therefore do not readily fit into either category. Examples of hard/soft metals and ligands as well as their classification are illustrated in **Table 2**. Also, most amino acids can interact with both "hard" and "soft" ligands. On the basis of the HSAB principle, metal ions preferentially bind to the functional groups of certain amino acids. The amino acids functionalities that often form the metal ion binding pocket include carboxyl (Asp and Glu), imidazole (His), indole (Trp), thiol (Cys), thioether (Met), and hydroxyl (Ser, Thr, and Tyr). The carbonyls

Table 2
Classification of hard and soft metal ions and ligands

Metal ions
Hard
Na^+, K^+, Mg^{2+}, Ca^{2+}, Mn^{2+}, Fe^{3+}
Borderline
Fe^{2+}, Co^{2+}, Cu^2, Zn^{2+}
Soft
Cu^{2+}, Cd^{2+}
Ligands
Hard
ROH, RO^-, R_2O, OH^-, Cl^-, NH_3, RNH_2
Borderline
$C_6H_5NH_2$, C_5H_5N, N_3^-, NO_2^-
Soft
R_2S, RSH, RS^-, SCN^-, R_3P, CN^-

of side chain amide groups of Asn and Gln, as well as the backbone carbonyl and amide nitrogen may also interact with metal ions *(37)*.

3. Molecular Recognition in Drug Design

3.1. Bioisosteres

Molecular recognition serves as the foundation for the bioisostere concept of drug design. Bioisosteric replacement, derived from Langmuir's isostere observation, is based on the idea that groups with comparable physical and chemical properties will likely produce similar biological effects *(38–40)*. By maintaining surface complementarity in terms of volume, polarity, and electronic influence, while making subtle changes to other physicochemical properties, compounds can be developed with similar bioactivities and more desirable characteristics. A bioisosteric approach can influence pharmacokinetic features (metabolic stability and bioavailability), selectivity, and can even simplify syntheses. However, trends seen for one biological system in terms of bioisosteric replacement will not necessarily translate to other biological targets. Classical isosteres, as described above, are generally divided into five groups, the first two of which are shown in **Table 3**. Nonclassical bioisosteres are groups of atoms shown to produce similar biological activities though they do not follow classical definitions pertaining to steric and electronic features, as well as atom number. These nonclassical isosteres are defined by the representative functional group, also included in **Table 3**. A detailed review by Patani and LaVoie describes the importance of bioisosterism in the drug design process *(41)*.

3.2. Computer-Aided Drug Design

Molecular modeling is a widely used tool for the rational design of new biologically active compounds. There are two main computational strategies for the design of new compounds, both of which are founded on molecular recognition factors. These are based on the presence or the absence of 3D structural information for the target protein, i.e., the direct structure-based drug design (SBDD) and the indirect ligand-based drug design (LBDD), respectively. Structure-based design relies on the availability of 3D structural information for the therapeutic target as determined by X-ray crystallography or NMR spectroscopy. Fragment-based and de novo design, as well as docking-based virtual screening make use of the receptor structure for the development of new, chemically diverse ligands *(42–44)*. Binding interactions are maximized by designing molecules that complement the residues within the putative binding site. In structure-based virtual screening, an effective alternative to high throughput screening (HTS), large

Table 3
Examples of classical and nonclassical isosteres

Classical isosteres
Univalent atoms and groups
CH_3, NH_2, OH, F, Cl
Cl, PH_2, SH
Br, i-Pr
I, t-Bu
Bivalent atoms and groups
$-CH_2-$ $-NH-$ $-O-$ $-S-$ $-Se-$
$-COCH_2R$ $-CONHR$ $-CO_2R$ $-COSR$
Nonclassical isosteres

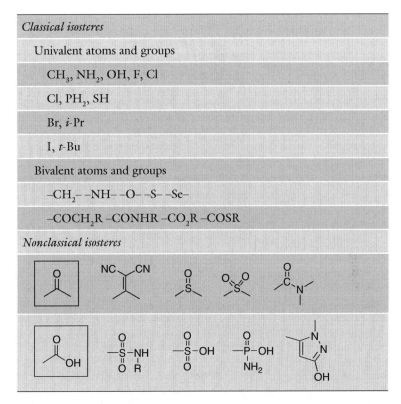

chemical databases can be docked into the active site of a receptor crystal structure and scored based on fit resulting in new, chemically diverse leads. Homology modeling, the generation of a 3D structure for a target protein based on the known structure of a homologous protein, is a useful approach in cases where detailed structural information is unavailable *(45)*. This technique is often used for generating the highly conserved 7-transmembrane domain structures of G-protein-coupled receptors (GPCRs) using templates of the only two available GPCR crystal structures of bovine rhodopsin *(46)* and more recently, the human β-2 adrenergic receptor *(47, 48)*.

Alternatively, ligand-based design has been a successful approach to the design of new chemical entities. In the absence of a 3D structure for the target molecule, QSAR studies *(49–52)* have been used to determine the physicochemical characteristics important for ligand binding, while additionally giving insight about the putative receptor binding site. QSAR analyses enable the mathematical determination of correlations between variables in a series of ligands using molecular descriptors. Pharmacophore modeling is a widely used variation of the QSAR technique in

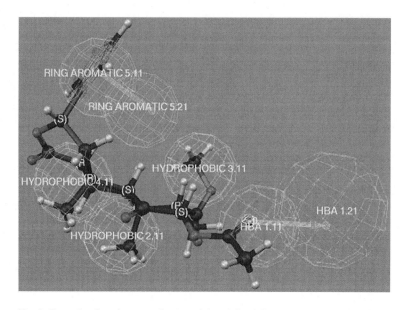

Fig. 5. Example of a pharmacophore model and fitted ligand containing six defining features.

which the chemical features essential for binding are represented as a 3D model; an example is shown in **Fig. 5**. These features consist of general chemical properties such as hydrogen bond acceptor/donor, hydrophobic, and positive and negative ionizable. These 3D models are very useful for the virtual screening of large databases *(53, 54)*. Critical assumptions must be made in ligand-based design which can greatly influence the resulting model, such as conformation, binding mode, and alignment, and thus should be done with caution.

3.3. Bivalent Ligand Design

It is widely known and of special interest that biological macromolecules can have more than one binding site and can also exist as oligomers. It has been suggested that ligands which can bind to two independent sites simultaneously would be more favorable, in terms of thermodynamics, than a monomeric ligand *(55)*. In the case of dimeric receptors, bivalent ligands may afford better affinity as well as selectivity. The bivalent serotonin reuptake inhibitors designed by Tamiz et al. supported this notion with a potency 8,000-fold greater than the monomeric counterpart *(56)*. It is still uncertain whether these ligands are bridging two separate monomers or rather two binding sites of the same receptor. The development of bivalent ligands has proven to be useful in pharmacology and molecular biology studies. The newly uncovered information will help to shed light on the details concerning molecular recognition of associated macromolecules and the multitude of cascades that they ultimately control.

4. Conclusions

Because molecular recognition is the basis for many scientific fields, a firm grasp of the underlying principles is crucial. The incomplete understanding of complex formation is a limiting factor for rational drug design. However, as our understanding of the molecular recognition process changes, so too does the current drug discovery paradigm. In the past, compounds found to have high affinity for the target were subsequently modified to optimize pharmacokinetic properties while often drastically affecting binding. However, as the details of molecular recognition are revealed, drug design is becoming more fluid with a simultaneous incorporation of necessary binding characteristics and pharmacokinetic [Absorption, Distribution, Metabolism, Excretion, Toxicity (ADMET)] considerations into the design of lead compounds. Furthermore, tools and techniques important in rational drug design are modified and updated as the molecular recognition process is elucidated.

References

1. Fischer, E. (1894). Einfluss der Configuration auf die Wirkung der Enzyme. *Ber. Dtsch. Chem. Ges.* **27**, 2985–2993

2. Koshland, D. (1958). Application of a theory of enzyme specificity to protein synthesis. *Proc. Natl. Acad. Sci. U.S.A.* **44**, 98

3. Gellman, S. H. (1997). Introduction: molecular recognition. *Chem Rev* **97**, 1231–1232

4. Lehn, J. M. (1993). Supramolecular chemistry. *Science* **260**, 1762–1763

5. Schneider, H. J. and Yatsimirski, A. (2002). *Principles and methods in supramolecular chemistry.* Chichester: Wiley

6. Steed, J. W. and Atwood, J. L. (2000). *Supramolecular chemistry.* Chichester: Wiley

7. Oshovsky, G. V., Reinhoudt, D. N. and Verboom, W. (2007). Supramolecular chemistry in water. *Angew. Chem. Int. Ed. Engl.* **46**, 2366–2393

8. Retrieved from http://nobelprize.org/nobel_prizes/chemistry/laureates/1987/

9. Chervenak, M. C. and Toone, E. J. (1994). A direct measure of the contribution of solvent reorganization to the enthalpy of ligand binding. *J. Am. Chem. Soc.* **116**, 10533–10539

10. Fisher, H. F. and Singh, N. (1995). Calorimetric methods for interpreting protein-ligand interactions. *Methods Enzymol* **259**, 194–221

11. Murphy, K. P., Xie, D., Thompson, K. S., Amzel, L. M., et al. (1994). Entropy in biological binding processes: estimation of translational entropy loss. *Proteins* **15**, 113–120

12. Gilli, P., Ferretti, V., Gilli, G., and Borea, P. A. (1994). Enthalpy-entropy compensation in drug receptor binding. *J. Phys. Chem.* **98**, 1515–1518

13. Brady, P. B. and Sharp, K. A. (1997). Entropy in protein folding and in protein-protein interactions. *Curr. Opin. Struct. Biol.* **7**, 215–221

14. Page, M. I. and Jencks, W. P. (1971). Entropic contributions to rate acceleration in enzymic and intramolecular reactions and the chelate effect. *Proc. Natl. Acad. Sci. U.S.A.* **68**, 1678–1683

15. Searle, M. S. and Williams, D. H. (1992). The cost of conformational order: entropy changes in molecular associations. *J. Am. Chem. Soc.* **114**, 10690–10697

16. Jeffrey, G. A. and Saenger, W. (1991). *Hydrogen bonding in biological structures.* Berlin: Springer

17. Frank, H. S. and Evans, M. W. (1945). Free volume and entropy in condensed systems III. *J. Chem. Phys.* **13**, 507

18. Nicholls, A., Mobley, D. L., Guthrie, J. P., Chodera, J. D., et al. (2008). Predicting small-molecule solvation free energies: an informal blind test fro computational chemistry. *J. Med. Chem.* **1**, 769–779

19. Poornima, C. S. and Dean, P. M. (1995). Conserved water molecules at the ligand-binding

sites of homologous proteins. *J. Comput. Aided Mol. Des.* **9**, 500–512

20. Amadasi, A., Surface, J. A., Spyrakis, F., Cozzini, P., et al. (2008). Robust classification of "relevant" water molecules in putative protein binding sites. *J. Med. Chem.* **51**, 1063–1067

21. Pauling, L. (1939). *The nature of the chemical bond and the structure of molecules and crystals.* Ithaca, NY: Cornell University Press

22. Platts, J. A., Howard, S. T. and Bracke, B. R. (1996). Directionality of hydrogen bonds to sulfur and oxygen. *J. Am. Chem. Soc.* **118**, 2726–2733

23. Warshel, A., Papazyan, A., Kollman, P. A., Cleland, W. W., et al. (1995). On low-barrier hydrogen bonds and enzyme catalysis. *Science* **269**, 102–106

24. Taylor, R. and Kennard, O. (1984). Hydrogen-bond geometry in organic crystals. *Acc. Chem. Res.* **17**, 320–326

25. Mcdonald, I. K. and Thornton, J. M. (1994). Satisfying hydrogen bonding potentials in proteins. *J. Mol. Biol.* **238**, 777–793

26. Vega, S., Kang, L.-W., Velazquez-Campoy, A., Kiso, Y., et al. (2004). A structural and thermodynamic escape mechanism from a drug resistant mutation of the HIV-1 protease. *Proteins* **55**, 594–602

27. Nishio, M. and Hirota, M. (1989). CH/π interaction: implication in organic chemistry. *Tetrahedron* **45**, 7201–7245

28. Sussman, J. L., Harel, M., Frolow, F., Oefner, C., et al. (1991). Atomic structure of acetylcholinesterase from *Torpedo californica*: a prototypic acetylcholine-binding protein. *Science* **253**, 872–879

29. Dougherty, D. A. (1996). Cation/π interactions in chemistry and biology: a new view of benzene, Phe, Tyr, and Trp. *Science* **271**, 163–168

30. Gallivan, J. P. and Dougherty, D. A. (1999). Cation/π interactions in structural biology. *Proc. Natl. Acad. Sci. U.S.A.* **96**, 9459–9464

31. Ma, J. C. and Dougherty, D. A. (1997). The cation/π interaction. *Chem. Rev.* **97**, 1303–1324

32. Blokzijl, W. and Engberts, J. (1993). Hydrophobic effects. Opinion and facts. *Angew. Chem. Int. Ed. Engl.* **32**, 1545–1579

33. Urzhumtsev, A., Tete-Favier, F., Mitschler, A., Barbanton, J., et al. (1997). A 'specificity' pocket inferred from the crystal structures of the complexes of aldose reductase with the pharmaceutically important inhibitors tolrestat and sorbinil. *Structure* **5**, 601–612

34. Mccurdy, C. R., Le Bourdonnec, B., Metzger, T. G., El Kouhen, R., et al. (2002). Naphthalene dicarboxaldehyde as an electrophilic fluorogenic moiety for affinity labeling: application to opioid receptor affinity labels with greatly improved fluorogenic properties. *J. Med. Chem.* **45**, 2887–2890

35. White, R. J., Margolis, P. S., Trias, J. and Yuan, Z. (2003). Targeting metalloenzymes: a strategy that works. *Curr. Opin. Pharmacol.* **3**, 502–507

36. Pearson, R. G. (1963). Hard and soft acids and bases. *J. Am. Chem. Soc.* **85**, 3533–3539

37. Sabat, M. (1990). Ternary metal ion-nucleic acid base protein complexes. In: Sigel, H. and Sigel, A. (Eds.), *Metal ions in biological systems.* New York: Marcel Dekker, pp. 521

38. Friedman, H. L. (1951). Influence of isosteric replacements upon biological activity. *N ASNRS* **206**, 295–358

39. Burger, A. (1991). Isosterism and bioisosterism in drug design. *Prog. Drug. Res.* **37**, 287–371

40. Langmuir, I. (1919). Isomorphism, isosterism, and covalence. *J. Am. Chem. Soc.* **41**, 1543–1559

41. Patani, G. A. and LaVoie, E. J. (1996). Bioisosterism: a rational approach in drug design. *Chem. Rev.* **96**, 3147–3176

42. Mohan, V., Gibbs, A. C., Cummings, M. D., Jaeger, E. P., et al. (2005). Docking: Successes and challenges. *Curr. Pharm. Des.* **11**, 323–333

43. Erickson, J. A., Jalaie, M., Robertson, D. H., Lewis, R. A., et al. (2004). Lessons in molecular recognition: The effects of ligand and protein flexibility on molecular docking accuracy. *J. Med. Chem.* **47**, 45–55

44. Xu, H. and Agrafiotis, D. K. (2002). Retrospect and prospect of virtual screening in drug discovery. *Curr. Top. Med. Chem.* **2**, 1305–1320

45. Patny, A., Desai, P. V. and Avery, M. A. (2006). Homology modeling of G-protein-coupled receptors and implications in drug design. *Curr. Med. Chem.* **13**, 1667–1691

46. Palczewski, K., Kumasaka, T., Hori, T., Behnke, C. A., et al. (2000). Crystal structure of rhodopsin: a G protein-coupled receptor. *Science* **289**, 739–745

47. Cherezov, V., Rosenbaum, D. M., Hanson, M. A., Rasmussen, S. G., et al. (2007). High-resolution crystal structure of an engineered human B2 adrenergic G protein-coupled receptor. *Science* **318**, 1258–1265

48. Rasmussen, S. G., Choi, H.-J., Rosenbaum, D. M., Kobilka, T. S., et al. (2007). Crystal structure of the human B2 adrenergic G protein-coupled receptor. *Nature* **450**, 383–387

49. Kubinyi, H. (2002). From narcosis to hyper-space: the history of QSAR. *Quant. Struct.-Act. Relat.* **21**, 348–356

50. Hansch, C. and Fujita, T. (1964). A method for the correlation of biological activity and chemical structure. *J. Am. Chem. Soc.* **86**, 1616–1626

51. Kubinyi, H. (1997). QSAR and 3D QSAR in drug design. Part 1: methodology. *Drug Discov. Today* **2**, 457–467

52. Kubinyi, H. (1997). QSAR and 3D QSAR in drug design. Part 2: applications and problems. *Drug Discov. Today* **2**, 538–546

53. Good, A. C., Mason, J. S. and Pickett, S. D. (2000). Pharmacophore pattern application in virtual screening, library design, and QSAR. In: Bohm, H.-J. and Schneider, G. (Eds.), *Virtual screening for bioactive molecules.* Weinheim: Wiley-VCH

54. Kurogi, Y. and Guner, O. F. (2001). Pharmacophore modeling and three-dimensional database searching for drug design using Catalyst. *Curr. Med. Chem.* **8**, 1035–1055

55. Portoghese, P. S. (1992). The role of concepts in structure-activity relationship studies of opioid ligands. *J. Med. Chem.* **35**, 1927–1937

56. Tamiz, A. P., Zhang, J., Zhang, M., Wang, C. Z., et al. (2000). Application of the bivalent ligand approach to the design of novel dimeric serotonin reuptake inhibitors. *J. Am. Chem. Soc.* **122**, 5393

Chapter 3

X-Ray Crystallography in Drug Discovery

Ana Luísa Carvalho, José Trincão, and Maria João Romão

Summary

Macromolecular X-ray crystallography is an important and powerful technique in drug discovery, used by pharmaceutical companies in the discovery process of new medicines. The detailed analysis of crystal structures of protein–ligand complexes allows the study of the specific interactions of a particular drug with its protein target at the atomic level. It is used to design and improve drugs. The starting point of these studies is the preparation of suitable crystals of complexes with potential ligands, which can be achieved by using different strategies described in this chapter. In addition, an introduction to X-ray crystallography is given, highlighting the fundamental steps necessary to determine the three-dimensional structure of protein–ligand complexes, as well as some of the tools and criteria to validate crystal structures available in databases.

Key words: X-ray crystallography, Protein structure, Crystallization, Protein–ligand complexes, Soaking, Cocrystallization, Drug design

1. Introduction

X-ray crystallography has been, historically, the standard method to obtain atomic resolution structures of macromolecules and is still the most powerful and common method to obtain protein structures as well as to design and optimize potential drugs. Although most of the techniques used in protein crystallography have evolved significantly in the recent years, in great part due to the exponential growth of the computing power available to crystallographers, it is still a difficult and time-consuming process. Manual interpretation of complex arrays of data and iterative electron density interpretation and model rebuilding take up most of the time spent in structure determination. Manual intervention

Ana Cecília A. Roque (ed.), *Ligand-Macromolecular Interactions in Drug Discovery: Methods and Protocols,*
Methods in Molecular Biology, vol. 572,
DOI 10.1007/978-1-60761-244-5_3, © Humana Press, a part of Springer Science+Business Media, LLC 2010

comes along with subjective interpretation of the data, leading to eventual errors in the structure and bottlenecks to the model refinement. However, many recent advances have made X-ray crystallography easier and faster. Most notably, tunable synchrotron X-ray sources, high speed CCD detectors, cryoprotection, as well as the ability to incorporate anomalous scatterers directly into the protein, have made crystal structure solution possible for a much wider range of targets than ever before.

Nuclear Magnetic Resonance (NMR), also a well-established technique for 3D structure determination, is limited by the size of the protein. In recent years, Cryo-Electron Microscopy (Cryo-EM) *(1, 2)* has become an increasingly powerful technique for structure elucidation but provides relatively low resolution data. It has been applied to the study of large macromolecular assemblies that are too large and/or too flexible to be solved by X-ray crystallography or NMR.

In this chapter, a general overview of protein crystallography is presented, focusing on the crystallization of proteins and ligand-protein complexes, the basic theory behind the method, the main steps involved in solving a crystal structure, and the criteria used to validate the structural models. **Figure 1** shows a flowchart of the main steps involved in a 3D structure determination by X-ray crystallography.

2. Crystallization of Protein–Ligand Complexes

The successful three-dimensional structure determination of proteins by X-ray crystallography is dependent on the availability of well-ordered single crystals. However, successful protein crystallization is very often the bottleneck of crystallography projects, in spite of high-throughput methodologies and considerable progress in the development of fully automated crystallization robots as well as in the diversity of crystallization screenings presently available.

In drug design, once the three-dimensional structure of the target apo-protein is available, crystallization experiments are necessary in order to obtain suitable crystals of the small molecule–protein complex *(3)*. However, even in those cases where crystallization conditions are well established for the ligand-free protein, the formation of crystals of the complex may not be straightforward, depending on the case under study. Furthermore, once crystals are obtained and the crystal structure is solved, one very often finds that the ligand did not bind to the protein at all. These negative results are a waste of time and effort. To decrease the failure rate in obtaining crystals of the complexed protein as much as possible,

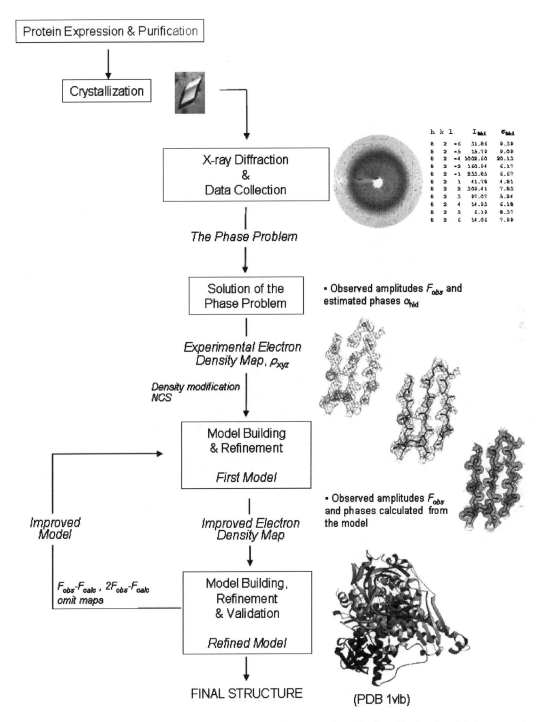

h	k	l	I_{hkl}	σ_{hkl}
8	2	-6	31.86	9.39
8	2	-5	15.79	9.09
8	2	-4	3002.60	20.13
8	2	-2	160.94	6.17
8	2	-1	233.85	6.67
8	2	1	41.78	4.81
8	2	2	309.41	7.83
8	2	3	97.07	5.24
8	2	4	14.93	6.18
8	2	5	6.19	8.37
8	2	6	14.06	7.99

Fig. 1. Main steps involved in a 3D structure determination by X-ray crystallography. Once the target protein is expressed and purified, it must be crystallized. Single crystals are necessary in order to measure diffraction data, either in-house or using a synchrotron X-ray source. These data are then processed and analyzed. The following steps involve a series of computational calculations. Once the phase problem (and therefore the 3D structure) is solved, the first (experimental) electron density map is calculated and model building can start by fitting protein atoms into the density. The structural model is refined and improved in an iterative way until convergence is reached. In the final steps of refinement and model building, several validation criteria are used that assess the quality of the model deposited on the Protein Data Bank (PDB).

several biophysical tools can be used prior to cocrystallization in order to assess the experimental conditions that will most likely lead to complex formation *(4)*. This preliminary analysis may increase the chances of obtaining crystals of the required complex, avoiding the waste of time spent in determining apo-structures.

2.1. Preliminary Analysis by Biophysical Methods

Many factors responsible for complex formation are strongly dependent on the protein under study. However, the use of the appropriate biophysical methods may help to confirm the binding of a particular ligand under defined experimental conditions, guiding the design of crystallization experiments that will eventually lead to suitable crystals of the desired complex.

There are several methods available to analyze ligand–protein binding in solution, some of which are described in detail in other chapters of this book. For example, Isothermal Titration Calorimetry (ITC), Surface Plasmon Resonance (SPR), NMR, Dynamic Light Scattering (DLS), Analytical Ultracentrifugation (AUC), and Differential Scanning Calorimetry (DSC) are some methods that can provide valuable information about protein–ligand complexes. The use of one or more of these methods may allow to confirm ligand–protein binding and to estimate binding affinities. The BindingDB (Database of Protein–Ligand Binding Affinities) is a public, Web-accessible database of measured binding affinities for biomolecules, genetically or chemically modified biomolecules, and synthetic compounds (*see* **Note 1**).

Some of these methods can study proteins in a nondestructive manner and in the same range of protein concentrations used for crystallization, such as ITC, SPR, NMR, and DLS. All these methods can be used prior to crystallization experiments, using the very same sample, confirming that complex formation has occurred.

Other methods involving thermal denaturation *(5–7)*, such as DSC or Differential Scanning Fluorimetry (DSF), allow monitoring of the shift in thermal stability upon ligand binding. In spite of their destructive nature, these techniques may also prove useful in determining the affinity of small molecules to the protein under study.

2.1.1. Isothermal Titration Calorimetry

ITC is a powerful technique that involves measuring the amount of heat released or absorbed upon ligand binding. It provides information about the stoichiometry and thermodynamic properties of protein–ligand interactions (association (K_a) and dissociation ($K_d = 1/K_a$) constants, enthalpy (ΔH), and entropy (ΔS)). ITC can be used to monitor the binding of small molecules to proteins under different experimental conditions. The molar concentration of protein required per ITC experiment, although dependent on the affinity, is approximately $10 < (K_a \times [\text{protein}]) < 100$ (*see* **Note 2**). The concentrations used are typically in the order of ≥ 0.5 mM for the ligand and 20–100 μM for the protein solution, which is within the range used for crystallization experiments. The nondestructive nature of the ITC experiment, and the information provided by this technique, makes it an ideal tool to use prior to cocrystallization.

2.1.2. Nuclear Magnetic Resonance

As with ITC, NMR is a technique also described in detail in this volume, focusing on its application to the monitoring of ligand binding to proteins. Two methods can be employed, either ligand-based or protein-based. In the first method, the ligand signals are analyzed as a function of protein binding, whereas the second analyzes changes in the protein spectrum upon ligand binding. In both cases, the technique is nondestructive and can be performed prior to crystallization.

2.1.3. Surface Plasmon Resonance

SPR is also a useful complementary technique in drug design. As with ITC, SPR allows the simultaneous determination of the affinity and stoichiometry of the ligand–protein complex, giving information on kinetic and thermodynamic data.

2.1.4. Dynamic Light Scattering

DLS analysis is another noninvasive technique that measures the change in the intensity of scattered light from a sample as a function of time. It monitors changes in size and protein aggregation as a function of varying solution conditions *(8, 9)*. It is generally accepted that monodisperse (uniform) samples are more likely to crystallize. DLS is frequently used to screen and monitor homogeneity prior to crystal growth as well as to study *crystallizability* of proteins based on aggregation analysis. The technique has established itself within the past 10 years, mainly in the field of protein crystallography, and has proved to be a valuable tool for the rapid screening of protein samples, which, if monodisperse, are expected to have a higher ability to form crystals.

Being a useful screening tool for identifying the formation of aggregates, it may help to identify ligand binding prior to crystallization assays in those cases where ligand binding induces protein oligomerization or large conformational changes *(4, 10)*.

2.1.5. Analytical Ultracentrifugation

AUC is a classical technique used to study the behavior of macromolecules in solution under the influence of a strong gravitational force. Like DLS, it is used to determine whether a protein sample is homogeneous or a mixture of forms (e.g., monomer/dimer or aggregates). AUC can quantify the shape and size of macromolecules in solution. One of its applications is to determine association and dissociation constants, solvation, stoichiometry, and ligand binding.

2.2. Soaking and Cocrystallization

Soaking and cocrystallization are the two most common methods used to prepare crystals of protein–ligand complexes (see review by McNae, 2005) *(11)*. The soaking method is the simplest one since it involves only incubation of crystals of the apo-protein with the ligand of interest. Cocrystallization consists in preparing crystals of the protein–ligand complex.

The success of soaking experiments is related to the properties of the protein crystal lattice, which is loosely packed and has a high solvent content, usually in the range of 30–70% *(12)*. The crystals have large solvent-filled channels (20–100Å) *(13)*, which provide good access for ligand molecules that may diffuse through the channels and bind to the protein molecules (**Fig. 2**).

Since protein crystals are fragile, it is usually necessary to stabilize them prior to handling, for cryo cooling and data collection (*see* **Subheading 3.1.3**) as well as for soaking experiments. The standard procedure is to pre-equilibrate the crystals in a convenient harvesting buffer, usually prepared by increasing the precipitant concentration. This allows stocking a large number of crystals that can later be used to soak the compounds of interest. In addition, one can perform chemical cross-linking of the crystals with glutaraldehyde *(14)* (*see* **Note 3**).

In practice, soaking times and inhibitor concentration vary considerably, which may be due to several factors related to: the diffusion rate of the ligand; the temperature and chemical environment of the soaking; the solvent used to dissolve the ligand; the presence of competing ligands, etc. [(see review by Danley *(15)*]. Furthermore, the experimental conditions of the soaking experiment (temperature, pH, salts, buffers, additives) may not be

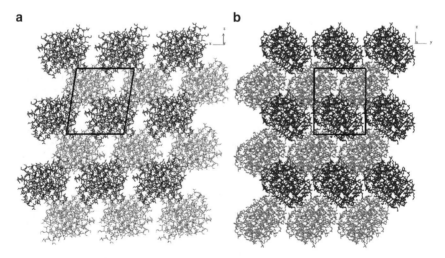

Fig. 2. (**a**) Representation of a protein crystal packing with three layers projected along the *b* axis. One unit cell is marked. These crystals have a solvent content of ~40% and the solvent channels are clearly visible in this lattice orientation. (**b**) Same representation but viewed along the *a* axis. In this orientation, the crystal packing is tighter with different crystal contacts. The picture was produced with program CCP4mg from the CCP4 suite of programs *(67)*.

optimal for ligand binding. In these cases, more than just increasing the soaking time, one needs to adjust other experimental parameters in order to maximize the desired ligand occupancy.

Sometimes, crystals of the protein bound to a natural ligand are obtained instead of the apo-protein. In these cases, the soaking experiment must be designed in order to replace the original ligand by the ligand of interest (*replacement soaking*) *(16)*. It is clear that, in these cases, the solubility and relative binding affinities of the two compounds must be carefully analyzed.

Ligand binding to a protein induces several kinds of conformational changes, which may lead to crystal cracking and/or dissolution. In addition, in part due to possible restrained conformational changes, it has been argued that the mode of binding in a soaked complex structure may not reflect the binding mode in solution *(17)*. In these cases, and whenever soaking is not successful, cocrystallization is usually the method of choice. In this method, the ligand is added to the target protein solution and the complex is formed prior to crystallization. However, due to differences in solubility and/or conformation of the complex, it is sometimes found that crystals do not form in the same conditions as the apo-protein crystals. In these cases, a full crystallization screening is necessary in order to find the appropriate crystallization conditions, including screening of the ligand-protein ratio that will lead to the formation of suitable crystals of the complex. Since homogeneity is critical for successful crystallization and formation of well-diffracting crystals, the presence of free protein may lead to difficulties in preparing crystals of the complex (heterogeneous sample). Therefore, the higher the ligand occupancy the better the chances of obtaining crystals of ligand-protein complexes.

2.3. Coexpression and/or Copurification of Protein and Ligand

In case of protein instability, it has been found that the addition of ligands may help to stabilize it and can be used for the optimization of protein expression and purification. The presence of ligands during protein expression often produces increased levels of protein expression thus leading to higher amounts of pure protein. In these cases, the protein-ligand complex, once purified, can be directly crystallized. Successful examples include the steroid nuclear receptors, where coexpression with the ligand was crucial for the structure determination of the complex binding domains of human progesterone receptor *(18)* and human androgen receptor *(19)*. Even when protein expression has been optimized, it is sometimes useful to add ligands throughout the purification steps, which can lead to stable protein–ligand preparations that can be subsequently crystallized (reviewed in *(3)*).

3. From Data Collection to the 3D Structure

3.1. X-Ray Diffraction and Experimental Data

3.1.1. X-Ray Diffraction

X-rays are scattered in all directions by the atoms present in a crystal, but only those waves that interfere constructively (according to Bragg's law (*see* **Note 4**)) give rise to a diffracted beam, registered as diffraction spots (reflections) on a detector. Each reflection results from the interference of the X-rays diffracted from all the atoms in the crystal at the same diffraction angle. Therefore, each reflection contains information from all atoms in the structure. How then does one extract information about the position of each individual atom from the experimental diffraction data?

The answer to this question involves a large number of calculations that allow reconstructing the image of the molecule from the diffraction pattern. In an iterative process (see flowchart in **Fig. 1**), a molecular picture of the protein molecules is determined as a contour map of the electron density $\rho(x, y, z)$ throughout the unit cell (*see* **Note 5**).

Each recorded reflection on the detector represents a scattered wave, characterized by its frequency (that of the X-rays), amplitude, and phase. Each wave can be mathematically described as a Fourier series by the so-called structure factor (F_{hkl}) equation. F_{hkl} has associated frequency, amplitude, and phase that can be formulated as a function of the electron density $\rho(x, y, z)$ of all atoms in the unit cell:

$$F_{hkl} = \int_{V} \rho(x, y, z) \, e^{[2\pi i(hx + ky + lz)]} \, dV$$

Since the Fourier transform is a reversible operation, the electron density can be described as a function of the structure factors:

$$\rho(x, y, z) = (1 / V) \sum_{h} \sum_{k} \sum_{l} |F_{hkl}| \, e^{2\pi i \alpha_{hkl}} \, e^{[-2\pi i(hx + ky + lz)]}.$$

where α_{hkl} is the phase angle of reflection *hkl*; $|F_{hkl}|$ is the structure factor amplitude; (x, y, z) are the fractional atomic coordinates in the unit cell; and V the volume of the unit cell.

In a diffraction experiment one determines intensities (proportional to $|F_{hkl}|^2$), but all the phase information α_{hkl} is lost. Therefore, the electron density distribution (and therefore the 3D structure) cannot be directly calculated, and phases must be determined indirectly. This is the phase problem in crystallography.

3.1.2. Cryoprotection

Once a crystal has been obtained, a number of decisions about data collection strategy have to be addressed. Due to the increasing availability for beam time, diffraction data are usually collected at a synchrotron source, using a CCD detector. In such high-intensity beams, crystal decay can occur due to radiation damage.

This increases crystal disorder, reducing the scattering intensity. In order to minimize the effects caused by radiation damage, Hope introduced a technique in which the mother liquor present in the crystal solvent channels is replaced by a cryoprotectant *(20)*. In these conditions, the crystal can be brought to very low temperatures (around 100 K), solidifying the cryo-solution in a state of amorphous glass, preserving the crystal's order and diffraction properties, and allowing for much easier crystal handling and storage. The formation of crystalline ice within the protein crystal must be avoided. Thus, the conditions for cryo-cooling the crystals have to be optimized prior to data collection. Usually a solution made up of the mother liquor with a small percentage increase in the precipitant concentration is supplemented with a viscous solution, such as PEG, ethylene glycol, sucrose, glycerol (solutions that replace the solvent in the crystal channels), or oils (which coat the crystal, avoiding dehydration) *(21)*. The crystal is then transferred into the cryoprotectant and then flash-cooled in liquid nitrogen, or directly in a stream of cold and dry nitrogen.

3.1.3. Data Collection and Processing

The optimal strategy of diffraction data collection varies considerably from crystal to crystal. Many factors influence data acquisition: the crystal unit cell parameters and symmetry, crystal orientation on the goniostat, mosaicity (*see* **Note 6**), susceptibility to radiation damage, resolution limits, as well as the type of experiment to be performed. Many programs are available to help delineate the best strategy for data collection, which analyze a few frames taken from the crystal and compute the best starting orientation and rotation range in order to maximize the data completeness for a given type of experiment. Current beamline/data-processing software includes strategy routines that are able to provide the optimal data collection parameters.

Processing of the diffraction images is then carried out. In this step, the diffraction pattern collected over a wide range of images/rotations is integrated into a list of indices (*hkl*) with their measured intensities and associated uncertainties. The programs start by autoindexing a few frames in order to predict the symmetry and unit cell parameters of the crystal, as well as the orientation of the crystal relative to the beam. Refinement of several parameters is then carried out, including crystal parameters (unit cell, orientation, and mosaic spread), detector parameters (position, orientation, distortion), and beam parameters (orientation and divergence). After this step, integration of all the images takes place. This involves prediction of the position of a given reflection on the images and estimation of its intensity (after subtraction of the X-ray background) and associated error estimate.

A number of physical factors can affect the intensities of the spots recorded on the detector. This leads to scaling differences between images. Data reduction takes advantage of the redundancy of symmetry-related reflections to put all observations on a common scale. The overall quality of the data can be assessed by the analysis of the agreement between equivalent reflections, highlighting eventual parts of the data that do not agree with the remainder. Data quality assessment is commonly achieved based on the internal consistency of the data, which can be measured as an R_{merge} factor (*see* **Note 7**) or as a correlation coefficient. Analysis of these factors against batch number helps detect outliers. Occasionally, individual observations are erroneous. These can be detected if the reflection has been measured many times, but not if it has been measured only once or twice. In these cases, probabilistic methods have to be used in order to identify outliers *(22)*.

3.2. From the Electron Density Map to a Structural Model

3.2.1. Solving the Phase Problem

There are several ways to recover the lost phase information. The most powerful technique today is the Multiwavelength Anomalous Dispersion (MAD) method. In this technique, X-rays of a particular wavelength are absorbed by the inner electrons of the heavy atoms in the crystal. The X-rays are re-emitted after a certain delay, inducing a phase shift in all of the reflections, known as the anomalous dispersion effect. Analysis of this perturbation, measured as very small differences between datasets collected at different wavelengths, allows the calculation of initial approximate phases. Other methods of experimental phase determination include Multiple Isomorphous Replacement (MIR), where heavy atoms are inserted into the structure (usually by synthesizing proteins with analogs or by soaking), and Single wavelength Anomalous Dispersion (SAD).

After confirming the initial location of the heavy atoms or anomalous scatterers (*see* **Note 8**), experimental phases can be calculated. Many programs are available to perform this task, most of them based on maximum likelihood methods (*see* **Note 9**). Some examples of these programs are: MLPHARE *(23)*, CNS *(24)*, SHARP/AutoSHARP *(25)*, SOLVE *(26)*. The latter two have the advantage of integrating heavy atom location, refinement, phasing, and, more recently, density modification and automated model building.

3.2.2. Density Improvement

The raw phases obtained are usually poor, making the following steps of structure solution extremely difficult. Experimental phases are the starting point for phase improvement by applying a variety of methods of density modification (**Fig. 3**). Solvent flattening, histogram matching, and noncrystallographic symmetry (NCS) averaging are the main techniques used to improve the phases (*see* **Note 10–12**).

Density modification is a cyclic procedure where the original phases are combined with the new phase estimates. This method will improve correct phases, but a bad map will not get better.

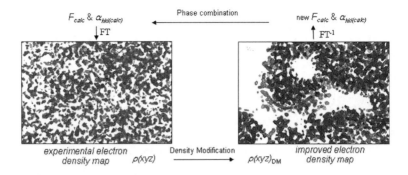

Fig. 3. Phase improvement by applying density modification methods (*see* **Notes 10–12**). The initial electron density map was of poor quality and no molecular boundaries were differentiated from the solvent. After density modification, the molecular contours are easily recognized and the electron density map is much improved. This is achieved in a cyclic procedure where density modification is followed by phase combination steps. In each cycle, phases are progressively improved, until convergence. The modified and improved electron density map allows starting model building.

3.2.3. Molecular Replacement

The method of molecular replacement (MR) is commonly used when a structure of a homologous protein is available. In this method, the homologous model is used to try to locate the protein of interest in two steps: first, the rotation of the molecule is assayed in order to have both molecules in the same orientation; second, a translation search is performed throughout the unit cell to try to find the translational orientation for the rotated model *(27)*. This method is particularly important for ligand-binding studies, as the nonligated molecular structure can be used as a search model for MR on a dataset of the protein with the ligand bound. Traditionally, this method required a large amount of user input. Recently, several programs were made available that significantly improved the probability of finding a correct solution (Amore *(28)*, EPMR *(29)*). Newer programs take advantage of better scoring functions, using maximum likelihood (Phaser *(30)*) or Bayesian methods (BALBES *(31)*) (*see* **Note 9**). This approach greatly improves the chances of finding a correct solution using the traditional search functions. These methods also perform a statistical treatment of simultaneous information from multiple search models using multivariate statistical analysis *(30)*. This not only allows the usage of information from various structures but also minimizes the introduction of bias.

3.2.4. Electron Density Map Interpretation

After the determined phases are as good as possible, an electron density map is calculated. If the initial phases are good, clear secondary structure features can be identified. Even if the interpretation is easy, model building is still a very laborious task. A three-dimensional electron density map is a rather daunting object to comprehend. Its quality is largely dependent on the quality of the initial phases, the quality of the measured data and

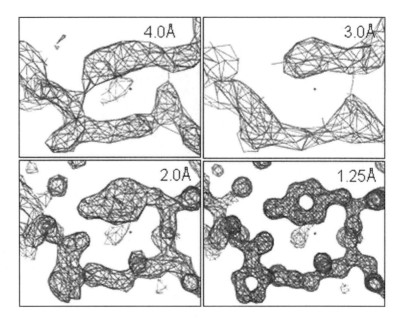

Fig. 4. Effect of the resolution on the diffraction data in $2F_{obs} - F_{calc}$ electron density maps calculated at increasing resolution. At 4.0Å resolution, the protein main chain is recognized; at 3.0Å resolution, the side chains are better resolved; at 2.0Å resolution, side chains, peptide bond planes, and carbonyls are well resolved; at 1.25Å resolution, individual atoms can be recognized and holes in the rings of Tyr and Pro are indicative of correct phases.

the resolution **(Fig. 4)**. The latter defines how much detail will be seen in the model and is highly dependent on the quality of the crystals (lack of order of the crystalline lattice, mosaicity (*see* **Note 6**), protein flexibility, heterogeneities, among others).

To get an overview of a density map, it needs to be simplified in some way. This is generally done through skeletonization. In skeletonization, a series of lines are drawn that run between the peaks of density, through regions of reasonably high density. This type of representation can help to clarify local features of the electron density and serves as a good scaffold for starting model building. After some landmarks in the density have been identified, the polypeptide chain can be fit by choosing plausible alpha-carbon positions from the bones. Alpha-carbons can be positioned at 3.8Å intervals, guided by the presence of the side-chain density branching off the main-chain. With most of the main-chain traced, one has to spot some sequence landmarks. Bulky residues such as tryptophans are very useful to identify local sequences. After fitting most of the main-chain and side-chains, refinement of the model can begin. When an atomic model has been built, it is likely that the phase angles calculated from it are more accurate than the experimental phases. These phases can either be used in conjunction with the measured amplitudes or combined with the experimental phases in order to calculate a new electron density map. Two types of maps can be calculated: the electron density difference map (usually referred to as $F_{obs} - F_{calc}$ or $F_o - F_c$ map) and the double difference map ($2F_{obs} - F_{calc}$ or $2F_o - F_c$ map). The $F_o - F_c$ map will have positive density where features exist which have not been adequately represented in the model and negative density where the model contains features not supported by the data. This map can be used to move the model from the negative density peaks into the positive ones. The use of phases calculated from the model inevitably creates electron density, which is biased by the model. In order to minimize this bias, the double difference $2F_o - F_c$ maps are used. These look like regular electron density maps of the protein, but with reduced bias from the model. Other common methods of reducing bias are the calculation of omit maps and sigmaA-weighted maps (*see* **Notes 13** and **14**).

3.2.5. Refinement

In general, the initial model of the protein contains errors and must be optimized to best fit the experimental data as well as the known chemical information. Moreover, the initial model is often incomplete. Thus, refinement must be carried out in order to generate improved phases, which will be used to calculate a more accurate electron density map, in an iterative way. For most protein structures there is a very poor observation-to-parameter ratio. For each atom, one refines its position, its temperature factor (a measure of how much an atom oscillates around the position specified by the model), and its occupancy (a measure of the fraction of molecules in which the atom is present throughout the crystal) (*see* **Notes 15** and **16**). At typical resolution ranges (~2.0–3.0Å), there will only be about one observation for each parameter. In order to increase the data-to-parameter ratio, additional information is incorporated by using restraints and constraints. Constraints are fixed conditions that cannot be changed (e.g., occupancy of atoms). Restraints allow variation within a certain limit. Examples of restraints include bond length, bond angle, and close contact restrictions. These are obtained from ideal values established from high-resolution structures of small molecules *(32–36)*.

Crystallographic observations are far from ideal (as well as the missing phase, the errors are fairly large) and it is easy to overfit the data by adding atoms that are not really present in the crystal structure, thus getting a misleading level of agreement with the observed data. Several methods have been developed in order to minimize the hurdles presented by over-fitting and model bias. Cross-validation in the form of an R_{free} value (*see* **Note 17**) is the most important method used to detect over-fitting *(37)*. Between 5 and 10% of the data are arbitrarily excluded from minimization of the R-value during refinement. After refinement, the R_{free} value is calculated like the R value but with the excluded reflections. The two values must not differ too much (ideally less than 5%).

Traditionally, data used to be fitted by a least-squares fit approach. The maximum likelihood method (*see* **Note 9**) is a more modern approach to the problem. It applies statistical assumptions and allows inclusion of more data and information, e.g., experimental phases. For macromolecules, maximum likelihood is more stable and leads to overall better results, often with reduced model bias. Maximum likelihood incorporates errors of the data and ensures the model is not built with higher accuracy than the data would permit, thus reducing over-fitting.

3.2.6. Water Picking

After building and refining the protein model, extra density can be interpreted. This is usually due to solvent molecules trapped in ordered positions around and in the protein interior. The higher the structure resolution, the better the water network can be defined. Water picking has become mostly automated, with programs such as Arp/wArp and COOT *(38)*. The user can define what sigma level of the electron density to be used as a threshold for water picking, define maximum and minimum distance for close contacts and maximum temperature factor value.

3.2.7. Validation

Validation of the macromolecular models is a crucial part of structure determination *(39)*. It is important both during structure refinement and at the final stages of data deposition in the Protein Data Bank (PDB) *(40)*. Structure refinement should be performed to convergence, i.e., all parameters are refined to their optimal value considering the quality of the data. The aim of validation is to estimate the errors associated with a given structure. Validation of a model is performed at three levels: quality of fitting of the data, agreement with known chemical parameters, and assessment of nonbonded contacts (solvent molecules and ligands). R/R_{free} factors are good indicators of how well the model fits the data (*see* **Note 17**), but other tools are available to assay this quality in more detail (e.g., average positional error, atomic B-factors). Assessment of stereochemical parameters complements the quality control of the model. This is independent of the experimental data. The most important tests available to evaluate the model of a protein are the Ramachandran plot, measure of the side-chain torsion angles, and analysis of close contacts. The solvent network must also be analyzed – water molecules should be forming hydrogen bonds to residues of the protein, maintaining a sensible distance to the surrounding atoms. All these tests can be performed by either standalone programs or Web servers (*see* **Note 1**), which can output highly detailed information that can help correct the model to its best final state.

4. Structure Determination and Interpretation of Protein–Ligand Complexes

Once the structure of a target protein is known, it may be used in the design of potential compounds that will alter its activity. Further crystallographic studies can demonstrate how compounds bind to their target sites and what changes are induced upon binding. After data collection and processing, structures are determined using the previously solved apo-structures, for which coordinates are available. Sensible starting coordinates of the protein structures are very important for model building and refinement.

If the presence of the ligand does not affect crystal packing and symmetry, one can directly calculate difference maps $(F_o - F_c)$ to find electron density for the ligand. In the case of nonisomorphism (altered crystallographic symmetry or protein packing), MR must be carried out (**Subheading 3.2.3**).

4.1. Ligand Identification and Fitting

4.1.1. The Binding Site

Usually, one does not expect to find significant differences between the native and the ligand-bound protein structures. Some differences may arise, particularly in solvent-exposed loop regions, but the most interesting regions are those where conformation changes occur only upon ligand binding. If a ligand enters the protein crystal and binds in an ordered way to the protein in most unit cells, it should be visible in a $F_o - F_c$ electron density map (**Subheading 3.2.4**). The whole difference electron density may not be due to the ligand, as the crystals may also contain any of the components present in purification, crystallization, or cryoprotection solutions.

The ligand of interest is expected to be found in the predicted binding site, but it is not uncommon to find it elsewhere, usually a cleft or a channel through the protein. The crystal structures of protein-ligand complexes usually show that the ligand binds in a preferred site, most commonly forming hydrogen bonds to surrounding amino acid residues. Nevertheless, ligands can undergo chemical changes or not bind with full occupancy, making interpretation of the electron density maps less straightforward. There is also the possibility that a mixture of ligands is used in the soaking or cocrystallization procedures (**Subheading 2.2**). Recently, methods to identify unknown electron density due to the presence of an unexpected ligand have been developed *(41)*. An unknown ligand found to bind to a macromolecule throughout purification and crystallization may turn out to play a crucial role in a drug discovery project.

4.1.2. Unknown Electron Density Interpretation

Two approaches have been used in order to interpret the electron density of an unknown ligand, both relying on ligand databases. One method tries to fit a model into the unknown density.

The second approach compares the unknown density with the calculated densities of ligands from the test set. Terwilliger and coworkers *(41)* have developed the ligand-fitting approach, applying it to many protein-ligand complexes found in the PDB using a large variety of ligands as diverse as the heme group, glycerol, ATP, cAMP, GTP, NAG. Each possible ligand is fitted to the density and ranked based on the correlation between calculated and observed density. Ligands with unique shapes usually generate the best results. In these cases, the highest score corresponds to the correct ligand identification and fitting (and is usually easily distinguishable from the next best scores). Unfortunately, this is not always true for densities where many ligands can fit into. The resolution and the quality of the electron density maps are crucial for the discrimination between candidate ligands. This approach is implemented in the PHENIX project *(42)*.

4.2. Crystallographic Model Building and Refinement

Generally, phasing and refinement of protein-ligand complex structures are straightforward if the quality of the diffraction data is reliable. The purpose of the refinement process is to produce a chemically realistic model of the protein–ligand complex. Structure solution is usually followed by rigid-body refinement of the best found solution and restrained refinement using maximum-likelihood statistics (*see* **Note 9**) using programs such as REF-MAC5 *(43)*, CNS *(24)*, or BUSTER-TNT *(44, 45)*. Ligand building can be performed using the same graphic programs used for model building, such as O *(46)* or COOT *(38)*. The latter incorporates routines that help to identify nonmodeled density and autofitting of the ligand.

During model building and refinement processes, sets of restraints have to be specified for the ligand entities, different from the restraints usually specified for the macromolecules. Due to the small size of the ligand molecules, calculation of ideal geometries is straightforward. However, factors like steric strain or unexpected environmental effects (such as pH) have to be considered. Ideal geometry should always be assumed unless obvious changes are visible in the ligand electron density. In most cases, restraints are calculated using the predicted ideal geometry.

4.2.1. Ligand Design

If the ligand is not present in the available databases (*see* **Note 1**), a starting model needs to be designed, with the corresponding set of restraints. In order to design a valid model for a novel ligand compound, knowledge of the compound's chemistry is essential and stereochemical restraints have to obey some general rules *(47, 48)*. Considering that ligands can be either very simple molecules or larger ones (such as oligosaccharides), designing a hetero-compound can imply specifying restraints not only for bond lengths or angles, planarity, and chirality, but also torsion angles. Ligands can even react with residues of the macromolecule and this is not predicted in the programs' libraries. Deviations from the target values are usually similar to those defined for macromolecules: 0.02Å for bond lengths and 2° for bond angles *(49)*.

The initial target values are usually obtained from a crystal structure of the compound, at high resolution, for which the coordinates are usually available from commercial databases such as the Cambridge Structural Database (CSD) or the Inorganic Crystal Structure Database (ICSD) (*see* Web links to the mentioned databases in **Note 1**). Recently, several Web servers, such as the Crystallography Open Database (COD), the NCI Open Database, and the Reciprocal Net, have been providing free access to the coordinates of a significant number of crystal structures of small molecules, although these are not as complete as the commercial databases. It is also possible to generate restraints from model structures.

Choosing a good set of restraints is crucial for the refinement process and is the best way to prevent errors in the ligand structures. Coordinates for structures of hetero-compounds that occur in the PDB are compiled and regularly updated in Web resources, such as the MSDChem, the HIC-Up, or the Ligand Depot *(50)* Web servers. These servers can generate dictionaries containing all the information needed by most refinement programs. In the cases where no structural information is available, theoretical models can be generated for the ligand of interest. Several programs, such as the PRODRG server *(51, 52)*, CORINA *(53)*, and AFITT *(54)* can be used to generate 3D coordinates from a 2D diagram representation of the compound of interest and calculate the subsequent restraint dictionary.

4.2.2. Ligand Fitting and Refinement

The ARP/wARP program *(55)* includes a module dedicated to ligand building, involving two main steps: identification of the ligand location and building of an atomic model *(56)*. The ligand location is easier for large ligands, while the building process requires high resolution and small to medium-size ligands. Other programs capable of building and positioning a ligand into the electron density, such as COOT or SOLVE/RESOLVE *(41)*, use different approaches and algorithms.

Strong restraints and constraints are advisable in the initial cycles of refinement. As refinement proceeds, these can be reduced to account for real differences between molecules. If NCS is present, constraints should be imposed. However, although the crystallization is usually performed with a large excess of ligand, there are cases where it is not present in all the molecules of the asymmetric unit (low occupancy) (*see* **Note 16**) (e.g., the Abl kinase-NVP-AFN941 complex reported in Levinson et al., 2006) *(57)*. The available programs are still limited in dealing with partial occupancy in ligands. The structure refinement process also involves refinement of individual B-factors (*see* **Note 15**).

Overall temperature factor and Wilson statistics (*see* **Note 18**) can indicate that a possible low-quality structure may originate from the diffraction data. This is usually reflected in high values for the refined individual B-factors, low occupancy of ligand atoms, and difficult convergence of *R*-factors. The electron density will resemble that of a low-resolution structure and, quite often, water molecules cannot be included. The electron density map may allow the identification of the ligand. However, if detailed protein–ligand interactions are to be inferred, then it is advisable to return to the data collection or even to the crystallization stages.

4.2.3. Validation

The same procedures of model validation applied to macromolecules apply to protein–ligand complexes. In these cases, one must not only check the model quality for the protein, but also for the ligand and its interactions with the protein. As for the protein, the ligand must be chemically correct and must contact the protein in a sensible way. Most of the programs used for protein validation can also analyze the quality of the ligand. The HETZE program *(58)* and ValLigURL *(49)* are two of the few validation programs that have been written specifically to analyze the geometrical parameters of hetero-compounds associated with PDB structures.

5. Notes

1. Useful Links

    BindingDB (Public Database of Protein–Ligand Binding Affinities): http://www.bindingdb.org/bind/index.jsp Public, Web-accessible database of measured binding affinities for biomolecules, genetically or chemically modified biomolecules, and synthetic compounds. The database currently contains data generated by ITC, enzyme inhibition, and receptor–ligand binding methods.

    Databases of macromolecular structures: Protein Data Bank (PDB) http://www.rcsb.org/pdb/

    Macromolecular Structure Database (MSD) http://www.ebi.ac.uk/pdbe/

    Databases of small molecules crystal structures:

    Cambridge Structural Database (CSD) http://www.ccdc.cam.ac.uk/products/csd (nonfree access)

    Inorganic Crystal Structure Database (ICSD) http://icsd.ill.edu/dif/icsd (nonfree access)

    Crystallography Open Database (COD) http://www.crystallography.net

    NCI Open Database http://cactus.nci.nih.gov/ncidb2/

Reciprocal Net http://www.reciprocalnet.org/ Databases of crystal structures of the small molecules found in the PDB and MSD: MSDChem http://www.ebi.ac.uk/msd-srv/msdchem/cgi-bin/cgi.pl HIC-Up http://xray.bmc.uu.se/hicup Ligand Depot http://ligand-depot.rutgers.edu/ Web resources to generate restraints from model structures: RESID http://www.ebi.ac.uk/RESID SWEET http://www.dkfz-heidelberg.de/spec/sweet2/doc/index.php Validation: EDS http://eds.bmc.uu.se/eds/ ValLigURL http://eds.bmc.uu.se/eds/valligurl.php

2. In ITC measurements, if for example $K_a = 10^5$ M ($K_d = 10\,\mu$M), the optimal protein concentration in the cell should be 100–1,000 μM, although for ligands with higher affinity constant the concentration can be much lower (*see* the Web link to the Binding Affinities Database in **Note 1**).

3. Crystal cross-linking with glutaraldehyde can be performed with 2–5 μl of 25% glutaraldehyde placed in a microbridge in a sealed reservoir. The glutaraldehyde vapor is then allowed to diffuse into the drop containing the crystals, for 30 min to 6 h, depending on the protein. Glutaraldehyde reacts with lysine residues. The number of lysines is one of the variables influencing cross-linking time. Note that amines will interfere with glutaraldehyde, thus ammonium sulfate or Tris buffer will have to be replaced before cross-linking is attempted.

4. *Bragg's law and resolution.* The diffraction phenomenon was described by Bragg as a reflection of the X-rays by imaginary planes in the crystal lattice *(59)*. Bragg's law states that diffraction only occurs when constructive interference of the scattered radiation occurs. For a planar interspacing *d* and an incident angle θ, constructive interference occurs when the path difference between the waves with wavelength λ is equal to an integral number *n*. The maximum θ angle corresponds to the minimum distance d_{min} in the crystal that can be resolved, and is called the *resolution of the diffraction pattern*: $d_{min} = \lambda/\sin\theta_{max}$ (*see* **Fig. 5**).

$$2d \sin\theta = n\lambda$$

5. *Unit cell.* Crystals are composed of ordered arrays of identical units distributed in the three-dimensional space. The smallest repeating unit is called *the unit cell* and is characterized by three cell axes *a*, *b*, and *c* and by three angles α, β, and γ. The entire crystal can be built by translations of the unit cell along the three axes (*see* **Fig. 6**).

6. *Mosaicity.* Intrinsic disorder of the crystal's unit cells. Low values indicate well-ordered crystals and, consequently, better diffraction.

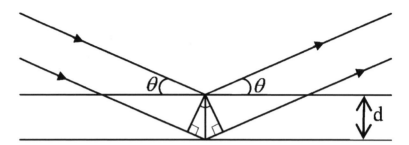

Fig. 5. Reflection of x-rays by two imaginary planes in a crystal lattice. d is the planar interspacing and θ is the incident angle.

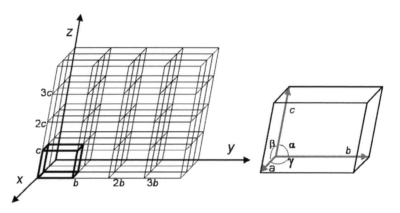

Fig. 6. The unit cell is characterized by three cell axes a, b, c and by three angles α, β and γ. Translations of the unit cell along the three axes build-up a crystal.

7. R_{merge}. A measure of agreement among multiple measurements of the same reflections. The different measurements belong to different frames or to different data sets. R_{merge} is calculated as follows (I_j is the jth intensity measurement of a reflection, and $<I>$ is the average intensity from multiple observations):

$$R_{merge} = \frac{\sum_A \sum_j \left| I_j - \langle I \rangle \right|}{\sum_A \sum_j I_j}$$

8. The *location of heavy atoms or anomalous scatterers* in MIR, MAD or SAD was traditionally performed by manual inspection of Patterson maps. The Patterson function (or Patterson map) is the Fourier transform of the intensities (squared amplitudes), and generates a map of the vectors between atoms. These maps are relatively simple to interpret for a small number of heavy atoms, but as this number increases, the map interpretation becomes extremely complicated. As a result, automated heavy-atom locating methods have been deve-loped, such as Solve *(26)* and CNS *(24)*,

which use Patterson-based searches. Other programs, such as Shelx-D *(60)* and Shake-and-Bake (SnB *(61)*) use the direct method approach. SnB uses a set of randomly positioned heavy atoms to refine phases, while Shelx-D derives starting phases by direct inspection of the Patterson map.

9. *Maximum likelihood.* Refinement criterion that evaluates the probability that the observations (the experimental data) will occur, given a certain model. The model fitting has to be performed so that the probability of the observed data is maximized. The likelihood (L) of a given model, in the presence of a set of observations, is the product of the probabilities (P) of all observations (F_{obs}) given the model (F_{calc}), i.e.,

$$L = \prod_{hkl} P\left(F_{obs}; F_{calc}\right)_{hkl} \quad \text{or}$$

$$\log L = \sum_{hkl} \log P\left(F_{obs}; F_{calc}\right)_{hkl}$$

(62). This function will have its maximum when F_{obs} and F_{calc} are equal.

10. *Solvent flattening* removes the negative electron density and sets the value of the electron density of solvent regions to a fixed value. Automatic methods are used to define a protein–solvent boundary *(63, 64)*. This method relies on the fact that protein crystals typically contain 30–70% solvent, forming channels through the crystal lattice.

11. *Histogram matching.* A density histogram is a probability distribution of values of the electron density sampled at regular intervals (grid points) throughout the three-dimensional map. Density histograms of protein structures have a characteristic form. The histogram matching method calculates the density histogram from the initial set of phases and modifies it so that it takes the form of an expected density histogram.

12. *NCS averaging.* When two or more copies of the same molecule are present in the asymmetric unit (*see* **Note 19**), NCS averaging can be used. This method averages the density of equivalent positions imposing the same value for each symmetrical molecule.

13. *Omit maps* are electron density maps that are usually calculated to elucidate specific zones of a protein structure. The calculated structure factors correspond to a model where certain atoms have been removed, so that possible local errors of the model do not contribute to the phases and, therefore, do not influence the resulting map *(65)*.

14. *SigmaA-weighted maps.* The use of model phases will intro-
    duce bias for the map to resemble the model more than it
    should. The use of a weighting scheme based on phase proba-
    bility profiles (calculated from the phases and figures-of-merit
    of a partial model) can produce a less model-biased electron
    density map *(66)*. This type of map can also be calculated
    using phases from several different sources and models.

15. *Temperature factor* or *B-factor.* Thermal vibration parameter
    that describes the degree of dynamic oscillation of an atom
    around a defined position. The thermal vibration parameter
    B_j for atom j is a local measure of the relative mobility of
    different regions of the molecule: $B_j = 8\{\Pi U_j^2\} = 79\{U_j^2\}$,
    U_j: mean deviation of atom j in relation to its average position.
    For example: $B_j \approx 80\text{Å}^2$ $U_j \approx 1\text{Å}$; $B_j \approx 20\text{Å}^2$ $U_j \approx 0.5\text{Å}$.
    The *B-factor* can also indicate errors in the model-building
    process. Wrongly placed atoms will exhibit higher B-factors,
    when compared to neighboring atoms.

16. *Occupancy* of an atom. Parameter that defines the fraction
    of asymmetric units where the atom is actually present in
    its mean position. It ranges from 0.0 to 1.0 and intermedi-
    ate values indicate that it does not occupy its position in all
    asymmetric units. It can be used to define alternate confor-
    mations of amino acid side chains.

17. *R* and *free R factor* (R_{free}). Validation criteria used as a meas-
    ure of the quality of the correctly oriented and positioned
    model(s). The difference between the experimental obser-
    vations and the refined model should be minimized during
    model building and refinement processes, according to the
    expression:

$$R = \frac{\left[\sum_{hkl} w_{hkl}\left(\left|F_{obs}\right|_{hkl} - k\left|F_{calc}\right|_{hkl}\right)\right]}{\left(\sum_{hkl}\left|F_{obs}\right|_{hkl}\right)},$$

where $\left|F_{obs}\right|_{hkl}$ is the measured amplitude for an individual
reflection *hkl*, $\left|F_{calc}\right|_{hkl}$ is the amplitude calculated from the
model for the same reflection, *k* is a factor that scales the calculated amplitudes to the observed data (independent of *h*,
k, and *l*). During refinement of a protein model, over-fitting
can be measured by using cross-validation in the form of a
free *R*-factor. A fraction of the data (5–10%) is excluded from
the refinement process, and used to calculate the model's
R-factor. A comparison of the two parameters can indicate
problems of model over-fitting. With high-quality data, *R*
and R_{free} should not differ by more than 5%.

18. *Wilson plot.* Plot of the observed intensities against resolution, used to determine an absolute scale and temperature factor.

19. *The asymmetric unit* is the smallest aggregate of molecules that can generate the whole unit cell by applying symmetry operations. In the case of multimeric proteins, the asymmetric unit can be composed of one or more molecules. In some cases the asymmetric unit can correspond to only a part of the functional unit (for example one $\alpha\beta$ dimer of hemoglobin). When the asymmetric unit contains more than one identical molecule, these can be related to each other by local symmetry operations – NCS.

Acknowledgments

The authors acknowledge Dr. Shabir Najmudin and Dr. Abhik Mukhopadhyay for critical reading of the manuscript.

References

1. Dror, O., Lasker, K., Nussinov, R., and Wolfson, H. (2007). EMatch: an efficient method for aligning atomic resolution subunits into intermediate-resolution cryo-EM maps of large macromolecular assemblies. *Acta Crystallogr. D Biol. Crystallogr.* **63**, 42–49

2. Sigworth, F. J. (2007). From cryo-EM, multiple protein structures in one shot. *Nat. Methods* **4**, 20–21

3. Hassell, A. M., An, G., Bledsoe, R. K., Bynum, J. M., Carter, H. L., III, Deng, S. J., Gampe, R. T., Grisard, T. E., Madauss, K. P., Nolte, R. T., Rocque, W. J., Wang, L., Weaver, K. L., Williams, S. P., Wisely, G. B., Xu, R., and Shewchuk, L. M. (2007). Crystallization of protein-ligand complexes. *Acta Crystallogr. D Biol. Crystallogr.* **63**, 72–79

4. Chung, C. W. (2007). The use of biophysical methods increases success in obtaining liganded crystal structures. *Acta Crystallogr. D Biol. Crystallogr.* **63**, 62–71

5. Niesen, F. H., Berglund, H., and Vedadi, M. (2007). The use of differential scanning fluorimetry to detect ligand interactions that promote protein stability. *Nat. Protoc.* **2**, 2212–2221

6. Senisterra, G.A. (2006). Screening for ligands using a generic and high-throughput light-scattering assay. *J. Biomol. Screen.* **11**, 940–948

7. Vedadi, M., Niesen, F. H., Allali-Hassani, A., Fedorov, O. Y., Finerty, P. J., Wasney, G. A., Yeung, R., Arrowsmith, C., Ball, L. J., Berglund, H., Hui, R., Marsden, B. D., Nordlund, P., Sundstrom, M., Weigelt, J., and Edwards, A. M. (2006). Chemical screening methods to identify ligands that promote protein stability, protein crystallization, and structure determination. *Proc. Natl. Acad. Sci. U.S.A.* **103**, 15835–15840

8. Brown, W., Ed. (1993). *Dynamic Light Scattering: The Method and Some Applications*, Vol. 49. Monographs on the Physics and Chemistry of Materials. Oxford University Press, USA

9. Schmitz, K. S. (1990). *An Introduction to Dynamic Light Scattering by Macromolecules.* Academic Press, Boston

10. Feng, B. Y., Shelat, A., Doman, T. N., Guy, R. K., and Shoichet, B. K. (2005). High-throughput assays for promiscuous inhibitors. *Nat. Chem. Biol.* **1**, 146–148

11. McNae, I. W., Kan, D., Kontopidis, G., Patterson, A., Taylor, P., Worrall, L., and Walkinshaw, M. D. (2005). Studying protein-ligand interactions using protein crystallography. *Crystallogr. Rev.* **11**, 61–71

12. Matthews, B. W. (1968). Solvent content of protein crystals. *J. Mol. Biol.* **33**, 491–497

13. Vilenchik, L. Z., Griffith, J. P., St Clair, N., Navia, M. A., and Margolin, A. L. (1998). Protein crystals as novel microporous materials. *J. Am. Chem. Soc.* **120**, 4290–4294

14. Lusty, C. J. (1999). A gentle vapor-diffusion technique for cross-linking of protein crystals for cryocrystallography. *J. Appl. Crystallogr.* **32**, 106–112

15. Danley, D. E. (2006). Crystallization to obtain protein-ligand complexes for structure-aided drug design. *Acta Crystallogr. D Biol. Crystallogr.* **62**, 569–575

16. Skarzynski, T. and Thorpe, J. (2006). Industrial perspective on X-ray data collection and analysis. *Acta Crystallogr. D Biol. Crystallogr.* **62**, 102–107

17. Hiller, N., Fritz-Wolf, K., Deponte, M., Wende, W., Zimmermann, H., and Becker, K. (2006). *Plasmodium falciparum* glutathione S-transferase – structural and mechanistic studies on ligand binding and enzyme inhibition. *Protein Sci.* **15**, 281–289

18. Williams, S. P. and Sigler, P. B. (1998). Atomic structure of progesterone complexed with its receptor. *Nature* **393**, 392–396

19. Matias, P. M., Donner, P., Coelho, R., Thomaz, M., Peixoto, C., Macedo, S., Otto, N., Joschko, S., Scholz, P., Wegg, A., Basler, S., Schafer, M., Egner, U., and Carrondo, M. A. (2000). Structural evidence for ligand specificity in the binding domain of the human androgen receptor. Implications for pathogenic gene mutations. *J. Biol. Chem.* **275**, 26164–26171

20. Hope, H. (1988). Cryocrystallography of biological macromolecules: a generally applicable method. *Acta Crystallogr B* **44(Pt 1)**, 22–26

21. Garman, E. F. and Schneider, T. R. (1997). Macromolecular cryocrystallography. *J. Appl. Crystallogr.* **30**, 211–237

22. Read, R. J. (1999). Detecting outliers in non-redundant diffraction data. *Acta Crystallogr. D Biol. Crystallogr.* **55**, 1759–1764

23. Otwinowsky, Z. (1991). *CCP4 Daresbury Study Weekend Proceedings*

24. Brunger, A. T., Adams, P. D., Clore, G. M., DeLano, W. L., Gros, P., Grosse-Kunstleve, R. W., Jiang, J. S., Kuszewski, J., Nilges, M., Pannu, N. S., Read, R. J., Rice, L. M., Simonson, T., and Warren, G. L. (1998). Crystallography & NMR system: a new software suite for macromolecular structure determination. *Acta Crystallogr. D Biol. Crystallogr.* **54**, 905–921

25. delaFortelle, E. and Bricogne, G. (1997). Maximum-likelihood heavy-atom parameter refinement for multiple isomorphous replacement and multiwavelength anomalous diffraction methods. *Meth. Enzymol.* **276**, 472–494

26. Terwilliger, T. C. and Berendzen, J. (1999). Automated MAD and MIR structure solution. *Acta Crystallogr. D Biol. Crystallogr.* **55**, 849–861

27. Tong, L. and Rossmann, M. G. (1990). The locked rotation function. *Acta Crystallogr. A* **46**, 783–792

28. Navaza, J. (1994). Amore – an automated package for molecular replacement. *Acta Crystallogr. A* **50**, 157–163

29. Kissinger, C. R., Gehlhaar, D. K., and Fogel, D. B. (1999). Rapid automated molecular replacement by evolutionary search. *Acta Crystallogr. D Biol. Crystallogr.* **55**, 484–491

30. Read, R. J. (2001). Pushing the boundaries of molecular replacement with maximum likelihood. *Acta Crystallogr. D Biol. Crystallogr.* **57**, 1373–1382

31. Long, F., Vagin, A. A., Young, P., and Murshudov, G. N. (2007). BALBES: a molecular-replacement pipeline. *Acta Crystallogr. D Biol. Crystallogr.* **64**, 125–132

32. Engh, R. A. and Huber, R. (1991). Accurate bond and angle parameters for X-ray protein-structure refinement. *Acta Crystallogr. A* **47**, 392–400

33. Engh, R. A. and Huber, R. (2001). Structure quality and target parameters. In: Rossmann, M. G., and Arnold, E. (eds.) *International Tables for Crystallography*, Vol. F. Kluwer, Dordrecht, The Netherlands, pp. 382–392

34. Hendrickson, W. A. (1985). Stereochemically restrained refinement of macromolecular structures. *Methods Enzymol.* **115**, 252–270

35. Parkinson, G., Vojtechovsky, J., Clowney, L., Brunger, A. T., and Berman, H. M. (1996). New parameters for the refinement of nucleic acid-containing structures. *Acta Crystallogr. D Biol. Crystallogr.* **52**, 57–64

36. Priestle, J. P. (2003). Improved dihedral-angle restraints for protein structure refinement. *J. Appl. Crystallogr.* **36**, 34–42

37. Brunger, A. T. (1992). *X-Plor Version 3.1: A System for X-ray Crystallography and NMR*, Yale University Press, New Haven

38. Emsley, P., and Cowtan, K. (2004). Coot: model-building tools for molecular graphics. *Acta Crystallogr. D Biol. Crystallogr.* **60**, 2126–2132

39. Kleywegt, G. J. (2000). Validation of protein crystal structures. *Acta Crystallogr. D Biol. Crystallogr.* **56**, 249–265

40. Berman, H. M., Westbrook, J., Feng, Z., Gilliland, G., Bhat, T. N., Weissig, H., Shindyalov, I. N., and Bourne, P. E. (2000).

The Protein Data Bank. *Nucleic Acids Res.* **28**, 235–242

41. Terwilliger, T. C., Klei, H., Adams, P. D., Moriarty, N. W., and Cohn, J. D. (2006). Automated ligand fitting by core-fragment fitting and extension into density. *Acta Crystallogr. D Biol. Crystallogr.* **62**, 915–922

42. Adams, P. D., Grosse-Kunstleve, R. W., Hung, L. W., Ioerger, T. R., McCoy, A. J., Moriarty, N. W., Read, R. J., Sacchettini, J. C., Sauter, N. K., and Terwilliger, T. C. (2002). PHENIX: building new software for automated crystallographic structure determination. *Acta Crystallogr. D Biol. Crystallogr.* **58**, 1948–1954

43. Murshudov, G. N., Vagin, A. A., and Dodson, E. J. (1997). Refinement of macromolecular structures by the maximum-likelihood method. *Acta Crystallogr. D Biol. Crystallogr.* **53**, 240–255

44. Blanc, E., Roversi, P., Vonrhein, C., Flensburg, C., Lea, S. M., and Bricogne, G. (2004). Refinement of severely incomplete structures with maximum likelihood in BUSTER-TNT. *Acta Crystallogr. D Biol. Crystallogr.* **60**, 2210–2221

45. Roversi, P., Blanc, E., Vonrhein, C., Evans, G., and Bricogne, G. (2000). Modelling prior distributions of atoms for macromolecular refinement and completion. *Acta Crystallogr. D Biol. Crystallogr.* **56**, 1316–1323

46. Jones, T. A., Zou, J. Y., Cowan, S. W., and Kjeldgaard, M. (1991). Improved methods for building protein models in electron-density maps and the location of errors in these models. *Acta Crystallogr. A* **47**, 110–119

47. Evans, P. R. (2007). An introduction to stereochemical restraints. *Acta Crystallogr. D Biol. Crystallogr.* **63**, 58–61

48. Kleywegt, G. J., Henrick, K., Dodson, E. J., and van Aalten, D. M. F. (2003). Pound-wise but penny-foolish: how well do micromolecules fare in macromolecular refinement? *Structure* **11**, 1051–1059

49. Kleywegt, G. J. (2007). Crystallographic refinement of ligand complexes. *Acta Crystallogr. D Biol. Crystallogr.* **63**, 94–100

50. Feng, Z. K., Chen, L., Maddula, H., Akcan, O., Oughtred, R., Berman, H. M., and Westbrook, J. (2004). Ligand Depot: a data warehouse for ligands bound to macromolecules. *Bioinformatics* **20**, 2153–2155

51. Schuttelkopf, A. W. and van Aalten, D. M. F. (2004). PRODRG: a tool for high-throughput crystallography of protein-ligand complexes. *Acta Crystallogr. D Biol. Crystallogr.* **60**, 1355–1363

52. vanAalten, D. M. F., Bywater, R., Findlay, J. B. C., Hendlich, M., Hooft, R. W. W., and Vriend, G. (1996). PRODRG, a program for generating molecular topologies and unique molecular descriptors from coordinates of small molecules. *J. Comput. Aided Mol. Des.* **10**, 255–262

53. Gasteiger, J., Rudolph, C., and Sadowski, J. (1992). Automatic Generation of 3D-Atomic Coordinates for Organic Molecules. *Tetrahedron Comput. Method.* **3**, 537–547

54. Peat, T. S., Christopher, J. A., and Newman, J. (2005). Tapping the protein data bank for crystallization information. *Acta Crystallogr. D Biol. Crystallogr.* **61**, 1662–1669

55. Perrakis, A., Morris, R., and Lamzin, V. S. (1999). Automated protein model building combined with iterative structure refinement. *Nat. Struct. Biol.* **6**, 458–463

56. Evrard, G. X., Langer, G. G., Perrakis, A., and Lamzin, V. S. (2007). Assessment of automatic ligand building in ARP/wARP. *Acta Crystallogr. D Biol. Crystallogr.* **63**, 108–117

57. Levinson, N. M., Kuchment, O., Shen, K., Young, M. A., Koldobskiy, M., Karplus, M., Cole, P. A., and Kuriyan, J. (2006). A Src-like inactive conformation in the Abl tyrosine kinase domain. *PLoS. Biol.* **4**, 753–767

58. Kleywegt, G. J. and Jones, T. A. (1998). Databases in protein crystallography. *Acta Crystallogr. D Biol. Crystallogr.* **54**, 1119–1131

59. Bragg, W. H. and Bragg, W. L. (1913). The reflection of X-rays by crystals. *Proc. R. soc. Lond. Ser. A-Contain. Pap. Math. Phys. Character* **88**, 428–428

60. Sheldrick, G. M. (1995). Structure solution by iterative peaklist optimization and tangent expansion in-space group P1. *Acta Crystallogr. B* **51**, 423–431

61. Weeks, C. M. and Miller, R. (1999). Optimizing Shake-and-Bake for proteins. *Acta Crystallogr. D Biol. Crystallogr.* **55**, 492–500

62. Ten Eyck, L. F. and Watenpaugh, K. D. (2006). Introduction to refinement. In: Rossmann, M. G., and Arnold, E. (eds.) *Crystallography of Biological Macromolecules*, Vol. F. Kluwer, Dordrecht, The Netherlands, pp. 369–374

63. Wang, B. C. (1985). Resolution of phase ambiguity in macromolecular crystallography. *Meth. Enzymol.* **115**, 90–112

64. Leslie, A. G. W. (1987). A reciprocal-space method for calculating a molecular envelope using the algorithm of Wang, B.C. *Acta Crystallogr. A* **43**, 134–136

65. Hodel, A., Kim, S. H., and Brunger, A. T. (1992). Model bias in macromolecular crystal-structures. *Acta Crystallogr. A* **48**, 851–858

66. Read, R. J. (1986). Improved Fourier coefficients for maps using phases from partial structures with error. *Acta Crystallogr. A* **42**, 140–149

67. Potterton, E., McNicholas, S., Krissinel, E., Cowtan, K., and Noble, M. (2002). The CCP4 molecular-graphics project. *Acta Crystallogr. D Biol. Crystallogr.* **58**, 1955–1957

Chapter 4

Virtual Screening of Compound Libraries

Nuno M.F.S.A. Cerqueira, Sérgio F. Sousa, Pedro A. Fernandes, and Maria João Ramos

Summary

During the last decade, Virtual Screening (VS) has definitively established itself as an important part of the drug discovery and development process. VS involves the selection of likely drug candidates from large libraries of chemical structures by using computational methodologies, but the generic definition of VS encompasses many different methodologies. This chapter provides an introduction to the field by reviewing a variety of important aspects, including the different types of virtual screening methods, and the several steps required for a successful virtual screening campaign within a state-of-the-art approach, from target selection to postfilter application. This analysis is further complemented with a small collection important VS success stories.

Key words: Drug design, Docking, Scoring, Filters, Compound libraries, Druggability, Protein flexibility, Computational chemistry.

1. Introduction

Virtual Screening (VS) methodologies involve the use of computational techniques capable of filtering chemical databases, the design/optimization of targeted combinatorial libraries, and the rapid assessment of large libraries of chemical structures, as a tool to guide the selection of likely drug candidates *(1)*. The main goal of virtual screening is to obtain hits of novel chemical structures that yield a unique pharmacological profile. Thus, the success of a virtual screening is defined in terms of finding interesting new scaffolds rather than many hits *(2)*. Because the cost of performing screens in silico is significantly less expensive than high-throughput screening (HTS) or combinatorial chemistry methods, virtual

Ana Cecília A. Roque (ed.), *Ligand-Macromolecular Interactions in Drug Discovery: Methods and Protocols,*
Methods in Molecular Biology, vol. 572,
DOI 10.1007/978-1-60761-244-5_4, © Humana Press, a part of Springer Science+Business Media, LLC 2010

screening can play an important role in drug design, limiting the number of compounds to be evaluated by HTS or combinatorial chemistry methods to a subset of molecules that are more likely to "hit" when screened.

2. General Types of Virtual Screening

The generic definition of virtual screening encompasses many different methodologies. Despite the difficulty in unambiguously grouping the variety of existing approaches, two main classes of virtual screening methods can be defined: ligand-based methods and receptor-based methods. These two general approaches and the principles associated are outlined below.

2.1. Ligand-Based Virtual Screening

Ligand-based methods aim to identify molecules sharing common features, both at the chemical and physical levels, with known binding compounds, grounded in the assumption that similar compounds can have similar effects (3). These methods normally discard all information related to the target and focus exclusively on the ligand (4). Within a lock-and-key paradigm, these approaches compare different keys, and neglect the lock. They are normally employed only as a stand-alone approach when no high-resolution structure or model of the target-binding pocket is available, but its concerted or parallel application, as a complement to receptor-based virtual screening, is gaining popularity (5). A strict requirement for ligand-based approaches is of course, the existence of one or more compounds already known to bind to the biological target, such as a patented structure or a compound previously described in the literature. In ligand-based protocols these active compounds are used as templates for the screening process, and are therefore of paramount importance.

The application of ligand-based methods traditionally relies on the use of computational descriptors, accounting for a variety of aspects such as the molecular structure, the physicochemical properties, and the pharmacophore features, to evaluate the relationship between a given active compound (or a set of active compounds) and a database of compounds. The representation of molecules as sets of computational descriptors is a cumbersome issue, as several thousands of different molecular descriptors, varying in complexity and sophistication, have been described in the literature (6–8). Furthermore, the choice of the number, type, and relative weight of the descriptors to use depends on the specific search problem and compound classes under consideration, with different combinations rendering either comparable or very different results (6).

One of the main limitations of traditional ligand-based virtual screening strategies is the limited diversity of the hits normally encountered. In fact, the underlying philosophy of ligand-based methods tends to bias the hits found towards the properties of the known ligands. Hence, novelty tends to be penalized, with the screening protocol ensuring only a very crude sampling of the chemical space. Another difficulty involving the application of ligand-based methods, concerns the notion of biological similarity between two pairs of compounds. Compounds that are structurally very unlike can behave alike when bound to the receptor. Grasping this notion of biological similarity, instead of the simple notion of molecular similarity, is one of the difficulties of current ligand-based approaches. Despite these potential pitfalls and the proprietary nature of the great majority of drug discovery projects, several successful applications *(6, 9–11)* have been reported in the literature, demonstrating the usefulness of ligand-based approaches.

2.2. Receptor-Based Virtual Screening

Receptor-based methods, also called structure-based methods, target the interaction between a ligand and the receptor, trying to single out ligands that bind strongly to the target protein from ligands that do not. The main requirement for a receptor-based virtual screening campaign is, of course, the existence of a 3D structure of the target. This can be a crystallographic X-ray structure, an NMR structure, or even a homology modelling structure. Receptor-based approaches are gaining considerable importance over ligand-based techniques, particularly as more and more 3D structures of target proteins are being determined and becoming available. Two fundamental types of receptor-based methods exist: protein–ligand docking and active-site-derived pharmacophore methods.

Protein–ligand docking methods basically take small-molecule structures from a database of existing compounds and explicitly dock them into the target protein. They comprise a particularly promising approach to find structural novel ligands, as the hits found will depend on the quality of the docked models and on the docked protocol rather than on a subjective opinion of the characteristics that the ligand is expected to have *(12)*. Protein–ligand docking methods involve a docking algorithm (which positions the ligand and searches for the ligand-binding pose) and a scoring function (which evaluates the ligand-binding ability). A wide range of docking algorithms, of different levels of sophistification, depending essentially on the extension at which the flexibility is included, has been defined in the literature *(13)*. The main problem in docking is that the scoring functions normally employed in ranking and evaluating ligand binding in virtual screening protocols are unable to fully capture all possible events that contribute to binding (or their relative weight).

Active-site-derived pharmacophore methods are based on a pharmacophoric model derived from the 3D structure of the receptor that tries to encompass all main structural features that could be of relevance for ligand binding, including all possible and essential points of interaction between the receptor and the ligand, and aspects such as the shape and size of the binding pocket, capturing the underlying ligand–receptor interaction pattern and general active-site topology. Active-site directed pharmacophore methods try to fit the active conformations of the molecules to the pharmacophore model of the receptor, discarding all the information of the target protein that is not directly related to the binding pocket *(14)*. The main difficulty of pharmacophore-based approaches lies in the way the conformational flexibility is handled.

Protein–ligand docking is the most expensive approach of all the virtual screening methods, but is also the most detailed, least biased, and most informative approach to assess which ligands might be good candidates for inhibition, ensuring a finer-grained compound sieving process. It overcomes many of the simplifications considered in pharmacophores. The progress in hardware that has characterized the last decades has rendered the application of these methods to larger and larger compound libraries not only a reality but even a standard procedure, making protein–ligand docking an extremely powerful and computational competitive approach to the virtual screening problem, and the number one choice. Alternative techniques are nevertheless still much employed, but normally within an integrated virtual screening strategy that makes of protein-ligand docking its main tool *(12, 15)*.

3. Steps for a Virtual Screening Campaign

In this section, the typical stages normally followed in a successful virtual screening campaign within a state-of-the-art approach are summarized. These include the definition of the target, the preparation of the compound library, the choice of appropriate filters and of the docking protocol, and the application of postfilters to limit the number of results from a virtual screening campaign, as depicted in **Fig. 1**.

3.1. Target

A successful structure-based virtual screening campaign starts necessarily with an accurate description of the target. The aspects related to the selection of the target, its druggability, structure, and flexibility are described in this section, together with the special cares required to ensure a proper description of all features of relevance for target–ligand binding.

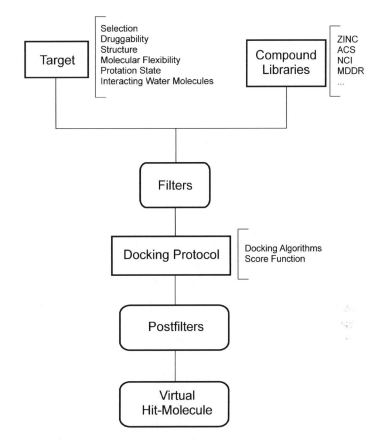

Fig. 1. Typical virtual screening scheme.

3.1.1. Target Selection

Target selection is at the basis of a virtual screening campaign and is the most fundamental step in the process of drug development. However, relating a given biological system, such as an enzyme, an intracellular receptor, or a cell surface receptor, to a specific disease aetiology is a thorny task. In fact, identifying the protein target with which a given molecule interacts is a traditional bottleneck in drug discovery efforts. Understanding its mechanisms of action, structure, and properties is an important prerequisite for the development of new drugs, with improved activity and specificity.

Traditionally, this process involved finding a chemical that had a favourable influence on a disease, using it as a tool to find out its cellular target, and then figuring by which means it could produce the observed effect. Recent strategies try to break down the disease process into the cellular and molecular levels using a variety of experimentally based techniques, together with the more recent genomics, proteomics, and bioinformatics approaches. Common strategies include the application of radioligand binding, DNA microarray, expressed sequence tags, and in silico methods. The function of the target, the disease pathway that is regulated by the target, and the importance of that specific disease pathway are carefully thought over. Only after the therapeutic target has been identified and validated the search for the appropriate compound can effectively start.

3.1.2. Druggability

An important question to take into consideration before embarking in a virtual screening campaign is the druggability of the chosen target. The selected target must be amenable to small-molecule intervention, i.e. druggable. Features such as active-site pocket size and geometry, surface complexity and roughness, shape and polarity need to be evaluated to predict whether a particular protein can bind with high affinity and specificity to small, drug-like compounds. Several indices to assist a differentiation between druggable and non-druggable binding pockets have been suggested by Hajduk et al. *(16)*. However, the available protocols to predict druggability are still rudimental and more developments are required.

3.1.3. Structure of the Target

The choice of the 3D structure of the target is also a key aspect to take into consideration before embarking in a virtual screening project. The choice normally relies on an X-ray crystallographic structure or an NMR structure, but successful virtual screening studies on homology models have also been reported in the literature *(17–19)*. The resolution of the diffraction data is a particularly important aspect to take into consideration when choosing a crystallographic structure. Other technical aspects to take into consideration include the population of the bound-ligand, B-factors next to the binding site, and the consistency of the hydrogen bond network *(15)*. NMR structures are normally limited to targets with a molecular mass of less than 20 kDa *(20)*, with the accuracy of the structure depending on the local distribution of the Overhauser effect distance information along the protein chain *(21)*. The use of a homology model is a riskier approach, usually only pursued when no X-ray of NMR structure of the target is available, and whose success strongly depends on the availability of quality 3D structures for related protein targets.

3.1.4. Molecular Flexibility

Protein targets have themselves internal degrees of freedom and can normally adopt several different conformational states. Properly dealing with molecular flexibility of the target significantly complicates the virtual screening problem. Standard docking programs can only consider the flexibility of the receptor up to a limited degree. In addition, the computational cost associated is normally very high, limiting its application to a relatively small number of compounds. Hence, it is not surprising that a significant parcel of the structure-based virtual screening studies has chosen to neglect molecular flexibility of the target. Recent approaches, however, try to account for this effect, at least up to some extent. Strategies to include the flexibility of the target in virtual screening are discussed in **Subheading 3.4**, within the context of docking algorithms.

3.1.5. Protonation States of the Active-Site Residues

Assigning the correct protonation state of the active-site residues can be a tricky task. In fact, the local electrostatic conditions can modulate by several orders of magnitude the K_a values of a given functional group. Polar hydrogens are of particular importance for hydrogen bonds, which can be of great significance in ligand-receptor interaction. Correctly assigning the protonation state of the residues surrounding the binding pocket is hence of extreme importance for a successful virtual screening campaign. Care should be taken in this step to ensure an accurate representation of the active-site characteristics, including visual inspection of the surrounding, and analysis of available experimental data on the ionisable groups.

3.1.6. Inclusion of Water Molecules in the Binding Site

The analysis of several thousands of crystallographic structures of ligand–protein complexes has revealed that in something like two-thirds of all cases, a water molecule is involved in ligand binding *(22)*. Hence, the classical approach of stripping the binding pocket of all water molecules can affect the ability to predict the binding of ligands whose coordination mode is stabilized by the presence of an explicit water molecule. Interstitial water molecules that might play an important role in ligand binding to the protein target should therefore be included as an integral part of the target structure. There is no perfect strategy to determine which active-site waters are important for ligand binding and which are not. The analysis of several X-ray or NMR structures for the presence of conserved water molecules at a specific position of the active-site can normally be taken as an indication.

3.2. The Compound Library

For a successful virtual screening campaign, it is normally desirable to have a database with the maximum degree of structural diversity, in order to maximize the chances of finding a usable hit. Nowadays, a large variety of small-molecule databases are available. Commonly used publicly available databases in virtual screening include ZINC (4.6 million compounds), the Available Chemical Directory (ACS, 4 million compounds), the National Cancer Institute compound database (NCI, 400,000 compounds), and the MDDR library (MDL, 110,000 compounds). Other alternative small-molecule collections include the Comprehensive Medicinal Chemistry Database and the Cambridge Structural Database *(23)*. All major pharmaceutical companies have also in-house corporate libraries, comprising several million compounds.

3.3. Filters

The molecular docking stage is extremely time-consuming if a large library of compounds is used. Consequently, the blind use of a large compound library is not generally pursued.

Whenever possible, undesirable compounds should be removed to avoid wasting computational time, and to ensure a more cost-effective approach. Several filters are normally applied to eliminate compounds containing reactive, toxic, or otherwise undesired groups, reducing the size of the initial database and lowering the computational cost of the virtual screening campaign, without compromising its quality.

Inorganic compounds are normally removed upfront, together with insoluble, reactive, and aggregating species. Commonly applied filters also include "drug-like" properties (or "lead-like"), the Lipinski's rule of five, and more general ADME (Absorption Distribution Metabolism and Excretion) features. Other target-independent filters include physical properties such as molecular weight, number of rotatable bonds, polar surface areas, or crude shape descriptors. Compounds containing specific chemical groups associated with poor chemical stability or toxicity are also eliminated from the database. Species with a marked propensity to bind multiple targets in virtual screening campaigns, the so-called promiscuous compounds, are normally also discarded as they tend to appear as false positives against a large number of targets, frequently without any biological meaning. Experimental information about known binders or the target properties can also be of aid in the filtering process. Ligand-based virtual screening techniques can be applied at this level, or even active-site-derived pharmacophore methodologies. The danger here lies in discarding novel compounds based on preconceived ideas or concepts *(15)*.

Globally, the filtering protocol employed normally reduces to number of compounds for the docking process down to 1–10% of the initial size of the compound library. It is this set of compounds that is subjected to atom-based molecular docking.

3.4. The Docking Protocol

Molecular docking tries to place the ligands in the most favourable configuration within the target and evaluates the strength of the ligand-target interactions. The selection of the appropriate docking tool and docking protocol ultimately determines the outcome of the virtual screening process. Several different docking programs, at different levels of sophistication are available. AutoDock *(24–26)*, GOLD *(27, 28)*, and FlexX *(29)* are the most popular docking programs, followed by DOCK *(30)*, and ICM *(31)*, as shown by a recent survey performed on the number of citations in the ISI Web of Science *(13)*, but the number of docking programs is high and ever increasing. The available methodologies differ on the docking algorithm considered, the scoring function used, the way the flexibility of the ligand and of the receptor is treated, and the CPU time associated. **Figure 2** summarizes the main steps involved in a standard docking protocol.

Docking methods treating the ligands as conformationally rigid are appreciably faster than flexible docking methods.

Fig. 2. Representation of the three main stages of a docking protocol. (**a**) Defining the binding region. (**b**) Docking ligand solutions to the pre-selected binding region. (**c**) Final solution (best scoring solution).

However, the neglect of the ligand flexibility may result in missed hits due to an inadequate representation of the ligand conformational space. Most docking tools currently available already include ligand flexibility but consider a rigid target. Ideally, however, the flexibility of the target should also be taken into account, and some approaches to partially include the flexibility of the target have been developed, albeit with a much larger computational cost. Flexible target docking is naturally the most computationally demanding step of all in a virtual screening campaign. Hence, it is advisable to limit its application to fewer candidates.

A common approach nowadays in virtual screening is to vary the sophistication of the docking method employed as the virtual screening process proceeds, i.e. a hierarchical docking protocol. Within a hierarchical docking protocol, a less demanding and less accurate method would be applied first (e.g. rigid docking), saving the more sophisticated and time-consuming approaches (flexible ligand and flexible target docking) for subsequent stages when less compounds are available.

3.5. Postfilters

At present, all docking and scoring methods have a tendency to generate a very significant number of false positives. At the end of the docking protocol, a selection of postfilters are normally also applied to limit the number of hits to be considered in subsequent drug development efforts. Visual inspection is naturally the number one option, but is unsuitable for large-scale applications, when too many hits have been identified. It is normally advisable at this stage to apply automatic filters, based on rules that can be adjusted taking into consideration the number of hits found. A common strategy is to re-apply some of the filters initially used to reduce the number of species in the compound library, but now following more rigorous criteria. Other common postfilter criteria include the final score, the proper balance between polar and apolar terms, diversity, and practical issues such as price, availability, and quality of the compounds.

4. Success Stories of Virtual Screening

Over the last decade, virtual screening has emerged as an important tool in the quest for novel drug-like compounds. In spite of its success, the importance of these methodologies is often masked due to the lack of results justifying their efficiency. This scenario occurs because most of the successful stories of virtual screening are unknown, and remain unpublished and not advertised by the pharmaceutical industry, as a strategy to protect their knowledge from their competitors. Nevertheless, some reports of successful virtual screening searches have been presented in the literature, illustrating the state of the art of this technology.

On of the most attractive targets for virtual screening techniques are metalloenzymes due to their participation in many diseases such as cancer, arthritis, glaucoma, and infection diseases (such as AIDS). These proteins are however challenging targets due to the covalent-like interactions between the metal centres and their ligands, and obviously due to the large electrostatic potential of the metals that is difficult to simulate by computational means (32). In spite of these difficulties, virtual screening techniques using the DOCK software have been successfully applied (33, 34) to several enzymes containing zinc (carbonic anhydrase II, matrix metalloproteinase-3, neural endopeptidase, Zn-β-lactamase, and Zn-dependent phosphotriesterase) (32, 35), nickel (peptide deformylase) (36), iron (cytochrome P450s) (37), and molybdenum (xanthine oxidase) (38).

Transmembrane receptors (G-protein-coupled receptors) represent possibly the most important class of target proteins for drug discovery (39). These receptors are involved in signalling from outside to inside the cell and their malfunction has been associated with diseases such as inflammation, hypertension, depression, obesity, and anxiety. Virtual screening approaches addressed to these receptors, namely to Serotonin 5-HT1A and 5-HT4, Dopamine D2 and NK1 receptor, have been successfully used to develop novel inhibitors (40). The DOCK software (41) was often used in these studies and in the end, new active compounds with new chemical entities were obtained with K_i values less than 5 μM.

In recent years, kinases have also become another major area of drug discovery and structure-based design, and hundreds of 3D structures for more than thirty different kinases are now available for virtual screening (42). Case studies on casein kinase II and protein tyrosine kinase have proven the potential of these techniques. Accordingly, the virtual screening of a large collection of compounds, containing around 400,000 structures, allowed the identification of a dozen of new hit compounds that belong to different chemical classes, with IC_{50} values between 10 and 200 nM. Some of these compounds are still the most potent and specific inhibitors of the CK2 kinase known today (43, 44).

Acetylcholinesterase is a serine hydrolase that belongs to the esterase family within higher eukaryotes *(45)*. This enzyme is responsible for the decomposition of acetylcholine into acetyl and choline fragments and the inhibition of this enzyme has been used strategically in the treatment of Alzheimer's patients. Recently, novel non-covalent inhibitors of acetylcholinesterase have been discovered by the use of virtual screening routines involving the ADAM & EVE software packages *(46)* and the ACD and Maybridge compound databases. Of the total hit list of 1,551 compounds, 13 compounds had IC_{50} values between 0.59 and 10 µM, with most of the core structures being very different from other known inhibitors and displaying high specificity for acetylcholinesterase.

The oestrogen receptor (ER) is a DNA-binding transcription factor that regulates gene expression *(47)*. There are two different forms of oestrogen receptors, usually referred to as α and β. Recently it was found that ERβ can be manipulated in order to prevent age-related neurodegeneration such as Alzheimer's disease. Consequently, several virtual screening trials have been addressed to this target, in particular using GOLD *(28)* and a compound library containing around 25,000 small molecules. From this process, five new hit compounds were isolated, showing IC_{50} values of 680 nM in ERβ-binding essays. Moreover, most of these compounds showed 100-fold binding selectivity to ERβ over ERα *(48)*, a very desirable property taking into consideration that these two different forms of oestrogen receptors have been shown to have distinct cellular distributions and can oppose each other's actions on some genes *(48)*.

5. Concluding Remarks

Although the term virtual screening has been coined only a decade ago, virtual screening is now a widespread lead identification method in the pharmaceutical industry. A myriad of different methods have been developed exploiting the growing library of target structures and assay data, as a basis for finding new lead structures. However, just as with many other promising technologies, it is important to have realistic expectations of the potential of virtual screening.

As any other young methodology, virtual screening has its own limitations. The confidence of any virtual screening algorithm will be always dependent on the type of properties that can be calculated, its relevance to the problem in hand, and of course the simplifications that must be adopted in order to make a good balance between accuracy and speed in order to maintain the algorithm attractive. Consequently, the output of virtual screening includes a significant number of compounds that are not truly bio-active (false positives), and omit others that are active (false negatives). Therefore, a good virtual screening method should therefore be capable of minimizing the number of both false positives and false negatives.

Improved explorations of conformational states for both ligand and receptor and more advanced scoring functions, with the use of more sophisticated models of solvation and a better balance of electrostatic and non-polar terms, will undoubtedly help virtual screening results in a near future. But the main challenge will be focused on the development of better physical models capable of describing the interaction of hundreds of thousands of possible ligands, in many thousands of possible receptor complexes.

Despite these limitations, virtual screening is still the best option available nowadays to explore a large chemical space in terms of cost effectiveness and commitment in time and material, as it allows access to a large number of possible ligands, most of them easily available for purchase and subsequent test. With the increasing number of targets identified by the HT genomics and proteomics, and improved methodologies capable of predicting better hit rates and better predictions of geometries, virtual screening methodologies will gather a preponderant role in drug design in a near future either as a complementary approach to HT screening or as standalone approach.

Acknowledgments

The authors would like to thank the FCT (Fundação para a Ciência e a Tecnologia) for financial support (POCI/QUI/61563/2004). Nuno M. F. S. A. Cerqueira and Sérgio F. Sousa also acknowledge the FCT for a post-doctoral scholarship (SFRH/BPD/27237/2007 and SFRH/BPD/34650/2007).

References

1. Chin D. N., Chuaqui C. E., and Singh J. (2004). Integration of virtual screening into the drug discovery process. *Min. Rev. Med. Chem.* **4**, 1053–1065

2. Brooijmans N., Kuntz I. D. (2003). Molecular recognition and docking algorithms. *Annu. Rev. Biophys. Biomol. Struct.* **32**, 335–373

3. Stahura F. L., Bajorath M. (2005). New methodologies for ligand-based virtual screening. *Curr. Pharm. Des.* **11**, 1189–1202

4. Lengauer T., Lemmen C., Rarey M., and Zimmermann M. (2004). Novel technologies for virtual screening. *Drug Discov. Today* **9**, 27–34

5. Schneider G., Bohm H. J. (2002). Virtual screening and fast automated docking methods. *Drug Discov. Today* **7**, 64–70

6. Bajorath J. (2001). Selected concepts and investigations in compound classification, molecular descriptor analysis, and virtual screening. *J. Chem. Inf. Comput. Sci.* **41**, 233–245

7. Livingstone D. J. (2000). The characterization of chemical structures using molecular

properties. A survey. *J. Chem. Inf. Comput. Sci.* **40**, 195–209

8. Xue L., Bajorath J. (2000). Molecular descriptors in chemoinformatics, computational combinatorial chemistry, and virtual screening. *Comb. Chem. High Throughput Screen.* **3**, 363–372

9. Bajorath J. (2002). Integration of virtual and high-throughput screening. *Nat. Rev. Drug. Discov.* **1**, 882–894

10. Engels M. F. M., Venkatarangan P. (2001). Smart screening: approaches to efficient HTS. *Curr. Opin. Drug Discov. Devel.* **4**, 275–283

11. Green D. V. (2003). Virtual screening of virtual libraries. *Prog. Med. Chem.* **41**, 61–97

12. Waszkowycz B., Perkins T. D. J., Sykes R. A., and Li J. (2001). Large-scale virtual screening for discovering leads in the postgenomic era. *IBM Syst. J.* **40**, 360–376

13. Sousa S. F., Fernandes P. A., and Ramos M. J. (2006). Protein–ligand docking: current status and future challenges. *Proteins* **65**, 15–26

14. Hou T. J., Xu X. J. (2004). Recent development and application of virtual screening in drug discovery: an overview. *Curr. Pharm. Des.* **10**, 1011–1033

15. Klebe G. (2006). Virtual ligand screening: strategies, perspectives and limitations. *Drug Discov. Today* **11**, 580–594

16. Hajduk P. J., Huth J. R., and Tse C. (2005). Predicting protein druggability. *Drug Discov. Today* **10**, 1675–1682

17. Evers A., Klabunde T. (2005). Structure-based drug discovery using GPCR homology mode-ling: successful virtual screening for antagonists of the alpha1A adrenergic receptor. *J. Med. Chem.* **48**, 1088–1097

18. Evers A., Klebe G. (2004). Successful virtual screening for a submicromolar antagonist of the neurokinin-1 receptor based on a ligand-supported homology model. *J. Med. Chem.* **47**, 5381–5392

19. Bissantz C., Bernard P., Hibert M., and Rognan D. (2003). Protein-based virtual screening of chemical databases. II. Are homology models of G-protein coupled receptors suitable targets? *Proteins* **50**, 5–25

20. Clore G. M., Gronenborn A. M. (1998). NMR structure determination of proteins and protein complexes larger than 20 kDa. *Curr. Opin. Chem. Biol.* **2**, 564–570

21. Wishart D. (2005). NMR spectroscopy and protein structure determination: applications to drug discovery and development. *Curr. Pharm. Biotechnol.* **6**, 105–120

22. Gunther J., Bergner A., Hendlich M., and Klebe G. (2003). Utilising structural knowledge in drug design strategies: applications using relibase. *J. Mol. Biol.* **326**, 621–636

23. Ghosh S., Nie A. H., An J., and Huang Z. W. (2006). Structure-based virtual screening of chemical libraries for drug discovery. *Curr. Opin. Chem. Biol.* **10**, 194–202

24. Goodsell D. S., Olson A. J. (1990). Automated docking of substrates to proteins by simulated annealing. *Proteins* **8**, 195–202

25. Morris G. M., Goodsell D. S., Huey R., and Olson A. J. (1996). Distributed automated docking of flexible ligands to proteins: parallel applications of AutoDock 2.4. *J. Comput. Aided Mol. Des.* **10**, 293–304

26. Morris G. M., Goodsell D. S., Halliday R. S., Huey R., Hart W. E., Belew R. K., and Olson A. J. (1998). Automated docking using a Lamarckian genetic algorithm and an empirical binding free energy function. *J. Comput. Chem.* **19**, 1639–1662

27. Jones G., Willett P., and Glen R. C. (1995). Molecular recognition of receptor sites using a genetic algorithm with a description of desolvation. *J. Mol. Biol.* **245**, 43–53

28. Jones G., Willett P., Glen R. C., Leach A. R., and Taylor R. (1997). Development and validation of a genetic algorithm for flexible docking. *J. Mol. Biol.* **267**, 727–748

29. Rarey M., Kramer B., Lengauer T., and Klebe G. (1996). A fast flexible docking method using an incremental construction algorithm. *J. Mol. Biol.* **261**, 470–489

30. Ewing T. J. A., Kuntz I. D. (1997). Critical evaluation of search algorithms for automated molecular docking and database screening. *J. Comput. Chem.* **18**, 1175–1189

31. Abagyan R., Totrov M., and Kuznetzov D. (1994). ICM – a new method for protein modeling and design: applications to docking and structure prediction from the distorted native conformation. *J. Comput. Chem.* **15**, 488–506

32. Irwin J. J., Raushel F. M., and Shoichet B. K. (2005). Virtual screening against metalloenzymes for inhibitors and substrates. *Biochemistry* **44**, 12316–12328

33. Stote R. H., Karplus M. (1995). Zinc binding in proteins and solution: a simple but accurate nonbonded representation. *Proteins* **23**, 12–31

34. Hoops S. C., Anderson K. W., and Merz K. M., Jr. (1991). Force Field Design for Metalloproteins. *J. Am. Chem. Soc.* **113**, 8262–8270

35. Irwin J. J., Shoichet B. K. (2005). ZINC – A free database of commercially available compounds for virtual screening. *J. Chem. Inf. Model.* **45**, 177–182

36. Wang Q., Zhang D. T., Wang J. W., Cai Z. T., and Xu W. R. (2006). Docking studies of

nickel-peptide deformylase (PDF) inhibitors: exploring the new binding pockets. *Biophys. Chem.* **122**, 43–49

37. Byvatov E., Baringhaus K. H., Schneider G., and Matter H. (2007). A virtual screening filter for identification of cytochrome P4502C9 (CYP2C9) inhibitors. *QSAR Comb. Sci.* **26**, 618–628

38. Gupta S., Rodrigues L. M., Esteves A. P., Oliveira-Campos A. M. F., Nascimento M. S. J., Nazareth N., Cidade H., Neves M. P., Fernandes F., Pinto M., et al. (2007). Synthesis of N-aryl-5-amino-4-cyanopyrazole derivatives as potent xanthine oxidase inhibitors. *Eur. J. Med. Chem.* doi:10.1016/j.ejmech.2007.06.002

39. Schneider G., Neidhar W., and Adam G. (2001). Integrating virtual screening methods to the quest for novel membrane protein ligands. *Curr. Med. Chem.* **1**, 99–112

40. Becker O. M., Marantz Y., Shacham S., Inbal B., Heifetz A., Kalid O., Bar-Haim S., Warshaviak D., Fichman M., and Noiman S. (2004). G protein-coupled receptors: in silico drug discovery in 3D. *Proc. Natl. Acad. Sci. U.S.A.* **101**, 11304–11309

41. Ewing T. J. A., Makino S., Skillman A. G., and Kuntz I. D. (2001). DOCK 4.0: search strategies for automated molecular docking of flexible molecule databases. *J. Comput. Aided Mol. Des.* **15**, 411–428

42. Muegge I., Enyedy I. J. (2004). Virtual screening for kinase targets. *Curr. Med. Chem.* **11**, 693–707

43. Wang R., Liu D., Lai L., and Tang Y. (1998). SCORE: a new empirical method for estimating the binding affinity of a protein–ligand complex. *J. Mol. Model.* **4**, 379–394

44. Vangrevelinghe E., Zimmermann K., Schoepfer J., Portmann R., Fabbro D., and Furet P. (2003). Discovery of a potent and selective protein kinase CK2 inhibitor by high-throughput docking. *J. Med. Chem.* **46**, 2656–2662

45. Holzgrabe U., Kapkova P., Alptuzun V., Scheiber J., and Kugelmann E. (2007). Targeting acetylcholinesterase to treat neurodegeneration. *Expert Opin. Ther. Targets* **11**, 161–179

46. Mizutani M. Y., Itai A. (2004). Efficient method for high-throughput virtual screening based on flexible docking: discovery of novel acetylcholinesterase inhibitors. *J. Med. Chem.* **47**, 4818–4828

47. Palmieri C., Cheng G. J., Saji S., Zelada-Hedman M., Warri A., Weihua Z., Van Noorden S., Wahlstrom T., Coombes R. C., Warner M., et al. (2002). Estrogen receptor beta in breast cancer. *Endocr. Relat. Cancer* **9**, 1–13

48. Zhao L. Q., Brinton R. D. (2005). Structure-based virtual screening for plant-based ER beta-selective ligands as potential preventative therapy against age-related neurodegenerative diseases. *J. Med. Chem.* **48**, 3463–3466

Part II

Protocols

Chapter 5

Combinatorial Chemistry and the Synthesis of Compound Libraries

Rolf Breinbauer and Matthias Mentel

Summary

Solid Phase Organic Synthesis (SPOS) has become a powerful tool for the preparation of compound libraries used for screening efforts in Chemical Biology. While different types of screening libraries have become commercially available through several vendors, the elaboration of a hit compound in a screening campaign to a useful chemical probe still requires the preparation of a focused library. In this report, protocols are given which allow the synthesis of a focused compound library on solid phase by the diversification of a central scaffold using reliable functionalization reactions. Diversity elements can be introduced by attaching building blocks via amidation, esterification, reductive amination, and Suzuki cross-coupling.

Key words: Combinatorial chemistry, Diversification, Libraries, Reductive amination, Amide coupling, Esterification, Suzuki cross-coupling, Scaffold, Solid phase synthesis

1. Introduction

Bruce Merrifield recognized in his Nobel prize-winning work that the synthesis of organic compounds can be simplified when a substrate is immobilized on an insoluble polymeric support ("resin") via a linker unit *(1)*. The substrate can now be easily modified by adding an excess of reagents and building blocks ensuring complete conversion. Consumed and excess reagents are then easily removed by washing the support and simple filtration. Finally, cleavage of the linker releases the desired product into the solution allowing its isolation in pure form (**Fig. 1**).

The application of solid phase methods for the synthesis of small molecule libraries has seen not only considerable progress

Ana Cecília A. Roque (ed.), *Ligand-Macromolecular Interactions in Drug Discovery: Methods and Protocols,*
Methods in Molecular Biology, vol. 572,
DOI 10.1007/978-1-60761-244-5_5, © Humana Press, a part of Springer Science + Business Media, LLC 2010

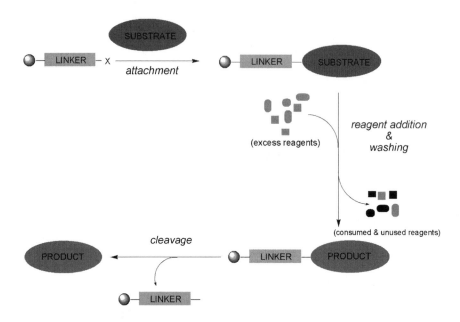

Fig. 1. A general scheme for solid phase organic synthesis.

over the last 15 years, but also several paradigm shifts. At the moment, an approach in which a scaffold with several points for diversification is prepared in solution and derivatized using solid phase methods is favored over pure solid phase or solution phase methods. This hybrid approach combines the advantages of both strategies. Scaffolds can be purchased from several chemical suppliers or be prepared in solution in one's own lab. The scaffold is attached to a polymeric matrix using a linker unit.

Polystyrene (PS) resins (cross-linked with 1–2% divinylbenzene (DVB), loading up to 2 mmol functional group/g) are the resins most commonly used for oligopeptide and small molecule solid phase synthesis. These resins swell in most organic solvents, but not in very polar solvents such as alcohols or water. If reactions need to be performed under polar conditions, the use of Tent-agel (PS-resin with grafted polyethyleneglycol chains), PEGA, Pepsyn, or Argogel is recommended, which are distinguished by polar, flexible polymeric chains.

A linker can be considered as a polymeric (or more general: immobilized) protecting group. It is a functional group responsible for the covalent attachment of the substrate to the support which will be cleaved from the product at the end of the synthesis sequence. The following features have to be considered, when choosing an appropriate linker: (1) the attachment of substrate should proceed in high yield, (2) the linker should be stable under all reaction conditions considered for the synthetic sequence ("orthogonality"), (3) the linker should be cleaved under mild conditions in quantitative yield without obscuring the integrity and purity of the product. More than 200 different linkers have been described exhibiting different degrees of orthogonality and releasing different functionalities at the product after cleavage *(2, 3)*. Those linkers which can be cleaved by acids are the most commonly used linkers.

For the protocols below we will discuss the use of a polystyrene resin to which a general scaffold can be attached using established protocols *(4)*. Once the scaffold has been immobilized, diversity can be created using reliable transformations that are applicable for a larger set of building blocks with different functional, electronic, and steric properties. Before a library can be synthesized, the reaction conditions for the used set of building blocks have to be validated. The best strategy is to try at least two model substrates exhibiting the most extreme properties of a class of building blocks. If for example in a reductive amination using benzaldehydes as substrates, a very electron-poor and an electron-rich representative produce the desired products in test reactions in similar and satisfying yields, one can expect that most other substrates with intermediate electronic properties will work equally well.

In **Fig. 2**, a hypothetical scaffold is presented to demonstrate the potential of the diversification reactions. Solid phase bound carboxylic acids can be transformed into amides using a well-established amidation procedure *(5)* taking advantage of the huge variety of commercially available amines (**Subheading 3.1**). Following along the same lines, **Subheading 3.2** describes the acylation of solid phase bound amines with a similarly large set of available carboxylic acids *(6)*. Similarly, solid phase bound carboxylic acids can be reacted with a generic set of alcohols to produce esters *(7)* (**Subheading 3.3**). One of the best diversification

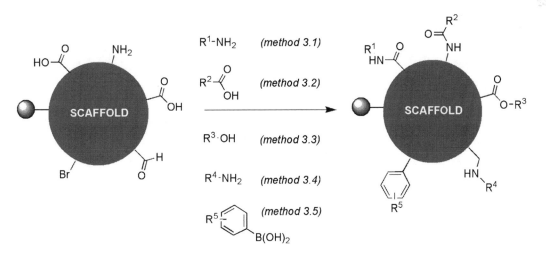

Fig. 2. Summary of the diversification reactions discussed in this chapter for the preparation of a focused library around a central scaffold using the building blocks which are readily available in great functional and structural diversity. Attention should be paid, that if some of these functional groups are present in the same molecule the use of protecting groups will be necessary.

reactions available is the reductive amination of solid phase bound aldehydes *(8)* (**Subheading 3.4**). Arylhalogenides can be coupled with aryl- or hetarylboronic acids via Pd-catalyzed Suzuki cross-coupling. The use of the very electron-rich S-PHOS-ligand *(9)* allows the use of almost any arylbromide, even with electron-rich character.

There are two excellent text books available in which further diversification reactions are summarized and experimental procedures given *(10, 11)*.

The release of the diversified compounds from the resin with acid-sensitive linkers will be initiated by addition of trifluoro-acetic acid, which has the advantage of being volatile enough, that an excess of this reagent can be removed in vacuo. Although the purities of the released compounds are often very high, it is highly recommended that compounds used for biological screening receive further purification by HPLC using reversed phase columns as minor impurities can obscure the results in biological screenings.

2. Materials

All chemical reagents have been purchased from Aldrich. Polystyrene resins with attached linker units are purchased from Novabiochem.

1. Dichloromethane (DCM).
2. *N,N'*-diisopropylcarbodiimide (DIC).
3. *N,N*-dimethylacetamide (DMA).
4. *N,N*-dimethylaminopyridine (DMAP).
5. *N,N*-dimethylformamide (DMF).
6. 1-hydroxybenzotriazole (HOBt).
7. 1-hydroxy-7-azabenzotriazole (HOAt).
8. Tetrahydrofuran (THF).

3. Methods

3.1. Diversification of a Solid Phase Bound Carboxylic Acid with a Set of Amine Building Blocks

1. A 100-mL round-bottom flask is charged with HOOC-functionalized polystyrene resin (loading: 2.50 mmol HOOC-groups). 10 mL dry *N,N*-dimethylacetamide (DMA) are added, and the resin allowed to swell for 10 min.

2. To this slurry were added 10 mL of a stock solution of 1 M HOBt in DMA (10 mmol HOBt) and the amine building block (5 mmol). The mixture is agitated with an orbital shaker for 5 min.

3. 10 mL of a stock solution of 1 M *N,N′*-diisopropylcarbodiimide (DIC) in DMA (10 mmol DIC) is added via syringe (*see* **Note 1**).

4. The flask is flushed with nitrogen, sealed with a stopper, and placed on an orbital shaker where it is shaken for 72 h at room temperature.

5. The suspension is transferred into a sintered glass funnel which sits on top of a filtering flask. The solution containing consumed and excess reagents is drained in vacuo.

6. The beads are subsequently washed with DMF (3 × 20 mL), MeOH (3 × 20 mL), and DCM (3 × 20 mL) (*see* **Note 2**).

7. The resin is dried in oil pump vacuum for 24 h. Then the resin is ready for further diversification steps. It should be stored in a closed flask in a refrigerator.

3.2. Diversification of a Solid Phase Bound Amine with a Set of Carboxylic Acid Building Blocks

1. A 25-mL round-bottom flask is charged with H_2N-functionalized polystyrene resin (loading: 0.60 mmol H_2N-groups) and so much DCM is added that the beads are swollen, but not significant amount of supernatant is formed.

2. In a 10-mL flask, 6 mmol carboxylic acid building block (10 equiv) and HOAt (0.815 g, 6 mmol, 10 equiv) are dissolved in 3 mL of DMF. DIC (0.91 mL, 6 mmol, 10 equiv) is added.

3. The reagent solution is added to the beads with a pipette. The flask is sealed and agitated on an orbital shaker for 6 h at room temperature.

4. The suspension is transferred into a sintered glass funnel which sits on top of a filtering flask. The solution containing consumed and excess reagents is drained in vacuo.

5. The beads are subsequently washed with DMF (5 × 5 mL) and DCM (5 × 5 mL) (*see* **Note 2**).

6. The resin is dried in oil pump vacuum for 24 h. Then the resin is ready for further diversification steps. It should be stored in a closed flask in a refrigerator.

3.3. Diversification of a Solid Phase Bound Carboxylic Acid with a Set of Alcohol Building Blocks

1. A 100-mL round-bottom flask is charged with HOOC-functionalized polystyrene resin (loading: 0.50 mmol HOOC-groups). 30 mL dry dichloromethane are added, and the resin is allowed to swell for 15 min.

2. The flask is immersed into an ice bath and allowed to equilibrate to 0°C (15 min).

3. 0.03 g (0.24 mmol) *N,N*-dimethylaminopyridine (DMAP) are added, followed by 2.0 mmol of alcohol-building block, and 0.19 mL (1.2 mmol) *N,N′*-diisopropylcarbodiimide (DIC).

4. The flask is sealed with a stopper and placed on an orbital shaker where it is shaken for 24 h at room temperature.

5. The suspension is transferred into a sintered glass funnel which sits on top of a filtering flask. The solution containing consumed and excess reagents is drained in vacuo.

6. The beads are subsequently washed with DCM (4 × 20 mL), THF (4 × 20 mL), DMF (2 × 20 mL), DMF/water [2 × (10/10 mL)], DCM/MeOH [2 × (10/10 mL)], and DCM (4 × 20 mL) (*see* **Note 2**).

7. The resin is dried in oil pump vacuum for 24 h. Then the resin is ready for further diversification steps. It should be stored in a closed flask in a refrigerator.

3.4. Diversification of a Solid Phase Bound Aldehyde with a Set of Aromatic or Electron Deficient Amines via Reductive Amination

1. A 50-mL round-bottom flask is charged with 1% HOAc in DMF (20 mL). The OHC-functionalized resin (0.30 mmol formyl groups) is added, followed by 0.64 g pulverized $NaBH(OAc)_3$ (3.00 mmol, 10 equiv).

2. Primary amine (3.00 mmol) is added.

3. The flask is sealed and the slurry agitated with an orbital shaker for 24 h.

4. The suspension is transferred into a sintered glass funnel which sits on top of a filtering flask. The solution containing consumed and excess reagents is drained in vacuo.

5. The beads are subsequently washed with MeOH (1 × 10 mL), DMF (7 × 20 mL), DCM (7 × 20 mL), MeOH (3 × 10 mL), and DCM (4 × 20 mL) (*see* **Note 2**).

6. The resin is dried in oil pump vacuum for 24 h. Then the resin is ready for further diversification steps. It should be stored in a closed flask in a refrigerator.

3.5. Diversification of a Solid Phase Bound Bromoarene with a Set of Boronic Acids via Suzuki Cross-Coupling

1. A 20-mL Schlenk tube with a magnetic stir bar is dried with a heat gun, while being evacuated by an oil pump to remove moisture. After cooling to room temperature, the Schlenk tube is filled with argon.

2. Under a stream of argon, the Schlenk tube is charged consecutively with bromoaryl-functionalized resin (0.191 mmol bromoaryl groups), 1.13 mmol (5.9 equiv) boronic acid, 0.475 g (2.24 mmol, 12 equiv) mortar powdered, anhydrous K_3PO_4, 0.010 g (0.045 mmol, 24 mol-%) $Pd(OAc)_2$, 0.074 g (0.179 mmol, 0.9 equiv) S-PHOS, and 6 mL absolute THF.

3. The Schlenk tube is degassed by applying vacuum and refilled with argon (repeated at least five times) (*see* **Note 3**).

4. The closed Schlenk tube is immersed in an oil bath and heated to 80°C. This temperature will be kept for 24 h. The beads are

carefully stirred with a magnetic stir bar to avoid mechanic abrasion of the beads.

5. After cooling to room temperature, the suspension is transferred into a sintered glass funnel which sits on top of a filtering flask. The solution containing consumed and excess reagents is drained in vacuo.

6. The beads are subsequently washed with THF/H$_2$O (1:1, 3 × 10 mL), THF/MeOH (1:1, 3 × 10 mL), MeOH (3 × 10 mL) (*see* **Note 2**), and dried under high vacuum.

7. The resin is dried in oil pump vacuum for 24 h. Then the resin is ready for further diversification steps. It should be stored in a closed flask in a refrigerator (*see* **Note 4**).

4. Notes

1. In SPOS, DIC is preferred over *N,N*′-dicyclohexylcarbodiimide as DIC produces a urea which is soluble in organic solvents and therefore can be removed form the resins easily by washing procedures.

2. It is important that in each washing step the interaction time between beads and solvent is sufficiently long to allow complete swelling of the beads.

3. Evacuation has to be performed very carefully to avoid loss of resin through bumping.

4. If the conversion of the reactions is not satisfactory, a double coupling strategy is recommended. Here, the supernatant of the reaction mixture after the first reaction step is removed via a Pasteur pipette and a fresh cocktail of reagents is applied and the reaction started again. Then the workup follows the line described for the single coupling reaction.

Acknowledgments

The research by the authors was supported by the Fonds der Chemischen Industrie, the Max-Planck-Society, Deutsche Forschungsgemeinschaft, University of Dortmund, University of Leipzig, Graz University of Technology, and BASF.

References

1. Merrifield, R. B. (1985). Solid-phase syntheses (Nobel lecture). *Angew. Chem.* **10**, 801–812

2. Guillier, F., Orain, D., and Bradley, M. (2000). Linkers and cleavage strategies in solid-phase organic synthesis and combinatorial chemistry. *Chem. Rev.* **100**, 2091–2157

3. James, I. W. (1999). Linkers for solid phase organic synthesis. *Tetrahedron* **55**, 4855–4946

4. Novabiochem Catalogue 2008/9, EMD Biosciences, http://www.emdbiosciences.com/

5. Hamper, B. C., Snyderman, D. M., Owen, T. J., Scates, A. M., Owsley, D. C., Kesselring, A. S. and Chott, R. C. (1999). High-throughput ¹H-NMR and HPLC-characterization of a 96-member substituted methylene malonamic acid library. *J. Comb. Chem.* **1**, 140–150

6. Garcia, J., Mata, E. G., Tice, C. M., Hormann, R. E., Nicolas, E., Albericio, F. and Michelotti, E. L. (2005). Evaluation of solution and solid-phase approaches to the synthesis of libraries of α,α-disubstituted-α-acylaminoketones. *J. Comb. Chem.* **7**, 843–863

7. Nad, S. and Breinbauer, R. (2005). Electroorganic synthesis of 2,5-dialkoxydihydrofurans and pyridazines on solid phase using polymer beads as supports. *Synthesis* **20**, 3654–3665

8. Boojamra, C. G., Burow, K. M., Thompson, L. A., and Ellman, J. A. (1997). Solid-phase synthesis of 1,4-benzodiazepine-2,5-diones. Library preparation and demonstration of synthesis generality. *J. Org. Chem.* **62**, 1240–1256

9. Barder, T. E., Walker, S. D., Martinelli, J. R. and Buchwald, S. L. (2005). New catalysts for Suzuki-Miyaura coupling processes: scope and studies of the effect of ligand structure. *J. Am. Chem. Soc.* **127**, 4685–4696

10. Czarnik, A. W. (ed.) (2001). *Solid Phase Organic Syntheses, Vol. 1.* Wiley-Interscience, New York

11. Dörwald, F. Z. (2002). *Organic Synthesis on Solid Phase. Supports, Linkers, Reactions, 2nd. Ed.* Wiley-VCH, Weinheim

Chapter 6

Ligand-Based Nuclear Magnetic Resonance Screening Techniques

Aldino Viegas, Anjos L. Macedo, and Eurico J. Cabrita

Summary

A critical step in the drug discovery process is the identification of high-affinity ligands for macromolecular targets, and, over the last 10 years, NMR spectroscopy has become a powerful tool in the pharmaceutical industry. Instrumental improvements in recent years have contributed significantly to this development. Digital recording, cryogenic probes, autosamplers, and higher magnetic fields shorten the time for data acquisition and improve the spectral quality. In addition, new experiments and pulse sequences make a vast amount of information available for the drug discovery process. All these techniques take advantage of the fact that upon complex formation between a target molecule and a ligand, significant perturbations can be observed in NMR-sensitive parameters of either the target or the ligand. These perturbations can be used qualitatively to detect ligand binding or quantitatively to assess the strength of the binding interaction. In addition, some of the techniques allow the identification of the ligand-binding site or which part of the ligand is responsible for interacting with the target.

In this chapter, we will use examples from our own research to illustrate how NMR experiments to characterize ligand binding may be used to both screen for novel compounds during the process of lead generation, and provide structural information useful for lead optimization during the latter stages of a discovery program.

Key words: Ct CBM11, Diffusion, DOSY, Drug design, Ligand binding, Screening, NOE, STD-NMR

1. Introduction

Nuclear Magnetic Resonance (NMR) spectroscopy is a unique tool to study molecular interactions in solution, and it became an essential technique to characterize events of molecular recognition. During the last few years, due to instrumental improvements in the development of higher magnetic fields and cryogenic probes, NMR became more sensitive. At the same time, the development

Ana Cecília A. Roque (ed.), *Ligand-Macromolecular Interactions in Drug Discovery: Methods and Protocols,*
Methods in Molecular Biology, vol. 572,
DOI 10.1007/978-1-60761-244-5_6, © Humana Press, a part of Springer Science+Business Media, LLC 2010

of new experiments and pulse sequences made NMR a unique technique to obtain information about the interactions of small ligands with biologically relevant macromolecules (proteins and/ or nucleic acids), key for drug discovery.

Depending on the NMR experiment that is chosen, it is possible to obtain information about the atoms' or ligands' functional groups, as well as amino acids or structural elements of proteins which are involved in the ligand–target interactions. Quantitative information about dissociation rate constants can also be obtained. This can be achieved by measuring different NMR parameters such as chemical shifts, coupling constants, signal intensities, and linewidths; and detecting their alterations by the addition of specific molecules to the target sample. In the case of small ligand-target complexes, the solution structure can be determined by NMR methods in a complementary way to X-ray crystallography.

Several reviews have been written about NMR-based screening methodologies and their contributions to different steps in the process of drug discovery, revealing the impact of the technique in the field (*1–5*).

NMR-based interaction studies can be followed by two different approaches, namely, by looking at the protein (target) spectrum and following the changes in chemical shift by ligand titration, or recording the spectra of a sample of ligand with small amounts of protein.

For the first approach, the one that uses techniques involving protein detection, the use of large amounts of labeled protein and the access to high magnetic fields is required. Chemical shift alterations due to ligand interaction can be easily followed by HSQC experiments with ^{15}N-labeled proteins, but total or partial resonance assignment needs to be done in order to identify the residues that are involved in the interaction. In this case, NMR is limited by the size of the protein. However, for the ligand detection techniques, spectra can be easily recorded at low magnetic field spectrometers, using samples with small amounts of nonlabeled protein, which makes this second approach more attractive. In this case, variations in peak intensities, relaxation behavior, diffusion or saturation transfer can be used to obtain information about the binding process.

In this chapter we will present the setup of ligand-detecting-based experiments using, as an example a 20-kDa noncatalytic protein from the cellulosome of *Clostridium thermocellum* (*Ct*) – family 11 Carbohydrate-Binding Module (*Ct*CBM11) (*6, 7*). This microorganism is able to perform a very important conversion of polysaccharides from plant cell wall into soluble sugars (*8–10*). The interaction between *Ct*CBM11 and a polysaccharide, cellohexaose, will be used to illustrate saturation transfer difference NMR, STD-NMR, and diffusion experiments that are used for the screening and epitope mapping of the ligand(s). **Subheading 3** will include a brief introduction to each of the techniques and a detailed description of the experimental setup.

2. Materials

1. Cellohexaose (Seikagaku Co.) 4 mM in D_2O (Euriso-top). Store at 4°C.

2. Laminarihexaose (Seikagaku Co.) 4 mM in D_2O (Euriso-top). Store at 4°C.

3. Store *Ct*CBM11 at 4°C.

4. TSP (3-(Trimethylsilyl)-Propionic acid-D4, sodium salt) (Euriso-top) 3 mg/ml. Store at room temperature.

5. 3-mm NMR tubes for 600 MHz (Wilmad LabGlass).

3. Methods

In this section two NMR-based screening techniques are described in detail: STD Spectroscopy in **Subheading 3.1**, and Diffusion Ordered Spectroscopy (DOSY) in **Subheading 3.2**. In spite of the fact that the examples presented here, from our own research, do not deal with the screening of drug candidates, the methodology described has been applied successfully in many examples in a drug design context (*see* **refs. *1–5***). The setup of the experiments as described here is general for any protein–ligand interaction study and independent of the context of the investigation. All NMR spectra were acquired in a Bruker Avance II spectrometer operating at a proton frequency of 600 MHz with a triple resonance cryogenic probehead.

3.1. Saturation Transfer Difference

STD-NMR spectroscopy can be used to characterize low-affinity interactions ($K_D \approx 10^{-8}–10^{-3}$ M) between small molecules and proteins, and it is based on the intermolecular transfer of magnetization (via spin diffusion) from the protein to the bound molecule through the Nuclear Overhauser Effect (NOE). This saturation transfer can be detected if there is a fast equilibrium between the bound and free state of the ligand in solution. The saturation transfer takes place only to molecules bound to the protein with a rate that depends on the protein mobility, ligand/protein complex lifetime, and geometry (*2, 11*).

STD involves selectively saturating a resonance that belongs to the receptor (a region of the spectrum that contains only resonances from the receptor such as 0 to −1 ppm must be found). If the ligand binds, saturation will propagate from the selected receptor protons to other protons of the receptor via spin diffusion, and ultimately the saturation is transferred to binding compounds by cross relaxation at the ligand-receptor interface. In order to obtain the data corresponding to the STD effect a control experiment must be performed by saturating a region of the spectrum that does not contain any signal. The difference spectrum yields only those resonances that have experienced saturation, namely those of the receptor and those of the compound that binds to the receptor. Since the receptor is present at very small concentrations its resonances should not be visible but, if needed, can be eliminated by a relaxation filter ($T_{1\rho}$). In order to improve the spectrum the solvent signal can also be suppressed (**Fig. 1**).

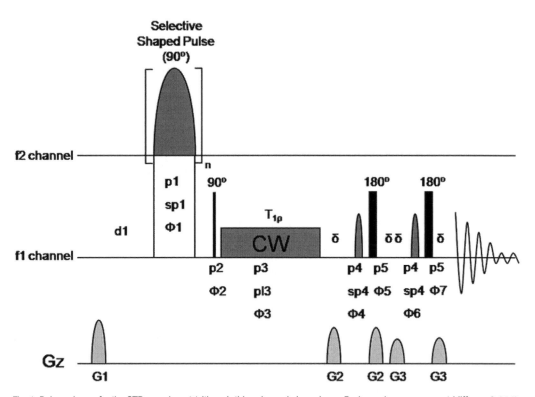

Fig. 1. Pulse scheme for the STD experiment (although this scheme is based on a Bruker pulse sequence, stddiffesgp.3 (*12*), the notation and numbering for pulses and delays differs considerably). *d1*: relaxation delay; *p1* and *sp1*: length and power for the 90° shaped pulse; *p2*: 90° ¹H transmitter pulse; *p3* and *pl3*: length and power for the spin-lock; *p4* and *sp4*: length and power for the 180° shaped pulse; *p5*: 180° ¹H transmitter pulse; δ: delay for homospoil/gradient recovery (200 μs); *G1*, *G2*, and *G3*: pulsed field gradients (gradient ratio 40:31:11 and length 1 ms). Phase cycle: $\Phi1 = x$; $\Phi2 = x, -x$; $\Phi3 = 8(y), 8(-y)$; $\Phi4 = 2(x), 2(y)$; $\Phi5 = 2(-x), 2(-y)$; $\Phi6 = 4(x), 4(y)$; $\Phi7 = 4(-x), 4(-y)$; aq = x, −x, −x, x, −x, x, x, −x.

3.1.1. Shaped Pulse Calibration

1. Prepare the NMR sample. In this example we have used (*Ct*CBM11) = 20 μM, in D_2O (total volume = 200 μL in a 3-mm tube).

2. Record a ^1H proton spectrum. Start with a regular ^1H spectrum (tune and match, lock, shim, and reference phase) and record any optimized parameters: (relaxation delay = 2 s, transmitter frequency offset = solvent ^1H frequency, time domain = 4 k, and optimized spectral width = 14 ppm in our example – **Fig. 5a**).

3. Calibrate the shaped pulse. Create a new experiment in order to calibrate the shaped pulse (you will have to load a pulse sequence with parameters corresponding to *selective excitation using a shaped pulse (12–14) (Fig. 2)*). The calibration of the shaped pulse involves the determination of the combination length/power level for a chosen pulse shape and flip angle and is described in detail in **Subheadings "Shaped Pulse for Solvent Suppression" and "Shaped Pulse for Protein Saturation"**.

Shaped Pulse for Solvent Suppression

1. Set the shape, length and flip angle. Set the shaped pulse for the solvent suppression (square-shaped pulse) with a length of 2 ms (p1); this will correspond to a 180° pulse (do not forget to set the offset to the water frequency).

2. Determine the power level. Depending on the manufacturer and Type of instrument that you have the calibration of the shaped pulses can be more or less automatic. Nevertheless, if you want or need to do it "by hand" set the shape and the length and start with a very low power and adjust the phase. Then repeat the experiment increasing the power until you

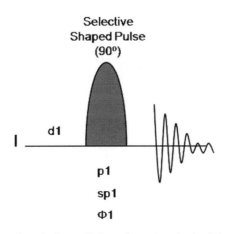

Fig. 2. Pulse scheme for selective excitation using a shaped pulse (this scheme is based on a Bruker pulse sequence, selzg *(12)*, but the notation and numbering for pulses and delays may differ considerably depending on the spectrometer). *d1*: relaxation delay; *p1* and *sp1*: length and power for the 90° shaped pulse. Phase cycle is: *Φ1* = x, −x, −x, x, y, −y, −y, y; aq = x, −x, −x, x, y, −y, −y, y.

get a spectrum with zero intensity of the solvent signal. (*see* **Note 1**).

Shaped Pulse for Protein Saturation

1. Set the shape, length and flip angle. Set the shaped pulse for the selective saturation of the protein (we have used E-Burp-1 *(15)*) in this example, but Gaussian-shaped pulses are the ones used more often) with a length of 100 ms (p1).

2. Determine the power level. Start with a very low power and adjust the phase. Then repeat the experiment increasing the power until you get a spectrum with zero intensity of the solvent signal – this is the 180° pulse. Now just divide the length by 2 (50 ms) and you have the 90° pulse (*see* **Note 1**).

3.1.2. Record a ¹H Spectrum with Water Suppression

1. Create a new experiment in order to calibrate the solvent suppression (pulse prog. *water suppression using excitation sculpting with gradients (12, 16)* (**Fig. 3**)) and update your parameters (relaxation delay = 2 s, transmitter frequency offset = solvent 1H frequency, time domain = 32k, and optimized spectral width = 14 ppm in our example).

2. Set the 180° shaped pulse for the solvent suppression: length 2.0 ms, square shaped and power level as determined according to Subheading "Shaped Pulse for Solvent Suppression".

3. Define the number of scans (NS) and dummy scans (8*n and 4, respectively) and acquire the spectrum (**Fig. 5b**).

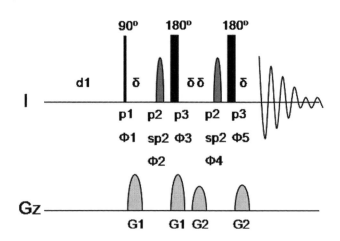

Fig. 3. Pulse scheme for water suppression using excitation sculpting with gradients (although this scheme is based on a Bruker pulse sequence (zgesgp *(12)*) the notation and numbering for pulses and delays differs considerably). *d1*: relaxation delay; *p1*: 90° ¹H transmitter pulse; *p2* and *sp2*: length and power for the 180° shaped pulse; *p3*: 180° ¹H transmitter pulse; δ: delay for homospoil/gradient recovery (200 μs); *G1* and *G2*: gradients (gradient ratio 31:11 and length 1 ms). Phase cycle: $\Phi1$ = x; $\Phi2$ = x, y; $\Phi3$ = −x, −y; $\Phi4$ = 2(x), 2(y); $\Phi5$ = 2(−x), 2(−y); aq = x, −x, −x, x.

(Note: to optimize the solvent suppression fine tune the frequency offset)

3.1.3. Setting up the Spin-Lock Filter (T₁ᵨ)

1. Create a new experiment in order to calibrate the spin-lock filter (**Fig. 4**) using the same parameters as before (relaxation delay = 2 s, transmitter frequency offset = solvent ¹H frequency, time domain = 32k, and optimized spectral width = 14 ppm in our example).

2. As before, set the 180° shaped pulse for the solvent suppression: length 2.0 ms, square shaped and power level as determined according to **Subheading "Shaped Pulse for Solvent Suppression"**.

3. The spin-lock filter is applied as a continuous-wave pulse with a fixed field (kHz) and adjustable length (p2 in **Fig. 4**). In this example we used a field of 2 kHz that, according to **Eq. 1**, corresponds to a length of 125 μs.

$$\gamma B_1 = \frac{1}{4p} \tag{1}$$

where γ is the magnetogyric ratio, B_1 is the radiofrequency field (in this case γB_1 = 2 kHz), and p is the length.

1. To determine the power level corresponding to the 2 kHz spin-lock you just have to proceed as if you were calibrating a 90° pulse with the length calculated previously (125 μs). This is the power level that should be used by the continuous-wave pulse, but its length will have to be optimized.

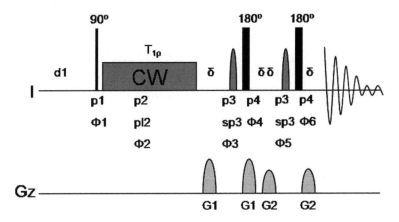

Fig. 4. Pulse scheme used for calibrating the spin-lock filter with water suppression using excitation sculpting with gradients (although this scheme is based on a Bruker pulse sequence, zgesgp *(12)*, the notation and numbering for pulses and delays differs considerably). *d1*: relaxation delay; *p1*: 90° ¹H transmitter pulse; *p2* and *pl2*: length and power for the spin-lock; *p3* and *sp3*: length and power for the 180° shaped pulse; *p4*: 180° ¹H transmitter pulse; *δ*: delay for homospoil/gradient recovery (200 μs); *G1* and *G2*: gradients (gradient ratio 31:11 and length 1 ms). Phase cycle: *Φ1* = x; *Φ2* = 8(y), 8(–y); *Φ3* = x, y; *Φ4* = –x, –y; *Φ5* = 2(x), 2(y); *Φ5* = 2(–x), 2(–y); aq = x, –x, –x, x.

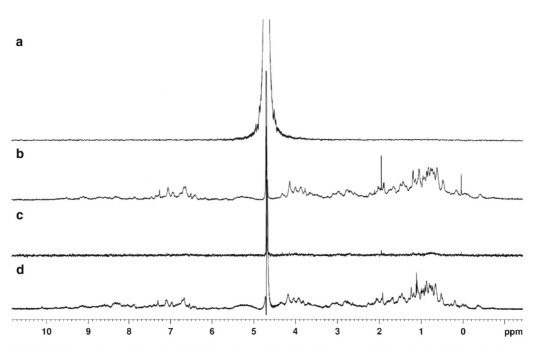

Fig. 5. (**a**) Reference ¹H spectrum of 20 μM *Ct*CBM11 sample in D₂0, displaying a very broad peak for the water signal. Due to the water signal and the small concentration we cannot see any signal from the protein. (**b**) ¹H spectrum of the same sample using water suppression via excitation sculpting with gradients. The few sharp resonances are due to low molecular weight impurities. (**c**) ¹H spectrum recorded with a T₁ₚ filter, consisting of a continuous-wave pulse with 2 kHz field and 125 μs length. Only the resonances from the low molecular weight impurities remain in the spectrum. (**d**) STD spectrum showing that, irradiating on resonance at 7.1 ppm for 2 s, the entire protein is saturated uniformly. We can also see that the impurities contained in the sample are effectively subtracted and do not give rise to signals in the difference spectrum. The spectra were obtained in a Bruker Avance II 600-MHz spectrometer, at 298K.

2. Set the spin-lock length to 50 ms (p2 in **Fig. 4**), define the NS and dummy scans (8*n and 4, respectively) and acquire the spectrum (**Fig. 5c**).

3. Optimize the spin-lock length to the minimum value necessary to suppress the protein signals; this will ensure the maximum sensitivity (*see* **Note 2**).

3.1.4. Setting up the Protein Saturation

1. Using the same sample and the optimized parameters create a new experiment. In order to set up the selective saturation of the protein we will use the pulse program corresponding to the STD experiment (pulse prog.: *STD with shaped pulse train for saturation, alternating between on and off resonance with water suppression using excitation sculpting with gradients and spin-lock to suppress protein signals* (*12, 16, 17*) (**Fig. 1**)).

2. Set the 180° shaped pulse for the solvent suppression, the 90° shaped pulse for the selective saturation of the protein, and turn off the T₁ₚ filter (the spin-lock filter has to be turned off because in order to calibrate the protein saturation the protein signals must be visible in the final spectrum).

3. Create the frequency list for the E-Burp-1 pulse train (*on resonance* – at a value where only resonances from the protein nuclei are located (at least 700 Hz away from any ligand signal)); usually values around –1 ppm or 11–12 ppm (if the ligand has no aromatic protons) work very well; *off resonance* – at a value far away from any protein or ligand resonance (for example 30 ppm). The offsets used in this example are 7.1 ppm as on-resonance frequency and 14 ppm as off-resonance frequency (the offsets can be determined from the spectra acquired in **Subheading 3.1.2**).

4. Define the saturation time. Depending on the pulse program syntax and spectrometer you either enter the total time of saturation or the number of loops (n) for repetition of the selective saturation pulse, having in mind that the total saturation time = number of loops × saturation pulse length. In our example we have used 2 s of saturation time, corresponding to 40 loops of the 50-ms length selective pulse.

5. Define the NS and dummy scans (16*n and 4, respectively) and acquire the spectrum.

6. Depending on your software you may obtain immediately the difference spectrum or the separate on- and off-resonance spectra. With the program used for this protocol we obtain a pseudo-2D spectrum where the rows correspond to the different saturation frequencies (on- and off-resonance). The difference spectrum is obtained after processing the individual rows.

7. Process the spectrum and extract the two rows.

8. Subtract the on-resonance from the off-resonance and the resulting spectrum is the STD spectrum. In this spectrum you should see all the signals of the protein meaning that the entire protein is saturated uniformly. In this case you can use the irradiation frequencies and saturation time to perform your STD experiment **(Fig. 5d)**. Otherwise you can try to choose another on-resonance frequency or a different saturation time.

3.1.5. The STD–NMR Experiment

1. Prepare the NMR sample. In this example we have added 30 μL of cellohexaose 4 mM to the *Ct*CBM11 (20 mM) NMR sample used in **Subheading 3.1.1** (CtCBM11)$_f$ ≈ 18 μM and (Cellohexaose)$_f$ ≈ 364 μM which yields approximately a 20-fold ligand excess.

2. Record a regular proton spectrum. Start with a regular ^1H spectrum to set up the shim again and to update your parameters namely the transmitter frequency offset (on the water signal).

3. Check your water suppression by performing an experiment with the parameters determined in **Subheading 3.1.2**. Update your spectral width if necessary and the offsets for the on- and off-resonance saturation pulse.

4. Create a new experiment number in order to perform the STD experiment (pulse prog.: the same as in **Subheading 3.1.4**).

5. Set the 180° shaped pulse for the solvent suppression, the 90° shaped pulse for the selective saturation of the protein, the $T_{1\rho}$ filter, the frequency list, and the saturation time.

6. Repeat the procedure described in **steps 3–6** in **Subheading 3.1.4**.

Figure 6 compares the STD result (B) with the proton spectra (A). The strongest signals correspond to the closest contact with the protein.

3.1.6. The STD Amplification Factor (Epitope Mapping)

This factor allows a better assessment of the absolute magnitude of the STD effect. The STD amplification factor is the fractional saturation of a given proton multiplied by the excess of the ligand over the protein. Usually, the saturation degree of an individual ligand proton reflects the proximity of this proton to the target surface. When creating an epitope map, the STD intensity relative to the reference is used to create the binding epitope maps; this is usually described by the STD factor A_{STD}, as shown by **Eq. 2**:

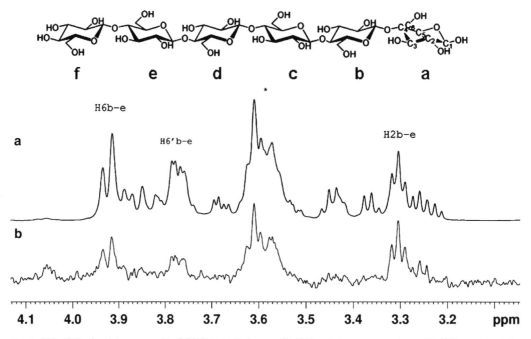

Fig. 6. STD-NMR of cellohexaose with *Ct*CBM11. (**a**) Reference ^{1}H NMR cellohexaose spectrum. (**b**) STD spectra of the solution of cellohexaose (364 μM) with the protein (18 μM). Protons H6b-e and H2b-e show the more intense signals, indicating that upon binding, these are the ones closer to the protein. In the region between 3.63 and 3.52 ppm, the signal overlap doesn't allow to determine the individual contributions of the protons to the binding. The spectra were obtained in a Bruker Avance II 600-MHz spectrometer, at 298K, with 256 scans and a spectral width of 6,410 Hz centered at 2,817 Hz. The protein saturation was achieved through the application of a train of E-Burp-1 pulses of 50 ms of pulse length in a total saturation time of 2 s. The on- and off-resonance saturation frequency switched between 7 and 14 ppm. The spectra were processed using the software TOPSPIN 2.0 from Bruker.

$$A_{STD} = \frac{I_0 - I_{sat}}{I_0} \times \text{ligand excess} \qquad (2)$$

where I_0 corresponds to the intensity measured in the reference spectrum and $I_0 - I_{sat}$ corresponds to the intensity measured in the STD spectra.

1. From integration of the STD spectra, calculate the STD amplification factors for all the signals in the STD spectrum (*see* **Subheading 3.1.5**).

2. Set the proton with the largest A_{STD} value to 100%.

In **Fig. 7** the structure of cellohexaose and the corresponding degrees of saturation, as determined according to this procedure, are presented. The degree of saturation was only determined for those protons whose resonances are not overlapped and of unequivocal assignment (H2 and H6 of the central glucose units). Note that the diastereotopic protons H6 have different saturation values which can be an indication of a very precise bound conformation.

3.1.7. STD Build Up (with Saturation Time)	To study the time course of STD epitope mapping, a build-up curve of STD factor against the saturation time was used (**Fig. 8**). The evolution of the A_{STD} against the saturation time allows determining the optimal conditions (saturation time (in this case) or ligand excess) for the STD-NMR experiment.

1. From the STD experiment performed in **Subheading 3.1.5** create a new set of experiments in which you will vary the saturation time. You only need to create a set of experiments where you will change the time delay corresponding to the saturation time. Depending on the pulse program syntax and spectro-meter you either enter the total saturation time or the number of loops (n) for the repetition of the selective saturation pulse (have in mind that the total saturation time = number of loops × saturation pulse length). In our example we

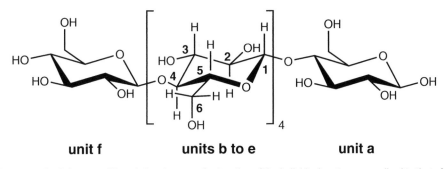

unit f   **units b to e**   **unit a**

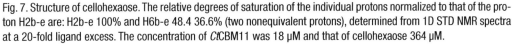

Fig. 7. Structure of cellohexaose. The relative degrees of saturation of the individual protons normalized to that of the proton H2b-e are: H2b-e 100% and H6b-e 48.4 36.6% (two nonequivalent protons), determined from 1D STD NMR spectra at a 20-fold ligand excess. The concentration of *Ct*CBM11 was 18 μM and that of cellohexaose 364 μM.

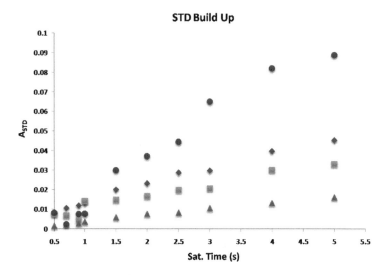

Fig. 8. STD build-up curve of protons H6b-e + βH6a + αH6a (*filled diamond*), H6′b-e + βH6′a + αH5a (*filled rectangle*), H5b-e + H3b + αH4a + βH3a + βH5a + H4b-e + αH3a (*filled triangle*), and H2b-e (*filled circle*). The data indicate that low saturation times (less than 1.5 s) is not favorable due to the low signal intensity of STD. Very long saturation times do not help either because they do not increase the STD sensitivity. The measurements were performed with 18 µM of *Ct*CBM11 and 364 µM of cellohexaose.

have used the follo-wing times corresponding to loops of for a selective pulse of 50 ms length: 0.5, 0.7, 0.9, 1.0, 1.5, 2.0, 2.5, 3.0, 4.0 and 5.0 s.

2. Process all the spectra as previously.

3. Take the last off-resonance spectrum as the reference one and integrate it.

4. Integrate all the STD spectra relatively to the reference one.

5. Plot the STD amplification factor vs. Saturation Time.

3.1.8. Screening by STD-NMR

The first application of the STD experiment was to screen a library of carbohydrate molecules for binding activity toward a carbohydrate-binding protein, wheat-germ agglutinin (WGA), which is of prime importance for the directed development of drugs *(17, 18)*.

To screen a library of possible ligands using the STD-NMR technique you just have to prepare your sample (target protein and the ligands) and proceed as previously. The combination of the STD technique with other standard 2D NMR experiments used for structural analysis allows the identification of the structure of the ligand that binds to the protein *(18)*. This methodology lies however outside the scope of this protocol.

3.2. Ligand Screening by Diffusion Ordered Spectroscopy

Self-diffusion is the random translational motion of molecules driven by their internal kinetic energy. Self-diffusion coefficients and the structural properties of a molecule are connected by the dependence of the self-diffusion coefficients on molecular size and shape; therefore, it is not surprising that the determination of molecular self-diffusion coefficients by pulsed field gradient (PFG) spin-echo NMR experiments has become a valuable methodology for studies of molecular interaction in solution. In the pharmaceutical industry, diffusion-based NMR techniques have been used in a wide range of applications, such as screening of chemical mixtures, determining the structures of bound ligands without physical separation, and measuring the diffusion coefficient of small metabolites in biofluids among others *(19, 20)*. The concept behind the application of diffusion NMR techniques for binding and screening studies is very simple and is based on the fact that the diffusion coefficient of a small molecule is altered upon binding to a large receptor. For this Type of studies it is usually sufficient to identify compounds that bind to a certain receptor from a mixture of nonbinding compounds, or to establish a relative binding affinity. The determination of association constants or size requires however the quantitative determination of the diffusion coefficients with precision and accuracy.

With DOSY it is possible to obtain the diffusion of individual compounds from a mixture separated in different rows, resembling a chromatographic separation *(21, 22)*.

From the extensive library of pulse programs to measure diffusion by NMR *(23)* we have chosen the stimulated echo using bipolar pulse gradients, incorporating water suppression by presaturation during the relaxation and diffusion delays (**Fig. 9**).

For this pulse sequence the relation between signal intensity and diffusion coefficient *D* is given by **Eq. 3**:

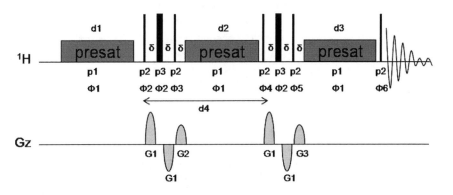

Fig. 9. (2D) Pulse scheme for diffusion measurement using stimulated echo and LED, bipolar gradient pulses for diffusion, two spoil gradients and with presaturation during relaxation delay (although this scheme is based on a Bruker pulse sequence, ledbpgppr2s *(12, 24)*, the notation and numbering for pulses and delays differs considerably). *d1*: relaxation delay; *d2*: delay calculated automatically from *d4*; *d3*: eddy-current delay; *d4*: diffusion time (big delta); *p1*: power level for presaturation; *p2*: 90° ^{1}H transmitter pulse; *p3*: 180° ^{1}H transmitter pulse; δ: delay for homospoil/gradient recovery (200 μs); *G1, G2,* and *G3*: gradients (gradient ratio 100: −17.13:−13.17 and length 1.1 ms for the encoding/decoding gradient *G1* (little delta * 0.5) and 600 μs for the spoil gradients *G2* and *G3*). Phase cycle is: $\Phi 1$ = x; $\Phi 2$ = x; $\Phi 3$ = 2(x), 2(-x); $\Phi 4$ = 4(x), 4(−x), 4(y), 4(−y); $\Phi 5$ = x, −x, x, −x, −x, x, −x, x, y, −y, y, −y, −y, y, −y, y; $\Phi 6$ = 4(x), 4(−x), 4(y), 4(−y); aq = x, −x, −x, x, −x, x, x, −y, y, y, −y, y, −y −y, y.

$$I = I_0\, e^{-D\gamma^2 g^2 \delta^2 (\Delta - \delta/3 - \tau/2)} \tag{3}$$

where γ is the magnetogyric ratio, g is the gradient strength, δ is the width of the gradient pulse, Δ is the diffusion period length, τ is the gradient recovery time, I is the signal intensity, and I_0 is the I when g is zero.

As mentioned earlier the experiment is based on the fact that the diffusion coefficient of a small molecule is altered upon binding to a large receptor. In this protocol we describe the experiment applied as a screening test to a mixture of two ligand candidates with very similar structure: one binding (cellohexaose) and the other nonbinding (laminarihexaose). As earlier the experiments are performed with diluted solutions in D_2O, so we start by setting-up the solvent suppression scheme.

3.2.1. Setting up the Solvent Suppression

1. Prepare four NMR samples in 3-mm tubes (*a*: cellohexaose, 40 µM in D_2O with TSP; *b*: Laminarihexaose, 40 µM in D_2O with TSP; *c*: mixture of cellohexaose and laminarihexaose, 40 µM in D_2O with TSP, *d*: mixture of cellohexaose, laminarihexaose, and *Ct*CBM11, 40 µM in D_2O with TSP).

2. Start with a regular ^1H spectrum in sample *a* (tune and match the probe head, lock, shim, and reference phase) and record any optimized parameters: (relaxation delay = 2 s, transmitter frequency offset = on the solvent ^1H frequency, spectral width = 20 ppm, time domain = 32k).

3. Create an experiment in order to calibrate the solvent suppression using presaturation (pulse prog.: *water suppression with presaturation (12)* **Fig. 10**); you will have to set up the power level and time for the presaturation and to fine tune the freq-uency offset for best results. Typically the presaturation time is set to 1 or 2 s and the power level and frequency offset optimized to obtain the best suppression.

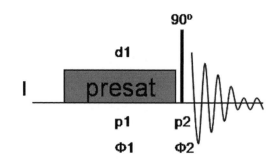

Fig. 10. Pulse scheme for water suppression with presaturation (although this scheme is based on a Bruker pulse sequence, zgpr *(12)*, the notation and numbering for pulses and delays differs considerably). *d1*: relaxation delay; *p1*: power level for presaturation; *p2*: 90° ^1H transmitter pulse. Phase cycle is: *Φ1* = x; *Φ2* = x, −x, −x, x, y, −y, −y, y; aq = x, −x, −x, x, y, −y, −y, y.

*3.2.2. Setting up
the Decay Function*

The diffusion experiment as will be described here corresponds to the acquisition of a series of 1D spectra with increasing gradient strength and fixed diffusion time. This is done to keep the timing constant throughout the whole experiment.

The analysis of the signal attenuation with increasing gradient strength, according to **Eq. 3**, allows to compute the diffusion coefficient and to construct the DOSY spectrum. It is however necessary to optimize the parameters that determine the decay function (gradient strength g, diffusion time Δ, and gradient length δ) for the sample in question. The optimization is easily done by running a few 1D measurements (1D version of the diffusion pulse programs *(12)*) or by running several diffusion experiments with only a few gradient increments.

1. Create a new experience, update the spectral parameters (frequency offset on the water signal and the correct power level for water presaturation), and select the diffusion sequence to use (in this example we have used *diffusion measurement using stimulated echo and LED, bipolar gradient pulses for diffusion, two spoil gradients and with presaturation during relaxation delay* – ledbpgppr2s *(12)*).

2. Set the diffusion time, Δ (*d4*), to 800 ms and the gradient length, δ (*G1*), to 1.1 ms. Your software should allow you to set up a gradient ramp corresponding to the several steps of the gradient strength increments to be used in the diffusion experiment. Start with 2% of the maximum gradient amplitude and build a linear ramp with eight steps until 95% of the maximum amplitude (*see* **Notes 3–5**).

3. Acquire the spectra with a minimum NS (8*n). As a result you will obtain a pseudo-2D spectrum. Fourier transform and phase the spectra in the direct dimension only (the other dimension corresponds to the increase in gradient strength).

4. Using your software, analyze the transformed spectra by examining the rows of the pseudo-2D spectrum. The first row should be used as your reference; note the change in signal intensity with increasing gradient strength from row to row. The signal decay should go down to 5% residual signal for the maximum gradient strength (i.e., 95% signal attenuation due to diffusion).

The observable residual signal intensity is, of course, dependent on the signal-to-noise ratio (S/N). For bad S/N you may have to increase the NS or go with less signal attenuation. The smallest signal to be detected (i.e., at highest gradient strength) has to be above the noise. If the signal intensity is already totally gone, reduce the gradient strength (*G1*). If the signal is still too big, you have to increase either the diffusion time Δ or the gradient length δ. Increasing δ is favorable, because it results in a bigger effect. δ^2 determines the signal attenuation, while Δ only affects the exponential decay function linearly.

5. Proceed in the same way for samples **b–d**.

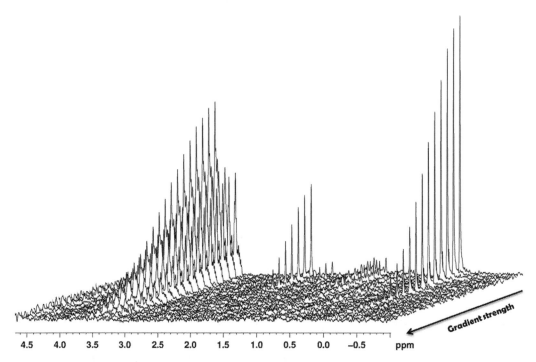

Fig. 11. Diffusion decay spectra of a mixture of cellohexaose and laminarihexaose, 40 μM in D₂O with TSP. The spectra were acquired in a Bruker Avance II 600-MHz spectrometer, at 298K, with 512 scans in 32 steps (only the first 22 steps are represented) and a spectral width of 12,376 Hz in the direct dimension centered in the solvent frequency. The length of the encoding/decoding gradient (*G1*) was 1.5 ms and 600 μs for the spoil gradients (*G2* and *G3*). The diffusion time was 400 ms. The spectra were processed using the software TOPSPIN 2.0 from Bruker.

Figure 11 shows the diffusion decay spectra of sample *c* (mixture of cellohexaose and laminarihexaose, 40 μM in D₂O with TSP) by varying the gradient strength from 2 to 95% in 22 steps.

3.2.3. The DOSY Experiment (Screening)

1. Create a new experiment from the dataset you used for the optimization. All parameters are already set correctly; only the new gradient ramp has to be generated.

2. To set the gradient ramp define the initial and the final values of the gradient ramp (2–95%), and the number of steps (i.e., number of increments), and the Type of ramp (linear, *l*, or squared, *q*); we have chosen a linear ramp with 32 steps.

3. Fourier transform and phase the spectra as in **step 3** from **Subheading 3.2.2**.

4. In order to process the DOSY spectrum you will have to input several diffusion-related processing parameters. The way you input these parameters will depend on the software package you are using. Start by processing the spectrum as a pseudo-2D spectrum **(step 3)** and correct the baseline in the direct dimension for all rows of the pseudo-2D matrix. Your software should have a module to analyze the decay of

the magnetization vs. gradient strength, according to **Eq. 3**, in order to compute the diffusion coefficients. For this module to work you will need to introduce the diffusion-related acquisition parameters relevant to the analysis of the decay. These are the diffusion time (Δ), gradient length (*G1*), gradient ramp (increments of gradient strength for each row of the pseudo-2D matrix), and the magnetogyric ratio of the observed nucleus (4257.70020 Hz/G for ^{1}H) *(20, 21, 24)*.

Typically the decays can be analyzed by monitoring the intensity or the integral of the signals. The Type of scaling used to construct the diffusion spectrum is usually logarithmic. In case you are using a Buker spectrometer proceed as in **Note 6**. As a result you should obtain a 2D DOSY spectrum with chemical shifts along the direct dimension axis and diffusion coefficients along the indirect dimension axis.

Alternatively you can choose to define integration regions and integrate all 1D spectra of the pseudo-2D matrix. With these integral values and the same diffusion-related experimental parameters described in **step 4** you can use a spreadsheet to determine the diffusion coefficients by fitting **Eq. 3** (or its logarithm) to the experimental values *(25)*.

5. Proceed in the same way for samples **b–d**.

Figure 12 compares the diffusion results from the mixture of sugars with (B) and without protein (A). From these results it is possible to say, only by direct observation of the DOSY spectra, that there is an interaction between cellohexaose and the protein (there was a change in the diffusion coefficient of cellohexaose).

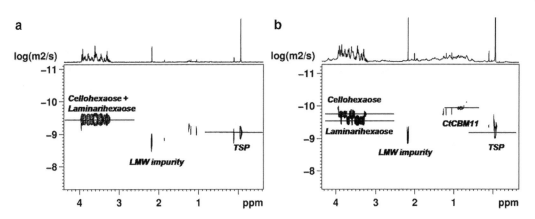

Fig. 12. (**a**) DOSY spectrum from the mixture of cellohexaose and laminarihexaose, 40 μM in D$_2$O with TSP, (**b**): DOSY spectrum from the mixture of cellohexaose, laminarihexaose, and *Ct*CBM11, 40 μM in D$_2$O with TSP. The spectra were acquired in a Bruker Avance II 600-MHz spectrometer, at 298K, with 512 scans in 32 steps and a spectral width of 12,376 Hz in the direct dimension centered in the solvent frequency. The length of the encoding/decoding gradient (*G1*) was 1.5 ms in (**a**) and 1.1 ms in (**b**) and 600 μs for the spoil gradients (*G2* and *G3*). The diffusion time was 400 ms in (**a**) and 800 ms in (**b**). The spectra were processed using the software TOPSPIN 2.0 from Bruker. *LMW*: low molecular weight.

This interaction can be quantified in terms of the association constant (K_a) using the following equations *(20)*:

$$K_a = \frac{[PL]}{[P][L]} \tag{4}$$

where K_a is the association constant; and (PL), (P), and (L) are the equilibrium concentrations of the protein-ligand complex, protein, and ligand, respectively.

$$f_{PL} = \frac{D_L - D_{obs}}{D_L - D_{PL}} \tag{5}$$

where f_{PL} is the molar fraction of the complex; and D_L, D_{obs}, and D_{PL} are the diffusion coefficients of the free ligand, the ligand when bound to the protein, and the protein when bound to the ligand, respectively, divided by the diffusion coefficient of the TSP (to account for viscosity changes *(20, 26)*). If it is assumed that D_{PL} is the same as the measurable diffusion of the free protein (D_p), then f_{PL} can be easily determined. Accounting for mass balance and combining **Eqs. 4** and **5** we get the expression for the association constant:

$$K_a = \frac{f_{PL}}{\left(1 - f_{PL}\right)\left([P]_0 - f_{PL}[L]_0\right)} \tag{6}$$

where $(P)_0$ and $(L)_0$ represent the total concentrations of protein and ligand, respectively.

As mentioned earlier it is possible to conclude that cellohexaose binds to *Ct*CBM11 whereas laminarihexaose does not (the diffusion coefficient of cellohexaose decreases when the protein is added to the mixture of sugars and the one from laminarihexaose remains the same). This way we demonstrate the ability of diffusion experiments to screen mixtures of compounds.

Using the equations given earlier we were able to calculate the association constant for the binding of cellohexaose to *Ct*CBM11: $K_a = 6.33 \times 10^4$ M^{-1}. This result is in agreement with previous studies *(6)*.

4. Notes

1. On a Bruker instrument use the "shape tool" (write *stdisp* on the command line) to calibrate the pulse power. After launching the shape tool, open the shape "Squa100.1000". In the

menu "*Analysis*" choose the option "*Integrate Shape*" and set the length of the pulse (2,000 μs), the total rotation (180°) and the width of the 90° hard pulse (p1). In the "*Results*" box click on "*update parameters*" and the correct power will be copied to your experiment. The same procedure can be repeated to calibrate for a variety of shaped pulses with the selection of the correct shape.

2. Depending on your sample you may want to have a broader spin-lock. Just calculate the power of a 90° pulse with the width that gives you the desired field according to **Eq. 1**. Use this power and adjust the spin-lock length (p2 in **Fig. 4**) for optimal results. Don't forget that the spin-lock filter acts as a relaxation filter used to suppress the residual protein signals – important for clarity where proteins and ligand signals overlap in 1D ^{1}H spectra – use the minimum length that allows you to suppress the protein signals in order to achieve maximum sensitivity in your STD experiment.

3. This procedure is preferable than acquiring just two spectra, one with 2% gradient amplitude and another with 95%, since it allows to observe the behavior of the decay in the eight steps.

4. It is important to start with a gradient strength bigger than 0, because you may get unwanted echoes when you don't apply a gradient.

5. It is recommended 95% to make sure that there is no non-linear behavior of the gradient amplifier at the end of the amplification range, but you may go up to 100%.

6. When acquisition finishes Type "*xf2*" and then "*abs2*" (set the limits ABSF1 and ABSF2 in F2 to 1,000.0 and –1,000.0 (outside the limits of the actual dataset), set the degree of the polynomial fitting function to 0 (ABSG(F2)) and execute "*abs2*"). Then Type "*setdiffparm*" to transfer some important parameters like Δ and δ into the appropriate parameters for the next processing (a wrong setting will obviously lead to wrong diffusion constants). Type "*dosy2d setup*" for calling the editor for the diffusion processing parameters. The only parameters you have to consider for a first trial processing step are the fitted intensity meaning (integral or intensity) and the Type of scaling in F1 (linear or logarithmic). Process the data by executing "*dosy2d*". Currently the software is not able to display a diffusion scale correctly; it can only display linear axes with ppm, Hz, or seconds as units. Even though you will see ppm on the F1 axis, you should read this as diffusion coefficient. If you have chosen a logarithmic scale just read the ppm values as log D values in m^2/s (i.e., –8.6 means $10^{-8.6} = 2.5 \times 10^{-9}$ m^2/s). In the case of a linear scale you will directly see the diffusion coefficient.

References

1. Salvatella, X. and Giralt, E. (2003) NMR-based methods and strategies for drug discovery. *Chem. Soc. Rev.* **32**, 365–372

2. Meyer, B. and Peters, T. (2003) NMR Spectroscopy techniques for screening and identifying ligand binding to protein receptors. *Angew. Chem. Int. Ed.* **42**, 864–890

3. Jahnke, W. and Widmer, H. (2004) Protein NMR in biomedical research. *Cell. Mol. Life Sci.* **61**, 580–599

4. Carlomagno, T. (2005) Ligand-target interactions: what can we learn from NMR? *Annu. Rev. Biophys. Biomol. Struct.* **34**, 245–266

5. Klages, J., Coles, M., and Kessler, H. (2006) NMR-based screening: a powerful tool in fragment-based drug discovery. *Mol. Biosyst.* **2**, 318–332

6. Carvalho, A. L., Goyal, A., Prates, J. A. M., Bolam, D. N., Gilbert, H. J., Pires, V. M. R., Ferreira, L. M. A., Planas, A., Romão, M. J., and Fontes, C. M. G. A. (2004) The family 11 carbohydrate-binding module of *Clostridium thermocellum* Lic26A-Cel5E accommodates beta-1,4- and beta-1,3–1,4-mixed linked glucans at a single binding site. *J. Biol. Chem.* **279**, 34785–34793

7. Viegas, A., Brás, N., Cerqueira, N., Fernandes, P. A., Prates, J. A. M., Fontes, C. M. G. A., Bruix, M., Romão, M. J., Carvalho, A. L., Ramos, M. J., Macedo, A. L., and Cabrita, E. J. (2008) Molecular determinants of ligand specificity in family 11 carbohydrate binding modules (CBM11): an NMR, X-ray crystallography and computational chemistry approach. *FEBS J.* **275**, 2524–2535

8. Béguin, P. and Lemaire, M. (1996) The cellulosome: an exocellular, multiprotein complex specialized in cellulose degradation. *Crit. Rev. Biochem. Mol. Biol.* **31**, 201–236

9. Demain, A. L., Newcomb, M., and Wu, J. H. D. (2005) Cellulase, clostridia, and ethanol. *Microbiol. Mol. Biol. Rev.* **69**, 124–154

10. Hashimoto, H. (2006) Recent structural studies of carbohydrate-binding modules. *Cell. Mol. Life Sci.* **63**, 2954–2967

11. Streiff, J. H., Juranic, N. O., Macura, S. I., Warner, D. O., Jones, K. A., and Perkins, W. J. (2004) Saturation transfer difference nuclear magnetic resonance spectroscopy as a method for screening proteins for anesthetic binding. *Mol. Pharm.* **66**, 929–935

12. Parella, T. (2004) Pulse Program Catalogue, NMR Guide 4.0, *Bruker BioSpin.*

13. Bauer, C., Freeman, R., Frenkiel, T., Keeler, J., and Shaka, A.J. (1984) Gaussian pulses. *J. Magn. Reson.* **58**, 442–457.

14. Kessler, H., Oschkinat, H., and Griesinger, C. (1986) Transformation of homonuclear two-dimensional NMR techniques into one-dimensional techniques using Gaussian pulses. *J. Magn. Reson.* **70**, 106–133

15. Cutting, B., Shelke, S. V., Dragic, Z., Wagner, B., Gathje, H., Kelm, S., and Ernst, B. (2007) Sensitivity enhancement in saturation transfer difference (STD) experiments through optimized excitation schemes. *Magn. Reson. Chem.* **45**, 720–724

16. Hwang, T. L. and Shaka, A. J. (1995) Water suppression that works – excitation sculpting using arbitrary wave-forms and pulsed-field gradients. *J. Magn. Reson. A* **112**, 275–279

17. Mayer, M. and Meyer, B. (1999) Characterization of ligand binding by saturation transfer difference NMR spectroscopy. *Angew. Chem. Int. Ed.* **38**, 1784–1788

18. Meyer, B. and Peters, T. (2003) NMR spectroscopy techniques for screening and identifying ligand binding to protein receptors. *Angew. Chem. Int. Ed.* **42**, 864–890

19. Yan, J. L., Kline, A. D., Mo, H. P., Zartler, E. R., and Shapiro, M. J. (2002) Epitope mapping of ligand-receptor interactions by diffusion NMR. *J. Am. Chem. Soc.* **124**, 9984–9985

20. Torsten, B., Cabrita, E. J., and Berger, S. (2005) Intermolecular interaction as investigated by NOE and diffusion studies. *Prog. Nucl. Mag. Res. Sp.* **46**, 159–196

21. Johnson, C. S. (1999) Diffusion ordered nuclear magnetic resonance spectroscopy: principles and applications. *Prog. Nucl. Mag. Res. Sp.* **34**, 203–256

22. Morris, K. F. and Johnson, C. S. (1992) Diffusion-ordered 2-dimensional nuclear-magnetic-resonance spectroscopy. *J. Am. Chem. Soc.* **114**, 3139–3141

23. Antalek, B. (2002) Using pulsed gradient spin echo NMR for chemical mixture analysis: how to obtain optimum results. *Concepts Magn. Reson.* **14**, 225–258

24. Wu, D. H., Chen, A. D., and Johnson, C. S. (1995) An improved diffusion-ordered spectro-scopy experiment incorporating bipolar-gradient pulses. *J. Magn. Reson. A* **115**, 260–264

25. Huo, R., Wehrens, R., van Duynhoven, J., and Buydens, L. M. C. (2003) Assessment of techniques for DOSY NMR data processing. *Anal. Chim. Acta* **490**, 231–251

26. Cabrita, E. J. and Berger, S. (2001) DOSY studies of hydrogen bond association: tetramethylsilane as a reference compound for diffusion studies. *Magn. Reson. Chem.* **66**, S142–S148

Chapter 7

Isothermal Titration Calorimetry and Differential Scanning Calorimetry

Geoff Holdgate

Summary

Isothermal titration [Holdgate (BioTechniques 31:164–184, 2001); Ward and Holdgate (Prog. Med. Chem. 38:309–376, 2001); O'Brien et al. (2001) Isothermal titration calorimetry of biomolecules. In: Harding, S. E. and Chowdhry, B. Z. (eds.), *Protein–Ligand Interactions: Hydrodynamics and Calorimetry, A Practical Approach*. Oxford University Press, Oxford, UK] and differential scanning calorimetry [Jelesarov and Bosshard (J. Mol. Recognit. 12:3–18, 1999); Privalov and Dragan (Biophys. Chem. 126:16–24, 2007); Cooper et al. (2001) Differential scanning microcalorimetry. In: Harding, S. E. and Chowdhry, B. Z. (eds.), *Protein-Ligand Interactions: Hydrodynamics and Calorimetry, A Practical Approach*. Oxford University Press, Oxford, UK] are valuable tools for characterising protein targets, and their interactions with ligands, during the drug discovery process. The parameters obtained from these techniques: ΔH, ΔG, ΔS, and ΔC_p, are properties of the entire system studied and may be composed of many contributions, including the binding reaction itself, conformational changes of the protein and/or ligand during complexation, changes in solvent organisation or other equilibria linked to the binding process. Dissecting and understanding these components, and how they contribute to binding interactions, is a critical step in the ability to design ligands that have high binding affinity for the target protein.

Key words: Isothermal titration calorimetry, Differential scanning calorimetry, Ligand binding, Protein stability, Thermodynamics, Drug discovery

1. Introduction

Both intermolecular ligand-binding reactions, and the intramolecular formation of the unique spatial orientation of folded macromolecules, involve molecular recognition processes, governed by thermodynamics. The thermodynamic profiles of these molecular recognition events can be complicated, involving different energetic transactions along the pathway from free partners to final complex.

Ana Cecília A. Roque (ed.), *Ligand-Macromolecular Interactions in Drug Discovery: Methods and Protocols*,
Methods in Molecular Biology, vol. 572,
DOI 10.1007/978-1-60761-244-5_7, © Humana Press, a part of Springer Science+Business Media, LLC 2010

The accessibility of high-sensitivity, commercially available calorimeters has established isothermal titration calorimetry (ITC) and differential scanning calorimetry (DSC) as the methods of choice for the study of the thermodynamics controlling these processes. Moreover, the combination of these techniques, which allow the direct measurement of the heat effects associated with these inter- and intramolecular recognition events under defined conditions, with structural data provides us with the tools to begin to understand how and why molecules interact. This knowledge has practical application, as an understanding of the forces governing protein stability and biomolecular interaction is key to driving drug discovery towards high affinity ligands showing the desired biological effects at specific targets required by the pharmaceutical industry.

The universality of the detection systems provides calorimetric methods with a distinct advantage over many other methods that can be applied toward this goal, in that the same instrumentation, methods, and protocols can often be used to study very different systems. This, coupled with high-sensitivity, rapid equilibration times and stable baselines, ensures that modern ITC and DSC instruments are important components of the wide range of biophysical and biochemical approaches that are available to be used within the course of drug discovery and development.

2. Materials

2.1. Dialysis

1. Suitable dialysis tubing or cassettes, with appropriate molecular weight cut-off (MWCO) membrane (*see* **Note 1**).
2. Dialysis buffer, which will serve as the buffer for the ITC/DSC experiment (*see* **Note 2**).
3. Dialysis vessel capable of containing 2–5 L of dialysis buffer. Retain the dialysis buffer following the dialysis.

2.2. Concentration Measurement

1. Spectrophotometer or plate-reader capable of carrying out scans between 230 and 350 nm.
2. Cuvette (such as spectrosil or suprasil quartz) or micro-plate (such as Corning UV micro-plates) suitable for use at wavelengths between 230 and 350 nm.

2.3. Preparation of ITC Samples

1. Protein (or nucleic acid) sample, preferably above a concentration of 5 µM, and more usually above 10 µM.
2. Dialysis buffer from the exhaustive dialysis; see earlier.
3. Ligand solution. Ideally the solid ligand should be dissolved in the remaining dialysis buffer to give a concentration typically around 5–20 times the protein concentration, or possibly higher if weak binding is suspected.

2.4. Preparation of DSC Samples

1. Protein (or nucleic acid) sample typically at or above a concentration of 5 µM.

2. Dialysis buffer from the exhaustive dialysis; see earlier.

3. If a ligand solution is to be used, it should be prepared as earlier for ligands for ITC at a concentration usually above the protein concentration.

2.5. ITC Instrumentation

The most widely used instruments are those originating from CSC (now part of the TA Instruments Division of Waters Corporation) and Microcal LLC (Now part of GE Healthcare). The Nano ITC from TA and the VP-ITC and iTC$_{200}$ from Microcal operate in similar principles, based on isothermal power compensation. The sample and reference cells are contained in an adiabatic chamber, with the sample cell being accessible for delivery of solution from the injection syringe. The reference cell is used as a thermal blank usually maintained filled with water or with 0.05%, w/v sodium azide to prevent bacterial growth. A feedback system provides a small, constant power to the reference cell and a small, variable power to the sample cell maintaining the temperature difference between the cells close to zero. Endothermic processes occurring in the sample cell lead to an increase in the applied power, and exothermic process lead to a decrease. The instrument monitors the power applied in this way over time, and integration with respect to time gives the amount of heat generated or absorbed, which can be related to the enthalpy of the reaction if the concentration of reactants is known.

The location of the ITC instrument should be considered carefully to minimise electrical interference through space or the power supply, as well as local heat fluctuations, which may affect the baseline quality (*see* **Note 3**).

2.6. DSC Instrumentation

Many manufacturers produce DSC or differential thermal analysis instruments that use small amounts of sample in either liquid or solid form. However, the majority of these instruments are unsuitable for biophysical or biochemical applications, which use relatively dilute samples in aqueous buffers. The reason is that the major source of the heat capacity in this situation is from the solvent. Thus, instruments that accurately compensate for these effects must be used for protein stability and/or ligand-binding studies using DSC. The two companies mentioned earlier in the ITC section also produce DSC instruments, the Nano DSC from TA and the VP-DSC from Microcal, which are capable of analysing dilute biological materials. There is also an automated version of the Microcal instrument (*7*). The principle of operation is similar to the ITC instruments, in that the instruments operate via a power compensation system. In the upscan mode, applying a constant power to the heaters raises the temperature of the adiabatic chamber. The power compensation system monitors the difference in temperature between the jacket and the cells and heats the cells to minimise this temperature difference. The difference in power applied to the sample and reference cells is a function of the heat capacity difference between the two cells and is monitored by the instrument as a function of temperature. The operation of the instrument is similar in the downscan mode, except that in this case the cells and jacket are cooled.

As with the ITC instrument, the location of the DSC instrument should be considered carefully in order to minimise through space or power supply electrical interference, or room-based heat fluctuations, which may affect the baseline quality (*see* **Note 3**).

2.7. Data Analysis for ITC

1. Personal Computer (PC) capable of supporting data analysis software.
2. Data analysis software package. Several software packages are suitable for analysing ITC data, for example Origin (supplied with Microcal instruments), BindWorks (supplied with TA instruments), Prism, Scientist, GraFit.
3. Suitable data for analysis.

2.8. Data Analysis for DSC

1. PC capable of supporting data analysis software.
2. Data analysis software package. Several software packages are suitable for analysing DSC data, for example Origin (supplied with Microcal instruments), CpCalc (supplied with TA instruments), Prism, Scientist, GraFit.
3. Suitable data for analysis.

2.9. ITC Instrument Cleaning and Maintenance

1. Buffer solution used for the ITC run.
2. Detergent solution, such as 5–10%, v/v Decon.
3. Purified water, such as double distilled or produced by a Milli-Q water purifier.

2.10. DSC Instrument Cleaning and Maintenance

1. Buffer solution used for the DSC run.
2. Detergent solution, such as 10%, v/v Decon.
3. Concentrated nitric acid.
4. Purified water, such as double distilled or produced by a Milli-Q water purifier.

3. Methods

3.1. Dialysis

Prior to ITC or DSC experiments protein samples should be equilibrated in an appropriate buffer.
1. The dialysis membrane is hydrated by submerging in dialysis buffer for up to 2 min in order to remove any glycerol and increase membrane flexibility.

2. Excess buffer is removed from the membrane without blotting.

3. Protein sample is added to the cassette or dialysis tubing using a syringe (*see* **Note 4**).

4. Air contained within the cassette or dialysis bag is removed without touching the membrane with the needle.

5. The dialysis bag is tied off and double clipped at both ends to ensure sample integrity. Dialysis cassettes are self-sealing via the silicon-covered injection ports.

6. The dialysis cassette or bag, containing sample, is added to the dialysis buffer.

7. Exhaustive dialysis is carried out using large dialysis buffer volumes and several changes of buffer (*see* **Note 5**).

8. Following dialysis the protein sample is removed from the dialysis cassette or bag.

9. The final dialysis buffer is kept for subsequent use in concentration measurement and sample preparation.

3.2. Concentration Measurement

1. Protein (or nucleic acid) sample is loaded into the cuvette or micro-plate (*see* **Note 6**).

2. The cuvette or micro-plate is placed into the spectrophotometer or plate reader and is scanned between the wavelengths of 230 and 350 nm.

3. Appropriate buffer blanks are subtracted from the protein spectrum. The buffer blank should be an aliquot of the saved dialysis buffer.

4. Usually protein concentration is measured by taking a reading of absorbance at 280 nm (260 nm for nucleic acids). Before this is done, the reading at this wavelength should be corrected for any light scattering problems. These effects can be identified by observing the spectrum in wavelength region 320–350 nm, where proteins (and nucleic acids) do not absorb.

5. The correction for scattering is applied. This is achieved by removing the non-zero value of absorbance, if it is constant between 320 and 350 nm. If this region is non-zero but sloping, then a correction should be made as follows:
the logarithm of the absorbance value is plotted against the logarithm of the wavelength. The straight line obtained is extrapolated to 280 nm (260 nm for nucleic acids), and the antilogarithm of the log absorbance value at the relevant wavelength is subtracted from the absorbance value obtained for the protein (or nucleic acid).

6. The concentration is then calculated using the Beer-Lambert law from the protein (or nucleic acid) extinction coefficient, which may be known, measured, or estimated (*see* **Note 7**).

7. The concentration of ligand should also be measured in this way if an extinction coefficient is available, rather than relying on the concentration estimated from the weighed solid (*see* **Note 8**).

3.3. Preparation of ITC Samples

1. The ligand solution is prepared by dissolving the solid ligand in the dialysis buffer saved from the dialysis (*see* **Note 9**).

2. The protein and ligand samples following the dialysis, dissolution, and subsequent concentration measurements are degassed. This is achieved by stirring the protein under vacuum at a temperature lower than the experimental temperature (*see* **Note 10**). Caution should be employed so that the protein solution does not boil or froth.

3. The protein sample is usually added to the calorimeter cell. The filling syringe is loaded with around 2.5 mL (for VP-ITC) of protein sample. To ensure that air bubbles are not added to the cell, any air bubbles contained within the syringe are dislodged from the sides of the syringe barrel, so that they accumulate near the plunger. The syringe is then inserted into the access tube until the tip of the syringe is only a small distance from the bottom of the cell. With the syringe in this position the protein sample is injected in a continual flow, until the cell is filled. Once the protein sample is observed in the access tube, several short pulses are applied to the plunger to administer further protein sample in short bursts. This helps to dislodge any air bubbles present inside the calorimetric cell.

4. The reference cell is filled with buffer or 0.05%, w/v sodium azide solution in the same way.

5. The ligand solution is usually added to the injection syringe. This is achieved by drawing the solution into the hole in the bottom of the paddle until the syringe is full, by means of the filling port. At this point the plunger is lowered to cover the filling port. To remove air bubbles, the syringe is purged and refilled. To minimise the first injection anomaly, the plunger is then moved down a short distance (*see* **Note 11**).

3.4. Preparation of DSC Samples

1. The protein sample, following the dialysis, and subsequent concentration measurement is degassed, as described earlier. Also degassed in the same way is a volume of dialysis buffer to be used in the reference cell.

2. The protein sample is then added to the calorimeter cell. The filling syringe is loaded with at least 600 µL (for VP-DSC) of protein sample. As with ITC, to ensure that air bubbles are not added to the cell, the syringe is checked visually and any air bubbles observed within the syringe are dislodged from the sides of the syringe barrel, to ensure that they accumulate near the plunger and will not be injected. The DSC cell is filled in a

similar manner to the ITC cell, with the syringe being inserted into the access tube so that the tip of the syringe is held a small distance from the bottom of the cell. Maintaining the syringe in this position, the protein sample is injected in a steady flow, until the cell is filled. Visual inspection of the access tube allows the observation of the sample emerging from the access tube. Whilst the syringe position is maintained, several small aliquots of protein sample are added in short, sharp bursts. This dislodges air bubbles potentially present inside the calorimetric cell.

3. The reference cell is filled in exactly the same way as the sample cell (*see* **Note 12**).

4. Once both cells have been filled in the manner described earlier, excess solution is removed from the access tube, so that the cells contain exactly the same volume of solution. This is achieved by drawing liquid from the cells by using a filling adjustment syringe, which contains a collar. The collar rests upon the top of the access tube during removal of the excess solution, ensuring that excess liquid is drawn from exactly the same position.

5. If positive pressure is to be applied to the cells, the cap is applied and screwed tightly down (*see* **Note 13**).

3.5. Operation of ITC

1. The filled injection syringe is carefully wiped dry, avoiding drawing any solution from the needle, and is then inserted carefully into the filled sample cell (*see* **Note 14**).

2. The settings for equilibrating the ITC run are input into the instrument control software in order for the thermal equilibration of the cells to be initiated (*see* **Note 15**).

3. The parameters for controlling the injection protocol are input into the software in order to achieve a spread of data throughout the isotherm, to allow efficient determination of the enthalpy, affinity, stoichiometry as well as a measure of the heat of dilution (*see* **Note 16**).

4. The experimental run is started and the instrument will equilibrate the cells and jacket, and will begin stirring. The system will then re-equilibrate with the stirring ongoing, since this will provide heat to the sample cell. Once the cells have equilibrated and a stable baseline is achieved, the instrument will, after an initial delay to allow a period of stable baseline data collection, begin the injection schedule.

5. The first few injections should be observed to verify that the injection parameters are suitable. If required, the injection protocol can be adjusted during the run to modify the volume or spacing of injections to allow for time-dependent or low enthalpy-binding interaction effects.

6. The ITC data are collected and stored in the filename specified.

7. The ITC cell, syringe, and filling syringes are then cleaned thoroughly before the next use.

3.5.1. Standard ITC

1. Under typical conditions ITC can be used to determine the enthalpy change, affinity, and stoichiometry of reaction. Having measured these parameters it is also possible to calculate the entropy change for the binding interaction. To carry out typical standard ITC experiments the concentrations of ligand and macromolecule should be set at concentrations that will give a final molar ratio of ligand to macromolecule of around 2–3:1, for a binding interaction having a stoichiometry of 1:1. The concentration ranges used should take into account the dilution of protein by the added ligand. Reasonable starting concentrations can be estimated based upon the expected amount of heat to be produced or absorbed by the binding interaction, along with a value of c in the range 5–500 (*see* **Note 17**).

2. The ITC run is conducted as described in **Subheading 3.5**.

3. Following the ligand into macromolecule titration, several control titrations should be carried out to correct for the heats of dilution and heat of mixing (*see* **Note 18**).

4. The sample cell should then be cleaned ready for the next experiment, as described in **Subheading 3.9**.

5. Data analysis, which usually involves the correction for the heats of dilution, adjustment for any concentration differences, and model fitting to determine parameter values, is carried out using the appropriate software, as described in **Subheading 3.7**.

3.5.2. Tight-Binding ITC

ITC can be used to determine the binding parameters for ligands binding tightly to the macromolecule, which would usually be out of the experimental range for the instrument. So called displacement experiments can be used to access these parameters:

1. A standard titration is carried out using a weak-binding ligand, as described in **Subheading 3.5.1**, which is known to compete with the compound of interest for the binding site on the macromolecule.

2. The binding parameters are determined by fitting an appropriate model to the data using the software supplied with the instrument, as described in **Subheading 3.7**.

3. For the displacement titration, the weak-binding ligand is added to the calorimeter cell, with the macromolecule. The concentrations should be determined, and the mixture degassed and added to the cell in the usual way. The concentration of added weak binder must be sufficiently high, in order to significantly reduce the apparent affinity of the high-affinity binder,

so that the apparent values fall within the measurable range. An appropriate concentration of weak binder may be estimated as described later (*see* **Note 19**).

4. The tight-binding ligand is degassed and added to the injection syringe in the usual way, described in **Subheading 3.3**, and is titrated into the mixture of macromolecule and weak-binding ligand.

5. Control titrations are carried out as appropriate.

6. The apparent parameters for the titration of the tight-binding ligand into the mixture of macromolecule plus weak-binding ligand are then evaluated using the data analysis software (*see* **Note 20**).

7. The syringe and cell are cleaned, ready for the next experiment.

3.5.3. Weak-Binding ITC

Weak-binding interactions can sometimes be studied directly by ITC, even if the c value is lower than around 10, which is a situation usually not recommended. In order to use ITC in this way, several criteria should be fulfilled:

(i) The binding isotherm should extend sufficiently so that a high proportion of the binding sites are saturated.

(ii) The binding stoichiometry should be known.

(iii) The data should be of sufficient signal:noise to allow rigorous data analysis.

(iv) Whilst determined K_d values are usually valid, the interpretation of ΔH values should be treated with caution.

1. The protein and ligand solutions are prepared in the usual way, as described in **Subheading 3.3**.

2. The concentration of both macromolecule and ligand solutions should be chosen carefully so that at least 80% saturation is achieved during the titration (*see* **Note 21**).

3. The injection protocol should also be optimised to ensure that the required final concentrations are met (*see* **Note 22**).

4. The ITC experiment is performed as discussed previously.

5. Control experiments are carried out, and the relevant corrections made.

6. Data analysis is undertaken using the fitting software.

7. The instrument is cleaned.

If these experiments do not yield the desired results, typically because the signal:noise is not sufficient for reliable data analysis, then the displacement experiments described earlier can be used instead (*see* **Note 23**).

Because of the universal nature of the detection system, ITC can be used to follow enzyme-catalysed reactions as well as binding interactions. This is because the heat evolved is proportional to the molar enthalpy of the reaction, and so is also proportional to the concentration of product formed (*see* **Note 24**).

1. In order to determine the apparent molar enthalpy it is necessary to convert all the added substrate to product. The enzyme and substrate solutions are prepared in the usual way. The enzyme solution is then added to the calorimetric cell. Concentrations between 1 nM and 10 µM are typically adequate.

2. The substrate solution is added to the syringe in the usual way, at a concentration typically in the range 10 µM–100 mM.

3. The injection protocol is set up to carry out a single injection in the range of 10–50 µL.

4. Data are usually collected for up to 30 min, or long enough for the baseline to return to its initial position following the measured heat pulse.

5. The data are analysed as described later to determine the apparent molar enthalpy (*see* **Note 25**).

6. A decision to carry out one of two different methods for measurement of the thermal power is made depending upon the estimate K_m of the substrate. If the K_m is expected to be greater than 10 µM, then the multiple injection method is preferred. Alternatively, if the K_m is expected to be lower than 10 µM, then the single injection method may be more appropriate. The multiple injection method is exemplified later, with the single injection method described in **Subheading 4** (*see* **Note 26**).

7. The calorimeter cell is filled with enzyme solution; suitable concentrations fall in the range 10 pM–1 mM, depending upon the activity.

8. The injection syringe is filled with substrate solution, at a concentration above K_m and in molar excess to the enzyme concentration. Typical concentrations range from 10 µM to 100 mM. It is important that the substrate concentration is not too high, as there should be a spread of concentrations both below and above K_m; thus after the first few injections, the substrate concentration in the cell should still be below K_m.

9. The injection protocol is set to deliver 15–30 injections each of around 2–20 µL of substrate. The spacing between each injection should allow enough time for the new thermal power baseline to be established, whilst ensuring that this baseline is linear and does not begin to decrease. Suggested interval times are around 2–3 min. As a rough guide, around only 5% of the substrate should be converted to product before the next injection takes place.

10. Injections may be made until there is no further change in the baseline power, which indicates that V_{max} has been attained.

11. After completion of the injection schedule, the data are analysed using the software provided with the instrument (*see* **Note 27**).

12. The power generated (dQ/dt) by each addition of substrate is determined by measuring the difference between the original baseline and the new baseline position following each injection.

13. The substrate concentration is determined after each injection, and the rate is calculated using the value of the apparent molar enthalpy, previously measured.

14. The Michaelis-Menten equation is then fitted to the data to determine the parameters K_m and V_{max}.

15. The instrument is thoroughly cleaned.

3.6. Operation of DSC

1. Once the cells are filled and the cap screwed down, the parameters are set for the scan as required. The starting temperature, final temperature, and scan rate are set, with typical values being around 20°C, 90°C, and 60°C/h, respectively. The data collection rate is also set, usually at 1 Hz. It is also possible to set the temperature and time period for which the instrument will equilibrate, both before and after each scan cycle. The number of scans required depends upon the application, and whether downscans and rescans are appropriate.

2. Usually buffer blank scans are carried out for the first few runs, before protein is added to the sample cell. This enables the thermal history of the instrument to be consistent for the buffer baseline scans and subsequent protein scans. The number of buffer scans required is a matter of choice and may decrease with practice and experience of what determines a "good" baseline scan (*see* **Note 28**).

3. Once the buffer scans have been carried out and a suitable baseline achieved, the cells are allowed to cool and the protein sample is added to the sample cell and scanned with the same parameter settings as for the baseline scans.

4. Depending upon the information required it might be useful to do repeat scans on the protein sample to investigate the reversibility and reproducibility of the unfolding process.

5. Once the cells have cooled following the series of protein scans, the sample is removed and it is good practice to check the sample for signs of precipitation or aggregation. These effects may manifest themselves as visible changes to the solution, for example changes in turbidity or the appearance of particles.

6. The cells should then be cleaned ready for the next experiment. The level of cleaning required depends upon the presence of any precipitation, aggregation, or protein adsorption to the inner surface of the cell, and these procedures are described in **Subheading 3.10**.

7. Data analysis, which includes subtraction of the collected buffer baseline, as well as concentration normalisation, and fitting to determine the experimental parameters, is carried out using suitable software, and is described in **Subheading 3.8**.

3.6.1. Standard DSC

The most frequent use of the DSC instrument is to characterise the thermal stability of apo-proteins, in terms of the melting temperature, T_m; the enthalpy of unfolding, ΔH; and the heat capacity change, ΔC_p (*see* **Note 29**).

1. The DSC sample and reference cells are filled with degassed buffer and the baseline scans are collected using the desired start, finish temperatures and scan rate, with typical values described earlier.

2. The DSC cells are allowed to cool and the sample cell is filled with protein solution, with typical sample concentrations of 5–20 µM, depending upon the system, data quality, and availability of sample. The buffer composition of the sample protein should be considered for optimising the DSC experiment (*see* **Note 30**).

3. The DSC scan is then repeated using the same experimental parameters as for **step 1**. Repeat scans or scans downward in temperature can be carried out to evaluate reversibility if desired.

4. The cells are allowed to cool, emptied, and cleaned.

5. Data analysis is carried out using the software to determine the relevant experimental parameters.

3.6.2. Ligand Binding by DSC

Le Chatelier's principle states that if a ligand binds to the folded state of the protein then this will stabilise the folded state, and so unfolding the protein will become more unfavourable. This allows DSC to be used to monitor the effects of ligand binding by following the effect on melting temperature in the absence and presence of ligand (*see* **Note 31**) (*8*).

1. The macromolecule and the ligand, which have been prepared in the usual way, are added to the DSC cell. The concentration of ligand used should be high enough to ensure that a good proportion of the protein is saturated with ligand.

2. The DSC run is carried out as usual, with the same parameters as the protein in the absence of ligand.

3. If required the concentration of ligand can be varied and eparate runs detailing a concentration response can be collected.

4. The data are analysed as described in **Subheading 3.8**.

3.7. Data Analysis for ITC

The following section provides an outline of the principles of data analysis for standard ITC-binding experiments. Data analysis is usually carried out by non-linear regression, using models supplied with the software, although other models may be fitted (*9*). The fitted curve and parameter values along with their standard errors or 95% confidence intervals are reported. It is also possible to generate residual plots in order to aid determination of the quality of fit.

1. For each peak in the ITC run the quality of the baseline and the shape of the peak are observed. If required the baseline is adjusted, so that it represents a smooth path, upon which the individual peaks are effectively superimposed. Markers should be set to define the integration period, which should incorporate the entire peak including the final transition back to baseline. Integrating large portions of baseline can, however, be avoided (*see* **Note 32**).

2. The initial concentrations of the macromolecule and the ligand as well as the relevant cell volume are entered into the software.

3. The control titrations are used to adjust the raw data to remove any additional heat effects present, other than that directly from the binding interaction itself (*see* **Note 33**).

4. Once the relevant controls have been subtracted, a suitable model is fitted to the data. Several models are supplied with the manufacturer's software, but other models can be used if appropriate to the system under study.

5. Initial estimates for the variables in the model are input and the non-linear regression procedure started (*see* **Note 34**).

6. Following convergence to the global minimum sum of squares, the software generates the best-fit line and reports back the parameter values associated with that fit.

3.8. Data Analysis for DSC

As with the ITC data analysis section the detailed analysis of DSC data will be dependent upon the type of sample, the application (straightforward stability measurement or ligand binding), and the models most appropriate for the analysis. Described as follows are the steps required for analysis of a standard DSC experiment.

1. The raw data files for the sample and a representative buffer scan are imported into the software.

2. The raw data files should be normalised with respect to scan rate. The software can do this automatically when the data files are read. The purpose of this normalisation is to convert from

the measured raw data values, which are collected as mcal/minute (1 cal = 4.184 J) to cal/degree. This allows the conversion of the data to differential heat capacity and is really just the raw data divided by the mean scan rate at each time point.

3. The reference data, which should also have been scan rate normalised, are then subtracted from the sample data.

4. The data are normalised with respect to concentration to produce differential excess heat capacity per mole (mcal/mole/degree). The software will usually have a functionality to do this automatically.

5. Before the data fitting can take place a baseline must be associated with the data. This is not the "instrumental" baseline, which has already been subtracted in **step 3**, but the "sample" baseline. This may be achieved in a number of ways (*see* **Note 35**).

6. The baseline may then be subtracted from the data, if there is no ΔC_p term required in the fitting. The baseline is required to be associated with the data if a model containing the ΔC_p term is fitted.

7. The model to be fitted to the data is chosen and the number of peaks in the data is input into the software and initial estimates provided (*see* **Note 36**).

8. The best fit line and the fitted parameter values are reported by the software.

3.9. ITC Instrument Cleaning and Maintenance

Baseline problems such as noise, drifts, spikes, and erratic or artefactual data are often due to ITC cells being contaminated with proteins or other reagents from previous runs. The ITC cells should be cleaned following the manufacturer's guidelines to ensure that problems due to dirty cells are minimised. The cell-cleaning protocol, for full cleaning, is given below (*see* **Note 37**):

1. The reference cell which is usually filled with water, buffer or 0.05%, w/v sodium azide requires no special cleaning. It should be rinsed every week using the filling syringe and refilled.

2. Sample cell cleaning following routine use can be achieved using the filling syringe to rinse the cell with 3–5 cell volumes of water, followed by 3–5 cell volumes of buffer.

3. To remove more obstinate proteinacious material, 5% v/v Decon solution should be added to the cell and left to incubate at room temperature for 15 min. The cells should then be rinsed thoroughly with water and then with buffer.

4. Sometimes a more vigorous cleaning is required. To achieve this the cell should be filled with 10%, v/v Decon solution, and the temperature raised to 65°C for 1 h. The cell should then be cooled, emptied, and rinsed with water.

5. The cell-cleaning apparatus should be used to flush detergent or water through the cell. The apparatus is connected to the vacuum pump, which draws solution into the apparatus and through the cell. This allows 500 mL of detergent or water to flush through the cell as required.

6. The injection syringe should be cleaned with detergent solution, water, and buffer as required using the same method, described earlier, for filling the syringe, by drawing solution through the filling port. Several syringe volumes should be sufficient. The syringe should then be rinsed with the ligand solution that is to be used for the next experiment.

7. The filling syringe should be cleaned by rinsing with detergent, water, and buffer.

3.10. DSC Instrument Cleaning and Maintenance

The sample cell of the DSC instrument should be cleaned routinely in order to avoid problems with cell filling, baseline stability, reproducibility, and artefactual thermal effects. The cell should be cleaned after each use and more stringently following issues such as protein precipitation (*see* **Note 37**).

A description of a full cleaning regime, assuming protein contamination not removed by the previous step, and thus requiring the subsequent steps, follows:

1. The cell should be rinsed with 3–5 cell volumes of buffer to be used in the next experiment.

2. The cell should then be rinsed with 3–5 cell volumes of detergent, such as Decon-90, at a concentration of 10%, v/v. The temperature of the cells may be raised to 50°C to improve the efficiency of the cleaning process. Following the detergent rinse the cell should be rinsed thoroughly with water and then buffer, and refilled.

3. The cell-cleaning device supplied (with the Microcal instruments) is inserted into the DSC cell. The device is connected to the vacuum pump and up to 500 mL of hot Decon solution (10%, v/v at 50°C) is passed through the cell. Following this, 500 mL (or more if required) of water is flowed through the cell. The cell is rinsed with buffer and refilled.

4. The cell is filled with detergent solution and, without being capped, the cells are raised to 65°C for 1 h. The cell is then rinsed with water using the cleaning device, rinsed with buffer, and refilled.

5. The cell is filled with concentrated (90%, v/v) nitric acid, left uncapped, and the temperature raised to 80°C for up to 24 h. The cell is allowed to cool, the acid removed, and then the cell is washed with 500 mL of water. The cell is then rinsed with buffer and refilled (*see* **Note 38**).

4. Notes

1. Several different MWCO membranes are available, for example the Pierce Slide-A-Lyzer cassettes have MWCO 3,500; 7,000; and 10,000 Da.

2. Many buffer systems are compatible with ITC and DSC instrument cells and so can often be chosen based on the requirements of the biological system. It should be remembered that protonation effects may contribute to the observed enthalpy change due to the enthalpy of ionisation of the buffer (*10, 11*). The large concentrations of injectant sometimes necessary in ITC experiments can often mean that effective buffering capacity must be considered. It may be prudent to choose relatively large buffer concentrations, at pH values close to the buffer pK_a, in order to ensure that injectant solution pH values are maintained. The use of reducing agent should be considered carefully as these can sometimes affect instrument baselines.

3. The location of the ITC and DSC instruments should be considered so as to minimise potential electrical interference. Whilst the instruments have been built with a recognition that the laboratory environment is filled with sources of electromagnetic radiation, careful planning of the location of the instruments can be beneficial for effective use, though space interference can occur in a variety of ways and can lead to different effects on the baseline noise. If problems do occur, it is helpful to try to identify the source of the problem, bearing in mind that the source of interfering radiation could be anywhere. It is sometimes useful to monitor the periodicity of the interference, if it is intermittent and to try to correlate this with other instruments operating nearby. Sometimes, moving the instrument to another location is the only solution.

Electrical interference through the power supply also may lead to a variety of effects on the baseline, often associated with increased noise or with spikes. The most common problem is usually voltage spikes, which the instruments can usually deal with if they are not too large. Severe fluctuations may be addressed using voltage stabilisers.

Preventing the instruments from fluctuations in room temperature is also recommended. For example, locating the instruments away from sources of heat, such as sunlight, or instruments that may have coolers or chillers is advised.

Location within a temperature-controlled room on a suitable, uninterruptible power supply is ideal.

4. It is good practice to draw some air into the syringe before loading the sample as the air can be used to expel sample from the dead volume of the syringe.

5. Volumes of up to 2 L with three changes of buffer dialysed for 2 h each is usually sufficient.

6. Strictly, for protein concentration measurements, the protein should be diluted into 6 M guanidine hydrochloride, in order to unfold the protein, before absorbance measurements are made. This helps to minimise the absorbance contribution from disulphide bonds. In practice, this often makes only around a 10% difference to the estimated concentration.

7. Extinction coefficients can be calculated online. For example, for proteins the ExPASy Proteomics tools website can be used: http://www.expasy.ch/tools/, or for single-stranded nucleic acids OligoCalc, the oligonucleotide properties calculator from Northwestern University can be used: http://www.basic.northwestern.edu/biotools/oligocalc.html.

8. The concentration of ligand may be determined by an appropriate analytical technique, for example determination of quantitative nitrogen content. A recent advance has been the introduction of charged aerosol detection, http://www.coronacad.com, which allows detection of many compounds, even those without chromophores following HPLC.

Even if the concentration of ligand is difficult to measure or estimate, this does not preclude it from being studied by ITC, as, if the stoichiometry of the interaction is known, then this can be fixed during the data fitting and the concentration of ligand determined from the fitting procedure.

9. By dissolving the ligand in the same buffer as the protein has been dialysed into, mismatches in buffer solution are reduced to a minimum. This is important, as mismatches in buffer can lead to heat effects upon mixing which may interfere with the signal from the binding interaction itself.

10. The Microcal instrument comes with the ThermoVac system which incorporates a stirred cooling block inside a vacuum chamber. Removal of dissolved gas helps to ensure that air bubbles do not occur in the calorimetric cell. These can become dislodged during the titration run, leading to interfering heat effects. Usually, a 5–7 min period of stirred degassing is sufficient, although longer periods may be required for samples that have to be defrosted. The cooling aids rapid equilibration of the instrument prior to the start of the run.

11. The first injection in an ITC titration often is smaller than the rest. Several explanations for this phenomenon have been proposed. However, backlash in the plunger screw mechanism movement appears to account adequately for the effect. Downward movement of the plunger, following the purge – refill, prior to the first injection, takes up the backlash and minimises the effect (*12*).

12. It is important that the reference cell is filled with exactly the same buffer to that which contains the protein sample. This is because the instrument measures the difference in heat energy uptake between the two cells, which is directly related to the apparent heat capacity of the solutions contained in the cells. Thus, the only difference in composition between the two solutions should be the presence of protein (or nucleic acid) in the sample cell.

13. The pressure inside the cell is volume dependent, and so the cells should contain the same volume to ensure that the pressure is consistent.

14. Caution should be employed to ensure that the syringe needle is not bent during the filling or insertion process, as this can lead to noisy baselines, spurious heat effects, and unreliable data.

15. The ITC instrument software controls the cell and jacket temperatures at the desired experimental temperature, the amount of power applied to the cells to maintain the cell temperature difference close to zero (which determines the baseline position), the syringe stirring speed, and the equilibration options. These parameters are set to minimise the equilibration time and to allow a baseline compatible with the expected heats of binding. The experimental temperature, often 25°C, but capable of being set at a wide range of temperatures, is input. The reference power can be modified for large exothermic or endothermic reactions, but a useful starting point is around 40 μJ/s (10 μcal/s). The initial delay, the time between the start of the run and the first injection, is set usually at around 300 s. The stirring speed is usually set at around 250–300 rpm, (around 0.5 g) for protein-ligand interactions, but may be increased for more viscous solutions, or decreased to avoid shearing of more delicate proteins.

16. The software allows the injection volume, the duration of the injection, the time spacing between injections, and the data collection rate to be set. Usually injections of at least 5 μL are made, typically of 10 s duration, with a spacing of 3–4 min between each injection. A data collection rate of 0.5–1 Hz is typically used. The number of injections can be varied, depending upon the relative concentrations of macromolecule and ligand, to ensure that a sufficient molar ratio of ligand to macromolecule is achieved, and binding reaction parameters determined.

The initial concentrations of macromolecule in the cell and ligand in the syringe are recorded and an appropriate filename supplied for data storage.

17. For an injection protocol of 20 injections, each averaging 20 µJ, the total heat produced in the calorimeter cell would be 400 µJ. The heat can be calculated from the following equation:

$$Q = \Delta H \,(\text{J/mol}) \times M_{tot} \,(\text{mol/L}) \times \text{cell volume}\,(\text{L})$$

So solving the equation for the earlier example gives:

$$400\mu J = 20,000(\text{J/mol}) \times M_{tot}(\text{mol/}L) \times 0.0014(\text{L})$$
$$\Rightarrow M_{tot} \approx 14\mu M$$

This gives a minimum concentration that will produce a measurable heat of interaction.

It is also useful to calculate the adimensional parameter, c (13), which is the ratio of the macromolecule concentration over the K_d value for the interaction:

$$c = \frac{M_{tot}}{K_d}$$

For standard ITC runs, it is usual practice for the value of c to be in the range 5–500.

In the earlier example, for an interaction with a K_d of 1 µM, the range of macromolecule concentrations giving a useful experimental window is 5–500 µM.

Combining this with the value with the estimated value from the heat produced thus gives a useable macromolecule concentration range of 14–500 µM.

Performing some simple calculations of this type prior to the experiment can help to ensure that the concentrations used are appropriate for the desired experimental outcome.

18. During the titration of ligand into macromolecule (A) there are heat effects arising from a number of processes: (1) the binding interaction itself; (2) the heat of ligand dilution; (3) the heat of protein dilution; (4) the heat of mixing. In order to correct the first of these for the unwanted effects of the others, it is necessary to carry out further control titrations. These are: (1) titration of ligand into buffer (B); (2) titration of buffer into macromolecule (C); (3) titration of buffer into buffer (D), which account for the heats of ligand dilution, protein dilution, and mixing, respectively.

The true heat associated with the binding interaction is then given by:

heat from ligand titration minus heat from ligand dilution minus heat from protein dilution plus heat of mixing, or A – B – C + D, from above.

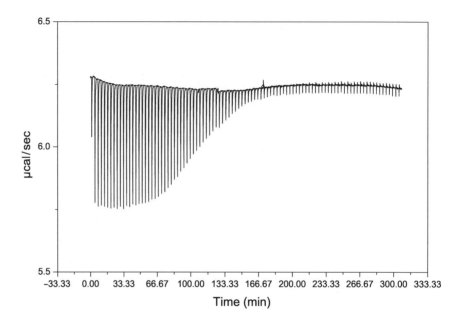

Fig. 1. Typical raw data from an ITC experiment.

The reason the heat of mixing is added back in is due to the fact that this effect is subtracted twice when the heats of dilution of the ligand and macromolecule are subtracted.

Often the heat of dilution of macromolecule and the heat of mixing are small compared to the potentially larger heat of dilution of the ligand (the protein is diluted only around 10–20% during the titration, whereas the ligand is diluted 200- to 300-fold during the early injections). This often means that the subtraction of only the heat of ligand dilution is sufficient to provide accurate data. The output, in terms or raw data, from a typical ITC experiment is shown in **Fig. 1**.

19. An approximation to the total concentration of weak-binding ligand, $[L]_t$, required to achieve a decrease in the apparent affinity of the tight-binding ligand by a factor of α, is given by:

$$[L]_t = K_L(\alpha - 1) + [M]_t$$

where K_L is the equilibrium dissociation constant for the weak-binding ligand and $[M]_t$ is the total macromolecule concentration.

20. Displacement or competition experiments involve titrating the compound of interest (A), into a mixture of macromolecule and a fixed concentration of competing ligand (L), the binding of which binding is known to be mutually exclusive with the compound of interest, and which has known K_d and H^o values. The presence of L will displace A from the complex with protein to a degree that is directly dependent upon the affinity and concentration of A and L. The apparent association constant for binding of A in the presence of L is given by:

$$K'_A = K_A \left(1 + \frac{[L]}{K_L}\right)$$

The apparent enthalpy is given by:

$$\Delta H'_A = \Delta H_A - \frac{\Delta H_L \cdot \frac{[L]}{K_L}}{\left(1 + \frac{[L]}{K_L}\right)}$$

where K_A, K_L are the equilibrium dissociation constants for A and L, respectively; K'_A is the apparent equilibrium constant for A in the presence of L; H_A and H_L are the enthalpy changes on binding for A and L, respectively; and H'_A is the apparent enthalpy change for the binding of A in the presence of L.

An exact analysis of competition ligand binding by displacement experiments has been described (*14*).

21. The total amount of ligand, X_t, that needs to be added in order to achieve the desired saturation, θ, can be determined from the following equation:

$$X_t = \frac{\theta(nM_t + K_d) - \theta^2 nM_t}{1 - \theta}$$

where n is the number of binding sites and M_t is the total macromolecule concentration.

22. As with any injection protocol in ITC experiments, the addition of ligand from the injection syringe displaces a portion of the macromolecule and ligand from the effective working volume of the cell, see **Fig. 2**. The corrected concentrations of macromolecule and ligand, respectively, are given by:

$$M_{t(i)} = M_t^0 \left(\frac{1 - \frac{\Delta V_{(i)}}{2V_0}}{1 + \frac{\Delta V(i)}{2V_0}} \right)$$

$$X_{t(i)} = \frac{\Delta V_{(i)} X_0}{V_0} \left(1 - \frac{\Delta V_{(i)}}{2V_0}\right)$$

where M_t^0 is the initial concentration of macromolecule, $V_{(i)}$ is the total sum of the added ligand volume following the *i*th injection, V_0 is the cell volume, and X_0 is the ligand concentration in the syringe.

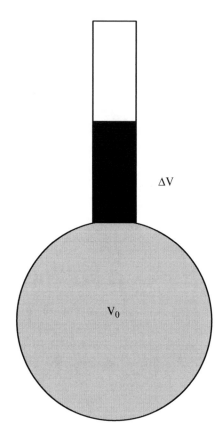

Fig. 2. Displacement of ligand from the calorimetric cell during titration.

The second equation can then be rearranged in order to express the total volume of ligand solution that is required to yield the desired concentration of ligand in the cell:

$$\Delta V = V_0 - \sqrt{\left(V_0^2 - \frac{2V_0^2 X_t}{X_0} \right)}$$

This equation can then be used in conjunction with the equation in **Note 21** to give a useful set of experimental conditions for the ITC experiment.

23. Displacement experiments, such as described earlier, can also be used to measure parameter values for weak-binding ligands (**15**). In this case the tight-binding ligand must be capable of having its binding parameters measured by the standard ITC process. The effect of repeating the titration of the tight-binding ligand in the presence of the weak-binding compound on the binding parameters of the tight-binding compound is assessed (*see* **Note 19**). In this case the concentration of weak-binding ligand added to the macromolecule

must be such that the apparent parameters for the tight-binding ligand remain in scope for determination by ITC.

24. The rate, v, of an enzyme-catalysed reaction measured in the calorimeter (*16–18*) is given by:

$$v = \frac{d[P]}{dt} = \frac{1}{V_0 \Delta H} \cdot \frac{dQ}{dt} = \frac{V_{max}[S]}{K_m + [S]}$$

where [P] is the concentration of product, V_0 is the volume of the calorimeter cell, H is the apparent molar enthalpy of the reaction, and dQ/dt is the heat flow or thermal power, V_{max} is the maximal rate for that enzyme concentration at saturating substrate, K_m is the Michaelis constant, and [S] is the substrate concentration.

In order to obtain a Michaelis-Menten plot it is necessary to determine both the power at different substrate concentrations and the total molar enthalpy. A typical plot showing the multiple injection method for enzyme kinetic reactions is shown in **Fig. 3**.

25. The apparent molar enthalpy is calculated by dividing the total heat generated during the reaction by the amount of product formed following total conversion of the substrate. Integration of the area under the peak with respect to time yields the total heat required for this calculation.

26. The single injection method involves a single continuous injection of substrate into the enzyme solution. Typical enzyme concentrations fall in the range 1–25 nM, with the substrate concentration being above the expected K_m and in molar excess compared to the enzyme concentration, typically ranging from 10 μM to 100 mM. The rate is calculated as

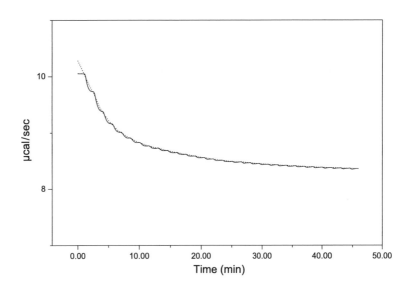

Fig. 3. Raw data from an enzyme kinetic assay using the multiple injection method.

shown earlier at any time, t, and the instantaneous substrate concentration at that time is determined from the equation:

$$[S]_t = [S]_0 - \frac{\int_0^t (\text{power})\mathrm{d}t}{V_0 \Delta H}$$

where $[S]_t$ is the substrate concentration at time t, $[S]_0$ is the initial substrate concentration, V_0 is the cell volume, and δH the apparent molar enthalpy.

As with any simple progress curve analysis, there is an assumption that there is no significant product inhibition. The output from a typical single injection experiment is shown in **Fig. 4**.

27. The Origin software provided with the Microcal instruments has options for analysing both the single injection method (described as method 1) and the multiple injection method (described as method 2), in the presence and absence of inhibitor.

28. The thermal history of the instrument really just refers to the fact that usually the first scan of a series of scans has been through a different process of heating and cooling than the others. For instance, for the first scan, the instrument will probably have been equilibrating at 25°C for the period before the scan took place. For subsequent scans, the instrument, assuming identical scan parameters, will have been cooled from the end of the first scan, equilibrated, and scanned. This process would be the same for any subsequent scans carried out. Thus, it is good practice for the buffer baseline scan to be at least the second scan of a series of

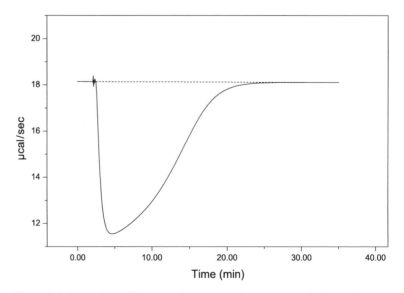

Fig. 4. Typical trace for an ITC enzyme kinetic run using the single injection method.

scans including those for the protein. The buffer scan is the measured response in the absence of sample and is used to correct the sample scan for any instrumental drifts associated with the scanning parameters and any ambient environment effects.

Good buffer baseline scans are usually those scans that are highly reproducible, both before and after the protein scans. However, with practice, good experimental technique and utilising the notion of thermal history, suitable baseline scans can be achieved in relatively few buffer scans. Typical DSC data, along with a good baseline trace, are shown in **Fig. 5**.

29. The reversible, two-state unfolding of proteins can be expressed by the following scheme:

$$N \quad U$$

The equilibrium constant, K_{eq}, can be defined at any temperature, and reflects the relative concentrations of N and U:

$$K_{eq} = \frac{[U]}{[N]}$$

The equilibrium constant is related to the Gibbs free energy change, G, by the following relationship:

$$\Delta G = -RT \ln K_{eq} = \Delta H - T\Delta S$$

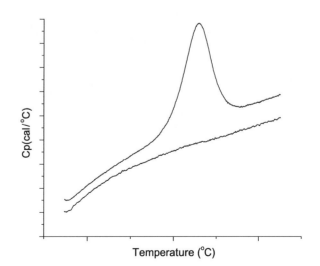

Fig. 5. Typical DSC data with an acceptable baseline scan.

The temperature at which the concentrations of D and N are equal, which defines the mid-point of the unfolding transition, is termed the melting temperature, T_m.

The change in the fraction of the protein unfolded, for a protein having $T_m = 60°C$, is shown in **Fig. 6**.

The DSC instrument measures the heat capacity, C_p, of the protein during the scan of temperature. The heat capacity is the amount of energy required to heat the sample by 1 K and is related to the enthalpy, ΔH, as follows:

$$\Delta H = \int_{T_N}^{T_D} C_p \, dT$$

The T_m is measured as the peak in the excess heat capacity of the protein, and from the earlier equation the enthalpy is calculated by integrating the excess heat capacity function. The ΔC_p is measured as the difference between the pre- and post-transition baselines.

30. Whilst most buffers are compatible with DSC (and ITC) experiments, caution should be used in the use of organic solvents and reducing agents. Organic solvents should be used at carefully matched concentrations in sample and reference (or cell and syringe) solutions. Reducing agents such as β-mercaptoethanol and dithiothreitol (DTT) are best

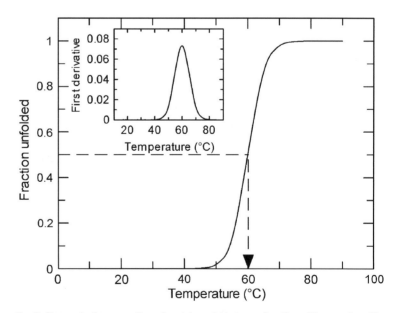

Fig. 6. Change in the proportion of protein unfolded as a function of temperature. The temperature at which 50% of the protein is unfolded is termed the melting temperature, T_m. Inset is the first derivative of the data, showing a peak at T_m.

avoided because of rapid oxidation leading to heat effects and/or heat degradation. If reducing agents are required for protein activity and stability, it is recommended that different reducing agents be investigated for suitability within DSC and ITC experiments. The use of tris(2-carboxyethyl) phosphine hydrochloride (TCEP), a reducing agent having several advantages over DTT, may be a suitable alternative.

31. The equilibria describing the unfolding and ligand-binding processes are shown earlier, in **Note 29** for unliganded protein, and as follows in the presence of ligand. The free-energy diagram is shown in **Fig.7**.

K_0, the equilibrium constant (K_{eq}) for the unfolding process in the absence of ligand is also given above. The ligand equilibrium association constant, K_L, for the ligand, L, binding to N is given by:

$$K_L = \frac{[NL]}{[N][L]}$$

It can be shown that these two constants can be combined to yield the value of K_L at the melting temperature, T_m:

$$K_{L(T_m)} = \frac{\left[K_{0(T_m)} - 1 \right]}{[L]_{(T_m)}}$$

This can be rewritten, allowing for the transposition of K_0 from T_0 (here, the melting temperature in the absence of ligand) to T_m (here, the melting temperature in the presence of ligand) to give:

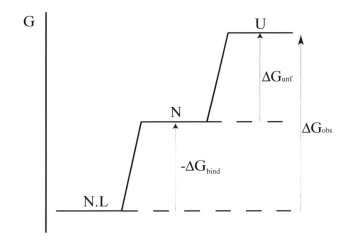

Fig. 7. Free energy diagram depicting protein unfolding with ligand binding to the native state.

$$K_{L(T_m)} = [\exp[(-\Delta H_{(T_0)} / R)(1/T_m - 1/T_0) +$$
$$(\Delta C_p / R)(\ln(T_m / T_0) + (T_0 / T_m) - 1)] - 1] / [L](T_m)$$

where $\Delta H_{(T0)}$ is the enthalpy of unfolding in the absence of ligand, R is the gas constant, and ΔC_p is the heat capacity change on protein unfolding.

The free ligand concentration, [L] at T_m, can be calculated from:

$$[L] = L_{tot} - \frac{nM_{tot}}{2}$$

where $L_{tot} \geq nM_{tot}$

In order to compare affinities ($K_{L(Trel)}$) for different ligands, extrapolation to a more relevant temperature (T_{rel}) must be undertaken, using the equation:

$$K_{L(T_{rel})} = K_{L(T_m)} \exp[(-\Delta H_{L(T_{rel})} / R)(1/T_{rel} - 1/T_m) +$$
$$(\Delta C_{PL} / R)(\ln(T_{rel} / T_m) + 1 - (T_{rel} / T_m))]$$

where $\Delta H_{L(T_{rel})}$ is the enthalpy of ligand binding at T_{rel}.

A typical plot showing DSC traces in the presence and absence of ligand is shown in **Fig.8**.

32. The diagram in **Fig.9** shows an example of the baseline and position of the integration markers typical for the analysis of an ITC injection peak.

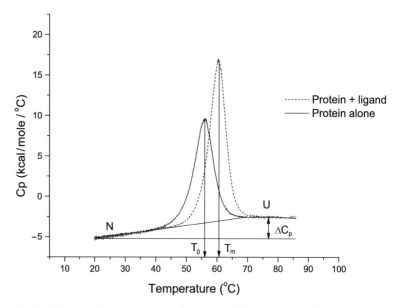

Fig. 8. DSC traces showing the effect of protein stabilisation by ligand binding.

33. It is sometimes possible to observe the injections at the end of the ITC titration, and where these are small and consistent, to use an average of these integrated peak areas as the background associated with the "heat of dilution", see **Fig.10**. It may also be possible to incorporate a constant

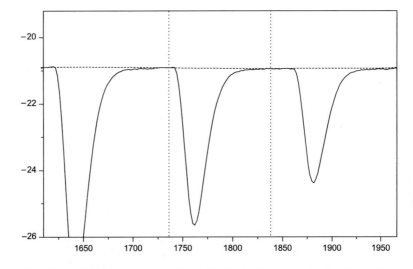

Fig. 9. Typical injection peak for an ITC run. The dashed line shows the baseline and the *dotted lines* indicate the position of the integration markers. The peak area between these two markers will be integrated by the software.

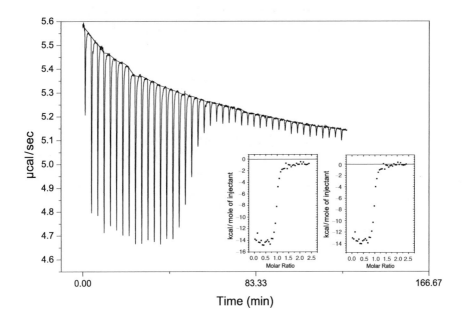

Fig. 10. ITC data showing the effect of removing an average of the last few "heat of dilution" injections. The raw data show that the injection heats at the end of the run are small and constant. The inset shows the effect of subtracting the average peak area of the last seven injections.

background parameter into the model used for data fitting to account for this heat effect. Sometimes it is not possible or not valid to remove a constant value for these background heat effects, for example if the heat of dilution of the ligand is concentration dependent. In these cases, the actual heats associated with the control experiment should be subtracted. Inaccurate estimates of the thermodynamic parameters associated with the binding reaction may be obtained if these controls are not properly dealt with.

34. The total heat content, Q, of the solution is given by:

$$Q = n\theta \, M_t \Delta H V_0$$

where n is the number of binding sites, θ is the fractional occupancy of the macromolecule, M_t is the total macromolecule concentration, ΔH is the apparent enthalpy of binding, and V_0 is the cell volume.

The fraction of the sites occupied by the ligand is given by:

$$\theta = \frac{[X]}{K_d + [X]}$$

where $[X]$ is the free ligand concentration given by:

$$[X] = Xt - n\theta M_t$$

where X_t is the total ligand concentration and the other parameters are as earlier.

Combining the three equations yields the following expression for the heat:

$$Q = \frac{nM_t \Delta H V_0}{2} \left[1 + \frac{X_t}{nM_t} + \frac{K_d}{nM_t} - \sqrt{\left(1 + \frac{X_t}{nM_t} + \frac{K_d}{nM_t}\right)^2 - \frac{4X_t}{nM_t}} \right]$$

The parameter of interest for data fitting however is the change in heat content associated with the completion of the $i - 1$ injection to the ith injection. As discussed in **Note 22** there must be a correction for the volume displaced from the calorimeter cell during the injection. The correction for the change in heat taking into account the contribution of the displaced volume to the heat before it is displaced from the calorimeter cell is:

$$\Delta Q_{(i)} = Q_{(i)} + \frac{\Delta V_{(i)}}{V_0} \left[\frac{Q_{(i)} + Q_{(i-1)}}{2} \right] - Q_{(i-1)}$$

The software uses the initial estimates to calculate $Q_{(i)}$ for each injection and compares this with the measured heat for each injection. A number of iterations occur where improvement in the parameter values takes place using a Levenberg-Marquardt approach. Once no further improvement in fit occurs, the final parameter values are returned for display.

35. The "sample" baseline should reflect the path of heat capacity baseline of the sample in the absence of the transition. Clearly, this is difficult to do, and various methods have been used to accomplish this. Four methods are included in the Origin software from Microcal: progress, linear, cubic, and step functions. The progress baseline option extrapolates the pre- and post-transition baselines and calculates the baseline in proportion to the estimated amounts of folded and unfolded protein present at a particular temperature. The linear and cubic functions interpolate the pre- and post-transition baselines. The step function allows a step between the pre- and post-transition baselines either at the peak of the transition or at the halfway point in terms of area under the curve. **Figure 11** shows the different baseline functions.

36. Whilst analysis of truly reversible processes using the models typically supplied with the manufacturer's software is valid, strictly speaking, thermodynamic analysis of processes involving aggregation of proteins, which therefore do not give reversible transitions, is not valid. However, these systems may be analysed to determine parameter values if the irreversible process is slow or occurs significantly after the unfolding transition. It can be useful to assess the scan rate dependence of the irreversible steps, in order to minimise the effect.

37. There are three stages to the cleaning of the ITC and DSC cells. The simplest is rinsing the cells with buffer or water with the filling syringe. More vigorous cleaning involves the use of the cell-cleaning device employing detergent solution followed by water. Finally the soaking method involves use of more harsh conditions, such as detergent or nitric acid (for the DSC) soaks at elevated temperatures. The cleaning regime should start with the mildest procedure and move towards the more vigorous approaches only after the previous approaches have failed. The steps described will illustrate the cleaning procedure assuming all steps are essential, but this should be applied on a case-by-case basis.

38. The use of nitric acid must be undertaken with caution, as although the cell is capable of withstanding exposure, the cell port and pressure transducer may be damaged and should be protected from this reagent. Compatibility of the cells with any cleaning reagent should be checked in the manufacturer's instrument manual.

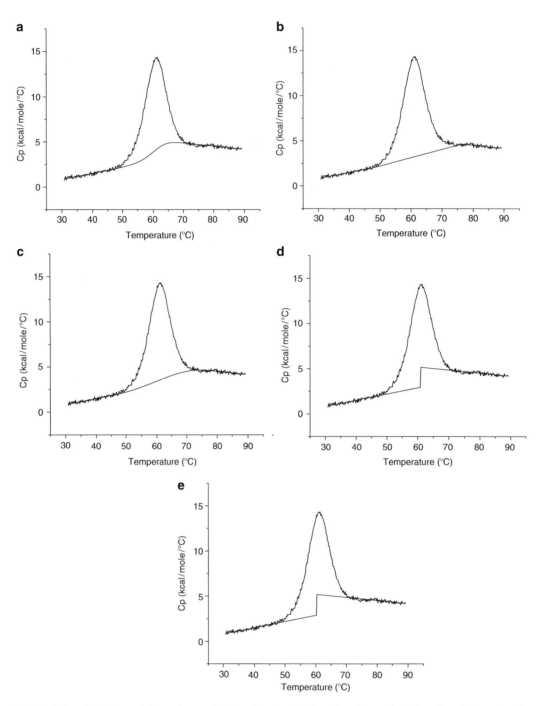

Fig. 11. Different DSC "sample" baselines. a: Progress baseline; b: Linear baseline; c: Cubic baseline; d: Step at peak baseline; e: Step at half area baseline.

Acknowledgments

The author would like to thank Dr. Gareth Davies for critically reading the manuscript.

References

1. Holdgate, G. A. (2001) Making cooldrugs hot: isothermal titration calorimetry as a tool to study binding energetics. *BioTechniques* **31**, 164–184

2. Ward, W. H. J. and Holdgate, G. A. (2001) Isothermal titration calorimetry in drug discovery. *Prog. Med. Chem.* **38**, 309–376

3. O'Brien, R., Chowdhry, B. Z. and Ladbury, J. E. (2001) Isothermal titration calorimetry of biomolecules, In: Harding, S. E. and Chowdhry, B. Z. (eds.) *Protein-Ligand Interactions: Hydrodynamics and Calorimetry, A Practical Approach.* Oxford University Press, Oxford, UK, pp. 263–286

4. Jelesarov, I. and Bosshard, H. R. (1999) Isothermal titration calorimetry and differential scanning calorimetry as complementary tools to investigate the energetics of biomolecular recognition. *J. Mol. Recognit.* **12**, 3–18

5. Privalov, P. and Dragan, A. I (2007) Microcalorimetry of biological macromolecules. *Biophys. Chem.* **126**, 16–24

6. Cooper, A., Nutley, M. and Wadood, A. (2001) Differential scanning microcaloriretry, In: Harding, S. E. and Chowdhry, B. Z. (eds.) *Protein-Ligand Interactions: Hydrodynamics and Calorimetry, A Practical Approach.* Oxford University Press, Oxford, UK, pp. 287–318

7. Plotnikov, V., Rochalski, A., Brandts, M., Brandts, J. F., Williston, S., Frasca, V. and Lin, L. (2002) An autosampling differential scanning calorimeter instrument for studying molecular interactions. *Assay Drug Dev. Technol.* **1**, 83–90

8. Sanchez-Ruiz, J. M. (2007) Ligand effects on protein thermodynamic stability. *Biophys. Chem.* **126**, 43–49

9. Buurma, N. J. and Haq, I. (2007) Advances in the analysis of isothermal titration calorime-

try data for ligand-DNA interactions. *Methods* **42**, 162–172

10. Fukada, H. and Takahashi, K. (1998) Enthalpy and heat capacity changes for the proton dissociation of various buffer components in 0.1 M potassium chloride. *Proteins* **33**, 159–166

11. Baker, B. M. and Murphy, K. P. (1996) Evaluation of linked protonation effects in protein binding reactions using isothermal titration calorimetry. *Biophys. J.* **71**, 2049–2055

12. Mizoue, L. S. and Tellinghuisen, J. (2003) The role of backlash in the 'first injection anomaly' in isothermal titration calorimetry. *Anal. Biochem.* **326**, 125–127

13. Wiseman, T., Williston, S., Brandts, J. F. and Lin, L.-N. (1989) Rapid measurement of binding constants and heats of binding using a new titration calorimeter. *Anal. Biochem.* **179**, 131–137

14. Sigurskjold, B. W. (2000) Exact analysis of competition ligand binding by displacement isothermal titration calorimetry. *Anal. Biochem.* **277**, 260–266

15. Zhang, Y.-L. and Zhang, Z.-Y. Low affinity binding determined by titration calorimetry using a high-affinity coupling ligand: a thermodynamic study of ligand binding to tyrosine phosphatase 1B. *Anal. Biochem.* **261**, 139–148

16. Williams, B. A. and Toone, E. J. (1993) Calorimetric evaluation of enzyme kinetic parameters. *J. Org. Chem.* **58**, 3507–3510

17. Todd, M. J. and Gomez, J. (2001) Enzyme kinetics determined using calorimetry: a general assay for enzyme activity? *Anal. Biochem.* **296**, 179–187

18. Bianconi, M. L. (2007) Calorimetry of enzyme-catalysed reactions. *Biophys. Chem.* **126**, 59–64

Chapter 8

Adaptive Combinatorial Design of Focused Compound Libraries

Gisbert Schneider and Andreas Schüller

Summary

Low-throughput screening for bioactive substances often represents the only way to discover new ligands of a drug target. This limits the number of compounds that can be tested for bioactivity. In such a situation, the design of small, focused compound libraries provides an alternative to the concept of large, maximally diverse screening collections. We present the technique of "adaptive" compound library design, which implements a simulated evolutionary process. Compound assembly and determination of bioactivity can be performed using computer-based methods (virtual screening), or in the laboratory. We show that there exists an optimal combination of the size of a screening library and the number of iterative screening rounds with the aim to keep experimental efforts at a minimum.

Key words: Bioinformatics, Diversity, Virtual screening, Combinatorial chemistry, Evolutionary algorithm

1. Introduction

Accurate structural information of validated drug targets and target–ligand complexes provide the basis for *receptor-based* (sometimes referred to as "structure-based") design of novel ligands that may eventually become lead structures and therapeutic agents in drug discovery *(1)*. High-resolution protein structures available from the Protein Data Bank (PDB) *(2)* represent the basis for rational, receptor-based molecular design. Knowledge about the shape of ligand binding sites and the distribution of functional groups in these pockets enables automated ligand docking and de novo design of selective low-molecular weight ligands *(3–5)*. For a large group of drug targets, namely G-protein-coupled receptors (GPCR)

Ana Cecília A. Roque (ed.), *Ligand-Macromolecular Interactions in Drug Discovery: Methods and Protocols*,
Methods in Molecular Biology, vol. 572,
DOI 10.1007/978-1-60761-244-5_8, © Humana Press, a part of Springer Science+Business Media, LLC 2010

and other integral membrane proteins like ion channels, reliable atomistic models are rare or unavailable. In these cases – and as a parallel drug discovery route to the receptor-based approach in general – a *ligand-based* design strategy can be followed. The ligand-based design of compound libraries requires known molecules exhibiting the desired properties or pharmacological activity as a starting point. These molecules are termed "seed structures" or "reference compounds." The aim is to identify novel substances having identical or very similar activity as the reference compounds *(6)*. Selected small sets of compounds ("focused libraries") can then be tested for activity in a laboratory.

Two complementary approaches are available for this task: virtual screening of compound libraries *(5, 7)* and de novo design of molecules from scratch *(8)*. In virtual screening studies, existing compound libraries are scanned to identify novel substances with favorable properties. Various complementary compound sources are accessible for virtual screening: databases of physically available compounds ("screening database," corporate compound repository) and virtual libraries including fully or partially enumerated combinatorial libraries. Commercial compound suppliers offer a total of approximately three million "druglike" agents which may be exploited for this purpose *(9, 10)*. To avoid complete screening of the plethora of compounds that are typically available from a compound repository, adaptive focused library design offers itself as a technique to select only a specialized, considerably smaller set of compounds with desired properties *(1)*.

De novo design, in contrast, creates new compounds from scratch. Following a fragment-based design concept, candidate compounds are constructed using a set of predefined building blocks and linkage rules *(11, 12)*.

Automated, computer-assisted de novo design as well as adaptive combinatorial library design is complementary to biochemical high-throughput screening (HTS) of screening databases *(7, 8)*. In theory, the search space for these design methods is given by all druglike compounds – a number estimated to be in the range of 10^{20}–10^{100} *(13)*. It is evident that exhaustive screening of such a large number of substances is by no means possible. Thus, efficient optimization algorithms are required to navigate through the vast chemical space toward regions populated by candidate compounds that exhibit a desired bioactivity (e.g., enzyme inhibition). Efficient scoring functions are required by optimization algorithms for an assessment of the quality of proposed candidate compounds. Consequently, the three basic components of an automated drug design approach are:

1. Method for virtual compound assembly or structure sampling

2. Optimization algorithm

3. Scoring function

With the advent of HTS and parallel chemical synthesis as standard tools in pharmaceutical drug research in the early 1990s, combinatorial libraries serving as basis for screening campaigns were desired to be large and maximally diverse *(14)*. Later, there has been a trend toward libraries that are focused on target families and even specific biological targets *(15, 16)*. Optimization algorithms have been successfully employed to perform "intelligent" sampling of screening libraries ever since *(1, 17)*. Such sampling techniques either explore the product space of molecular libraries by "cherry picking" compounds of interest, e.g., on basis of chemical similarity to a reference structure, or they explore the reactant space of combinatorial libraries by choosing only limited combinations of starting materials. Both concepts result in a specifically problem-tailored focused compound library.

Automated de novo design dates back to the early 1990s as well. Here, new molecules are constructed, rather than sampled from a set of existing structures. Consequently, de novo design is not restricted to available screening repositories. Some strategies proved to be more effective than others; for instance, the use of molecular fragments or building blocks for structure assembly instead of exclusive atom-wise construction *(8, 11–13)*, and the construction of chimeras or new molecules from the assembly of drug-derived building blocks *(18)*. The necessity for multi-dimensional optimization in drug discovery projects has been realized and taken up by de novo design software by consideration of pharmacokinetic properties and synthetic accessibility *(19)*. While *positive design* restricts the design and optimization process to regions of chemical space that contain the desired molecules, *negative design* defines selection criteria that help prevent adverse molecular properties (e.g., very high lipophilicity) and unwanted structures (e.g., molecules containing reactive groups). Experience shows, however, that de novo design will rarely yield novel lead structures with nanomolar activity, high target selectivity, and an acceptable pharmacokinetic profile in the first place. Yet, what can be expected is an increased hit rate in a tailored focused library compared to screening of an arbitrary compound collection. De novo design should primarily be seen as an idea generator for medicinal chemistry *(1, 8)*.

2. The Algorithmic Concept of Adaptive Evolutionary Design

During the drug design process (as well as in automated de novo design), molecules typically grow from smaller predecessors *(20)*. Considering all possible variations, we speak of a combinatorial search space, as we add, delete, or replace building blocks attached to the molecular core structure in a combinatorial fashion. Numerous strategies exist for searching this space. The idea is, of course, to do this in some systematic way with a high probability of finding the globally best solution(s). One such strategy is evolutionary compound design *(21, 22)*, which can be formalized by an evolutionary algorithm (EA) (**Fig. 1**). EAs are intuitive, as the iterative process of variation and selection is similar to the work of a project team.

EAs are population-based optimization strategies that simulate the evolutionary pressure of selection and the evolutionary operators "mutation" and "crossover." In case of adaptive drug design, individuals are represented by candidate compounds. Application of the mutation and crossover operators leads to the variation of individuals *(23)*. A mutation event introduces new information into a population. The crossover operator, on the other hand, exploits the information that is already present in a population. This is achieved by recombining the information of two individuals of the population. A scoring function – in the context of evolutionary optimization sometimes termed "fitness" function – determines which individuals are selected for breeding new compounds. Basically, an EA contains an iterative cycle of variation and selection ("generation loop") until a termination criterion is reached. The termination criterion can be given by a budget of single point assays or syntheses.

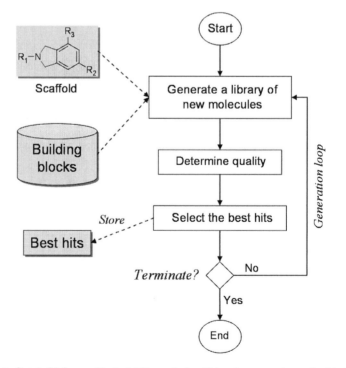

Fig. 1. Simple EA for combinatorial library design. This scheme can be realized in the laboratory by iterative synthesize-and-test cycles, or in the computer.

A common pattern of drug design programs that consider ease of synthesis is to assemble molecular building blocks by virtual reaction schemes. Suitable building blocks can be obtained by virtual *retro*-synthesis of drug molecules or from physically available building block collections **(Fig. 2)** *(24)*. When obtaining building blocks by virtual *retro*-synthesis, the same set of reaction is then used to assemble new candidate compounds *(25)*. It is reasonable to assume that the designed compounds will have some degree of "drug-likeness" and contain only few awkward or undesired structural elements *(18)*. Ideally, virtual structure assembly is guided by simulated organic synthesis steps so that a synthesis route can be proposed for every generated structure. Many of the advanced programs automatically analyze generalized synthetic routes and pick potential synthons from databases of available molecular building blocks *(26, 27)*.

The flowchart in **Fig. 1** represents a general scheme for iterative compound screening. A search will be nothing but a random

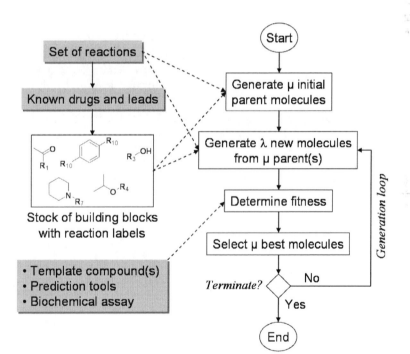

Fig. 2. Flowchart of the (μ, λ) Evolution Strategy for ligand- and fragment-based molecular design *(21, 23, 25)*. First, a stock of molecular fragments is generated by virtual *retro*-synthesis of known drugs and lead structures, for example, by the RECAP fragmentation rules *(24)* (*left*). In an adaptive process, generations of new compounds are bred using mutation and crossover operators together with the stock of building blocks (*right*). One or more template compounds and various prediction tools may be used for fitness calculation. Alternatively, "fitness" is determined by a biochemical assay. Parent molecules of each new generation are selected from the offspring. For generation of offspring, the reactions used for *retro*-synthesis of the parents and forward synthesis of new compounds are randomly selected from the set of reactions.

walk unless an *adaptive* parameter is included, that is, some kind of "memory." A straightforward way to implement such a memory system is to keep record of the best compounds designed, and breed new compounds ("offspring") from the best structures found in the population. As a consequence, with each new cycle, the search will be increasingly directed toward an "activity island" in chemical space. This idea is depicted in **Fig. 2**. A formal description of a simple EA with adaptive memory is the (μ, λ) Evolution Strategy *(28, 29)*. The (μ, λ) notation indicates the size of the population (that is, the number of compounds in each generation) by λ, and μ is the number of parents (best of a generation) that are used for breeding offspring. This population-based optimization approach was shown to be able to cope with noisy fitness landscapes (note that biological activity assays, measurements, and predictions of bioactivity are usually erroneous) *(30, 31)*, and many variations of the idea of self-adaptive optimization have been developed *(32, 33)*.

To appreciate the impact of an adaptive search strategy, one can compare different optimization techniques. **Figure 3** shows optimization results that were obtained for the example of finding trypsin inhibitors in a library of products of a multicomponent Ugi-type reaction *(34, 35)* (**Scheme 1**). This is an example of combinatorial library design. In this exercise, the budget for

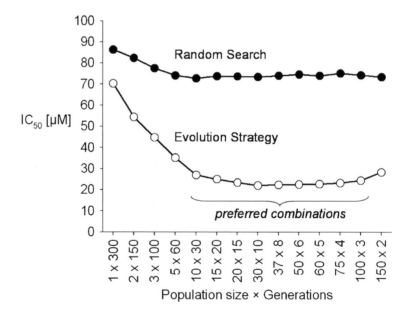

Fig. 3. Course of optimization of a library containing three-component Ugi reaction products. The fitness function was the experimentally determined IC_{50} value indicating trypsin inhibition (average values shown). The budget for synthesis and testing was fixed (300 compounds). The optimization was started from up to four seed structures in the initial population of molecules. An Evolution Strategy (*open circles*) clearly outperformed simple random searching (*filled circles*). Different combinations of population size and generations (synthesize-and-test steps) lead to different results.

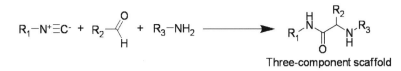

Scheme 1. Three-component Ugi-type condensation.

synthesis and testing was 300 compounds. One may ask whether 300 compounds should be tested in a single screening run or in consecutive screens with a smaller population size. **Figure 3** clearly demonstrates that there exist preferred combinations of population size and the number of screening cycles *(31, 34, 36)*. It is noteworthy that systematic optimization is not possible without adaptation, as can be seen from the average IC_{50} values yielded by a random search compared to the Evolution Strategy **(Fig. 3)**.

3. Design of a Focused Combinatorial Library

3.1. Preparation of Scaffold and Building Blocks

For virtual library enumeration, the constant parts of the molecules (scaffold) and the variable parts (side-chains) must be prepared so that virtual compounds can be assembled by the computer. Different software suites require different formats. Thus, no general rule can be given here. As a guideline, SMILES notation *(37)* is readable by many of these programs. For usage in our own publicly available library enumeration software SmiLib (http://gecco.org.chemie.uni-frankfurt.de/smilib/) *(38)* the molecular building blocks are prepared as shown in **Fig. 4** (examples only; for a detailed instruction, see the SmiLib web site).

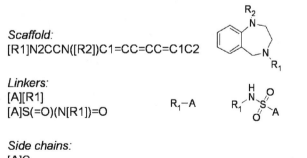

Scaffold:
[R1]N2CCN([R2])C1=CC=CC=C1C2

Linkers:
[A][R1]
[A]S(=O)(N[R1])=O

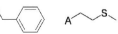

Side chains:
[A]C
[A]CC1=CC=CC=C1   A−CH₃
S(CC[A])C

Fig. 4. Examples of string representations (SMILES) of a scaffold, two linkers, and three side-chains for processing by the software SmiLib *(38)*. "A" represents an attachment site of a building block or a linker to the scaffold, and "R$_x$" denotes a variable side-chain of a scaffold or a linker.

SMILES can be used for virtual molecule assembly from scaffolds, linkers, and side-chains by string concatenation. According to this scheme, side-chains are attached to the scaffold via linker groups. Various chemical reactions can be considered by using different linker types and building block collections. It also simplifies virtual compound library enumeration since the connecting functional groups (linkers) may be completely left out in the set of building blocks. An advantage of this concept of a combinatorial library is its simplicity and ease of implementation yielding very short computing times. However, realistic chemical reactions cannot be modeled deliberately.

For the application example, the *R*-groups of the scaffold shown in **Scheme 1** were decorated with 44 aldehyde-derived building blocks, 15 amine-derived building blocks, and 24 isonitrile-derived building blocks, respectively, yielding the full 15,840 (44 × 15 × 24)-member combinatorial library *(34, 35)*.

3.2. Selection of a Fitness function

Automated computer-based library design requires a fitness function. This function can be seen as a substitute for the biochemical activity assay. Numerous possibilities have been suggested, from generic drug-likeness estimators to target-specific docking scores. We often employ a similarity index for this purpose. The similarity function computes a numeric value indicating some kind of pharmacophoric similarity between a candidate compound and one or multiple reference compounds with known activity. The reason for this choice is that pharmacophoric descriptions of molecules, expressed either as topological pharmacophores computed from the molecular graph or three-dimensional pharmacophores, allow for scaffold-hopping from the chemotype of the reference molecules to the scaffold of the combinatorial library *(39, 40)*. For the sake of fast computing, and to avoid potential pitfalls with the generation of three-dimensional conformers, we typically use the CATS similarity as a first, coarse-grained fitness function *(22)*. There are many alternatives available, and ideally the appropriate search space coordinates should be selected for each optimization problem. The goal is to find a descriptor that allows for systematic optimization in a "smooth" fitness landscape which lacks "activity cliffs," and where small changes in molecular structure result in small changes in the function (bioactivity) of the resulting molecule *(41–43)*.

The computational protocol of CATS assigns each atom of a molecule to zero, one, or two potential pharmacophoric points (PPP) CATS considers the PPPs of hydrogen-bond donor (D), hydrogen-bond acceptor (A), lipophilic (L), positively charged or ionizable (P), and negatively charged or ionizable (N). A distance matrix with entries d_{ij} holds the shortest topological path (expressed as number of bonds) between PPP i and PPP j for all PPP pairs. A mapping function combines the information of the distance matrix and the PPP assignments to obtain a pharmacophoric correlation vector. The mapping function counts the frequencies of all possible PPP pair distances and sums them in ten bins covering the distances of 0–9 bonds. The 15 pairing combinations of PPPs (DD, DA, DP, DN, DL, AA, AP, AN, AL, PP, PN, PL, NN, NL, LL) and ten different distances (bins) yield a 150-dimensional correlation vector. The final step of the CATS descriptor calculation is the scaling of the vector *(44)*. In many studies, CATS descriptors have proven their applicability to de novo design, compound library optimization, and similarity searching *(45)*. A web interface is available for computing CATS similarities for library sizes up to 100 compounds (http://gecco.org.chemie.uni-frankfurt.de/cats_light/).

3.3. Performing the Optimization

One will usually begin library design with one or a few representative reference compounds as the initial parent molecules. The first library generation will be structurally diverse, aiming at a broad coverage of the search space. The actives found in the first generation are the parents for the second round. In **Fig. 5**, the "evolution" from reference molecule **1** is shown for our example of optimizing a focused Ugi-type product library toward trypsin inhibitors. In a (5, 30) Evolution Strategy, compound **1** was selected as one of the parents of the first generation. It then generated six "children," two of which were selected as parents for the next generation. From these parent structures, only compound **2** generated successful offspring in subsequent synthesize-and-test cycles. Molecules **3–5** are parents that were selected and produced the most active variants. It is noteworthy that the example of trypsin inhibitor design is characterized by a peculiarity: The benzamidine moiety found in parents **1–6** is a preferred fragment binding to the S_1 substrate recognition pocket of the enzyme (**Fig. 6**). Almost any Ugi-type product containing this fragment in side-chain position R_3 (**Scheme 1**) exhibits some trypsin binding. Therefore, all successful designs contained this specific building block. A similar observation had been made for Ugi-type products that were evolved to block thrombin activity *(46, 47)*.

An example of the successful design of a combinatorial library focused toward binding to human cannabinoid receptor 1 (hCB-1), a GPCR, is shown in **Scheme 2** *(48)*. The scaffold was identified by adaptive computer-based design. Then, real libraries were synthesized and tested. In this case study, the library size was 80 compounds, and eight products showed binding to hCB-1 (10% hit rate).

Summarizing, these examples indicate the potential of adaptive evolutionary approaches for the rational design of focused compound libraries with appropriate hit rates. Through the use of virtual fragment-based synthesis, chemically tractable ligands, which are ready for small array synthesis offering adaptive feedback cycles, can be obtained.

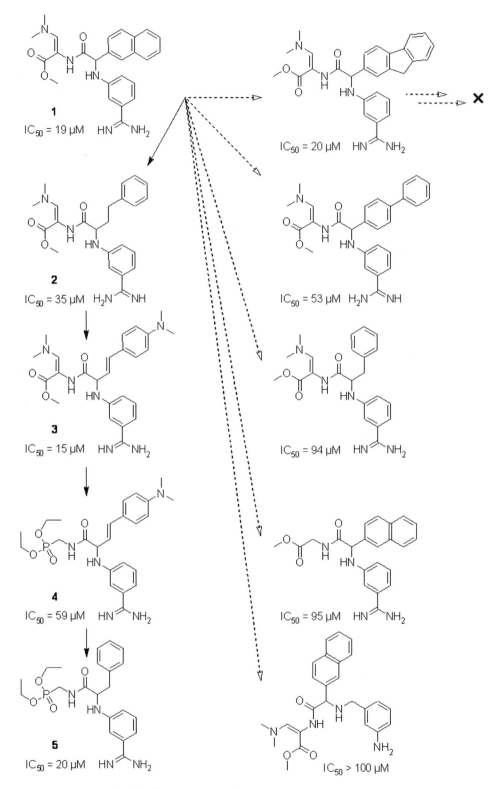

Fig. 5. Parent molecules of the first five generations of an evolutionary design run (*left column*) following the strategy shown in **Fig.1**. IC$_{50}$ values indicate trypsin inhibition. On the *right*, the best offspring of reference molecule **1** are shown. These compounds did not breed successfully and their descendants died out in the course of library optimization.

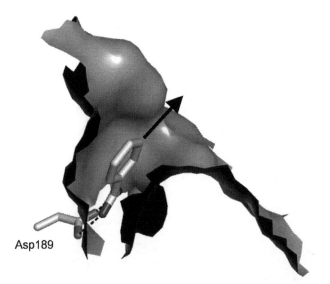

Fig. 6. Benzamidine bound to the S_1 substrate recognition pocket of trypsin (PDB file 2oxs *(49)*, 1.3 Å resolution). Only a part of the solvent-accessible pocket surface is shown. Benzamidine binds via hydrogen bridges (*dotted lines*) to Asp189 at the bottom of the pocket. The *arrow* indicates a preferred direction for elongation of the fragment. This structurally motivated design concept is mirrored in the automatically generated compounds (see text).

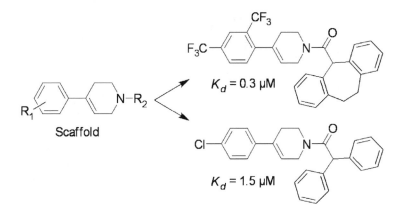

Scheme 2. Combinatorial scaffold and two active derivatives that were found to bind to human CB-1 receptor. The molecules were obtained by evolutionary library design *(48)*.

Acknowledgments

The authors would like to thank Dr. L. Weber for providing the Ugi-type compound database. This work was supported by the Beilstein-Institut zur Förderung der Chemischen Wissenschaften, the DFG Sonderforschungsbereich 579 (project A11.2), and the Fonds der Chemischen Industrie.

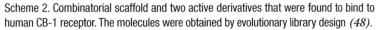

References

1. Schneider, G. and Baringhaus, K.-H. (2008) *Molecular Design – Concepts and Applications.* Wiley-VCH: Weinheim, New York

2. Berman, H. M., Westbrook, J., Feng, Z., Gilliland, G., Bhat, T. N., Weissig, H., Shindyalov, I. N., and Bourne, P. E. (2000) The Protein Data Bank. *Nucleic Acids Res.* **28**, 235–242

3. Rester, U. (2006) Dock around the clock - current status of small molecule docking and scoring. *QSAR Comb. Sci.* **25**, 605–615

4. Cavasotto, C. N. and Orry, A. J. (2007) Ligand docking and structure-based virtual screening in drug discovery. *Curr. Top. Med. Chem.* **7**, 1006–1014

5. Schneider, G. and Böhm, H.-J. (2002) Virtual screening and fast automated docking methods. *Drug Discov. Today* **7**, 64–70

6. Eckert, H. and Bajorath, J. (2007) Molecular similarity analysis in virtual screening: foundations, limitations and novel approaches. *Drug Discov. Today.* **12**, 225–233

7. Bleicher, H. K., Böhm, H.-J., Müller, K., and Alanine, A. I. (2003) Hit and lead generation: beyond high-throughput screening. *Nat. Rev. Drug Discov.* **2**, 369–378

8. Schneider, G. and Fechner, U. (2005) Computer-based *de novo* design of drug-like molecules. *Nat. Rev. Drug Discov.* **4**, 649–663

9. Irwin, J. J. and Shoichet, B. K. (2005) ZINC – a free database of commercially available compounds for virtual screening. *J. Chem. Inf. Model.* **45**, 177–182

10. Schuffenhauer, A., Popov, M., Schopfer, U., Acklin, P., Stanek, J., and Jacoby, E. (2004) Molecular diversity management strategies for building and enhancement of diverse and focused lead discovery compound screening collections. *Comb. Chem. High Throughput Screen.* **7**, 771–781

11. Honma, T. (2003) Recent advances in *de novo* design strategy for practical lead identification. *Med. Res. Rev.* **23**, 606–632

12. Erlanson, D. A. (2007) Fragment-based lead discovery: a chemical update. *Curr. Opin. Biotechnol.* **17**, 643–652

13. Mauser, H. and Stahl, M. (2007) Chemical fragment spaces for *de novo* design. *J. Chem. Inf. Model.* **47**, 318–324

14. Drewry, D. H. and Young, S. S. (1999) Approaches to the design of combinatorial libraries. *Chemom. Intell. Lab. Syst.* **48**, 1–20

15. Valler, M. J. and Green, D. (2000) Diversity screening versus focussed screening in drug discovery. *Drug. Discov. Today* **5**, 286–293

16. Schneider, G. (2002) Trends in virtual combinatorial library design. *Curr. Med. Chem.* **9**, 2095–2101

17. Schneider, G. and So, S.-S. (2002) *Adaptive Systems in Drug Design.* Landes-Bioscience: Georgetown

18. Siegel, M. G. and Vieth, M. (2007) Drugs in other drugs: a new look at drugs as fragments. *Drug Discov. Today.* **12**, 71–79

19. Segall, M. D., Beresford, A. P., Gola, J. M., Hawksley, D., and Tarbit, M. H. (2006) Focus on success: using a probabilistic approach to achieve an optimal balance of compound properties in drug discovery. *Expert Opin. Drug Metab. Toxicol.* **2**, 325–337

20. Carr, R.A., Congreve, M., Murray, C. W., and Rees, D. C. (2005) Fragment-based lead discovery: leads by design. *Drug Discov. Today* **10**, 987–992

21. Schneider, G., Lee, M.-L., Stahl, M., and Schneider, P. (2000) *De novo* design of molecular architectures by evolutionary assembly of drug-derived building blocks. *J. Comput. Aided Mol. Des.* **14**, 487–494

22. Schneider, G., Clement-Chomienne, O., Hilfiger, L., Schneider, P., Kirsch, S., Böhm, H.-J., and Neidhart, W. (2000) Virtual screening for bioactive molecules by evolutionary de novo design. *Angew. Chem. Int. Ed. Engl.* **39**, 4130–4133

23. Fechner, U. and Schneider, G. (2007) Flux (2): comparison of molecular mutation and crossover operators for ligand-based *de novo* design. *J. Chem. Inf. Model.* **47**, 656–667

24. Lewell, X. Q., Judd, D. B., Watson, S. P., and Hann, M. M. (1998) RECAP – retrosynthetic combinatorial analysis procedure: a powerful new technique for identifying privileged molecular fragments with useful applications in combinatorial chemistry. *J. Chem. Inf. Comput. Sci.* **38**, 511–522

25. Fechner, U. and Schneider, G. (2006) Flux (1): a virtual synthesis scheme for fragment-based *de novo* design. *J. Chem. Inf. Model.* **46**, 699–707

26. Gillett, V. J., Myatt, G., Zsoldos, Z., and Johnson, A. P. (1995) SPROUT, HIPPO and CAESA: tools for *de novo* structure generation and estimation of synthetic accessibility. *Perspect. Drug Discov. Des.* **3**, 34–50

27. Vinkers, H. M., de Jonge, M. R., Daeyaert, F. F., Heeres, J., Koymans, L. M., van Lenthe, J. H., Lewi, P. J., Timmerman, H., van Aken, K., and Janssen, P. A. (2003) SYNOPSIS:

SYNthesize and Optimize System In Silico. *J. Med. Chem.* **46**, 2765–2773

28. Rechenberg, I. (1994) *Evolutionnstrategie '94.* Fommann-Holzboog: Stuttgart

29. Schneider, G., Schrödl, W., Wallukat, G., Nissen, E., Rönspeck, G., Müller, J., Wrede, P., and Kunze, R. (1998) Peptide design by artificial neural networks and computer-based evolutionary search. *Proc. Natl. Acad. Sci. U.S.A.* **95**, 12179–12184

30. Arnold, D. V. and Beyer, H. G. (2003) On the benefits of populations for noisy optimization. *Evol. Comput.* **11**, 111–127

31. Schneider, G., Schuchhardt, J., and Wrede, P. (1996) Evolutionary optimization in multimodal search space. *Biol. Cybern.* **74**, 203–207

32. Hansen, N., Ostermeier, A. (2001) Completely derandomized self-adaptation in evolution strategies. *Evol. Comput.* **9**, 159–195

33. Arnold, D. V. and Beyer, H. G. (2006) Optimum tracking with evolution strategies. *Evol. Comput.* **14**, 291–308

34. Illgen, K., Enderle, T., Broger, C., Weber, L. (2000) Simulated molecular evolution in a full combinatorial library. *Chem. Biol.* **7**, 433–441

35. Schüller, A., Fechner, U., Renner, S., Franke, L., Weber, L., and Schneider, G. (2006) A pseudo-ligand approach to virtual screening. *Comb. Chem. High Throughput Screen.* **9**, 359–364

36. Schüller, A., and Schneider, G. (2008) Identification of hits and lead structure candidates with limited resources by adaptive optimization. *J. Chem. Inf. Model.* **48**, 1473–1491

37. Weininger, D. (1988) SMILES, a chemical language and information system. 1. Introduction to methodology and encoding rules. *J. Chem. Inf. Comput. Sci.* **28**, 31–36

38. Schüller, A., Hähnke, V., and Schneider, G. (2007) SmiLib v2.0: a Java-based tool for rapid combinatorial library enumeration. *QSAR Comb. Sci.* **26**, 407–410

39. Güner, O. (Ed.) (2000) *Pharmacophore Perception, Development, and Use for Drug Design.* International University Line: La Jolla

40. Schneider, G., Schneider, P., and Renner, S. (2006) Scaffold-hopping: how far can you jump? *QSAR Comb. Sci.* **25**, 1162–1171

41. Johnson, M. A. and Maggiora, G. M. (Eds.) (1990) *Concepts and Applications of Molecular Similarity.* Wiley: New York, p. 393

42. Peltason, L. and Bajorath, J. (2007) SAR Index: quantifying the nature of structure-activity relationships. *J. Med. Chem.* **50**, 5571–5578

43. Bull, J. J., Meyers, L. A., and Lachmann, M. (2005) Quasispecies made simple. *PLoS Comput. Biol.* **1**, e61

44. Fechner, U. and Schneider, G. (2004) Evaluation of distance metrics for ligand-based similarity searching. *Chembiochem* **5**, 538–540

45. Renner, S., Fechner, U., and Schneider, G. (2005) Alignment-free pharmacophore patterns – a correlation-vector approach. In: Langer, T. and Hoffmann, E. (Eds.) *Pharmacophores and Pharmacophore Searches.* Wiley-VCH: Weinheim, New York, pp. 49–79

46. Weber, L., Wallbaum, S., Broger, C., and Gubernator, K. (1995) Optimization of the biological activity of combinatorial compound libraries by a genetic algorithm. *Angew. Chem. Int. Ed. Engl.* **34**, 2280–2282

47. Weber, L. (2005) Current status of virtual combinatorial library design. *QSAR Comb. Sci.* **24**, 809–823

48. Rogers-Evans, M., Alanine, A. I., Bleicher, K. H., Kube, D., and Schneider, G. (2004) Identification of novel cannabinoid receptor ligands via evolutionary *de novo* design and rapid parallel synthesis. *QSAR Comb. Sci.* **23**, 426–430

49. Alok, A., Sinha, M., Singh, N., Sharma, S., Kaur, P., and Singh, T. P. (2007) Crystal structure of the trypsin complex with benzamidine at high temperature (35 C). PDB entry 2oxs, *unpublished*

Chapter 9

Chemical Microarrays: A New Tool for Discovery Enzyme Inhibitors

Shuguang Liang, Wei Xu, Kurumi Y. Horiuchi, Yuan Wang, and Haiching Ma

Summary

Enzymes, the catalytic proteins, are playing pivotal roles in regulating basic cell functions. Drugs that inhibit enzyme activities cover varying aspects of diseases and offer potential cures. One of the major technologies used in the drug discovery industry for finding the enzyme inhibitors is high-throughput screening, which is facing a daunting challenge due to the fast-growing numbers of drug targets arising from genomic and proteomic research and the large chemical libraries generated from high-throughput synthesis. Chemical microarray, as a new technology, could be an excellent alternative for traditional well-based screening, since the technology can screen more compounds against more targets in parallel with a minimum amount of materials, reducing cost and increasing productivity. In this chapter, we have introduced the basic techniques and applications of chemical microarrays, and how to use them routinely for identifying enzyme inhibitors with functional-based assays. Sample assays for kinases, proteases, histone deacetylases, and phosphatases are demonstrated.

Key words: Chemical microarray, High-throughput screening, Aerosol deposition, Kinase, Protease, Histone deacetylase, Phosphatase

1. Introduction

During the past decade, biological studies involving genomics and proteomics have identified thousands of more potential drug targets, simultaneously with chemical libraries, produced by parallel combinatorial synthesis techniques, and have supplied more compounds that are in need of screening. Therefore, one of the challenges facing lead identification through the practice of high-throughput screening (HTS) is to find more efficient ways to

Ana Cecília A. Roque (ed.), *Ligand-Macromolecular Interactions in Drug Discovery: Methods and Protocols*,
Methods in Molecular Biology, vol. 572,
DOI 10.1007/978-1-60761-244-5_9, © Humana Press, a part of Springer Science+Business Media, LLC 2010

identify new lead compounds. The new methods should be faster, economical, and more quantitative. The chemical microarray, derived from genomic microarray technologies, is one good example of such a method *(1)*. Conventional microtiter plate-based assays have progressed from the 96-well format to the 384-well system, or even the 1,536-well (or more) platform or microfluidic systems *(2)*. However, the cost associated with replacing old systems by the HTS/precision automations is substantial.

The widely adopted DNA microarray technology in genomic research has given new impetus to miniaturization, providing unprecedented possibilities and capabilities for parallel functional analysis of hundreds or even thousands of bio-entities. Dr. Stuart Schreiber and colleagues from Harvard University have expanded this technology for studying the interaction between chemical compounds with biological targets and have created the first chemical microarray technology, also called small molecule microarray (SMM), by immobilizing chemical compounds on the surface of microarrays *(3–5)*. This technology is a very useful tool for studying the binding effect of chemicals to biological targets both in purified form and in cell lysates. However, SMM requires the whole library to be synthesized with a common functional group that can be automatically reacted with the functional group on the activated slides to immobilize each compound *(6)*; therefore, this technology is not suitable to general chemical libraries. To avoid the potential complicated immobilization process, scientists from Abbott Laboratories have developed a new form of chemical microarray, Micro Arrayed Compound Screening (µARCS), by spotting and drying the compounds onto a polystyrene sheet with conventional well-plate size *(7–9)*. To determine how a group of compounds will inhibit special biological targets, such as enzymes, a thin layer agarose gel containing the enzymes will be laid on top of the polystyrene sheet and the dried compounds will be gradually diffused into the agarose gel to interact with the enzymes. Another thin layer agarose gel containing the substrate will be laid on top of the first agarose gel to start the reaction *(8)*. One of the major problems of µARCS is that the rate and capability of resolubilization and diffusion of different classes of dry compounds will complicate the dynamic range of arrayed compounds vs. bio-available compounds. This is a particular problem for in vitro screening campaigns in which fast enzyme reactions, weak inhibitors, weak receptor binders, hydrophobic compounds, and complex natural products may be involved. Differential transfer rates in HTS will detect inhibitors at different kinetic time points, resulting in noncomparable inhibition profiles. This would be a major challenge for IC_{50} profiling of a group of compounds against the same target under the same reaction conditions.

Conventional enzyme inhibition assay involves those compounds that are always in solution to freely access their biological targets. A microarray platform is similar to this format, and will not only produce an assay system comparable to regular well-based reaction, but also capable of using any commercial chemical libraries. By using a unique aerosol deposition technology, Dr. Scott Diamond and colleagues from the University of Pennsylvania have invented a microarray platform that is close to this ideal format, which can perform solution-phase chemical compound microarrays without immobilizing the compound *(10)*; it was later commercialized as DiscoveryDot™ technology by Reaction Biology Corporation for different enzyme targets and applications *(1, 11–14)*. In this format, each compound is individually arrayed on a glass surface with a reaction buffer containing a low concentration of glycerol to prevent evaporation. Therefore, as in conventional well-based screening, the chemical compounds and biological targets are always in reaction solutions, which are activated with biological targets by delivering analytes using aerosol deposition technology *(1, 10–15)*. This technology is further used for proteomic studies by using arrayed peptide library *(16, 17)*. In this chapter, since both SMM and µARCS platforms cannot be used for general chemical library screening, only the solution from chemical microarray technology DiscoveryDot™ will be further discussed.

2. Materials

2.1. Creating Chemical Microarray

1. *Microarrayers:* OmniGrid™ series, (GeneMachines, CA), Molecular Dynamics Gen III (GE, former Amersham Biosciences, NJ), and Nanoprint™ series (TeleChem International, Inc., Sunnyvale, CA).

2. *Microarray pins:* ArrayIt™ stealth microspotting pins (TeleChem International, Inc., CA); Tungsten microarray pins from Point Technologies, Inc. (San Diego, CA).

3. *Scanner:* GenePix laser scanner series (Molecular Devices, Sunnyvale, CA). SpotLight Microarray Scanner (TeleChem International, Inc., Sunnyvale, CA).

4. *Microarray analysis software:* Acurity and GenePix software from Molecular Devices, (Sunnyvale, CA). ArrayPro from Media Cybernetics, Inc. (Bethesda, MD).

5. Aerosol deposition machine, NanoSprayer™, for activating solution-phased microarray reactions, is manufactured and sold by Reaction Biology Corp. (Malvern, PA).

6. *Microarray slide*: Polylysine slides and Protein slides are from Fullmoon Biosystems (Sunnyvale, CA), Epoxy slide are from TeleChem International, Sunnyvale CA), streptavidin slides are manufactured internally or purchased from Xenopore (Hawthorne, NJ).

7. *General materials*: Microarray pin solution, SpotBot®, from TeleChem International, Inc.; DMSO, phosphate buffered serine (PBS) buffer, and other general chemicals are from Sigma. 384-well microtiter plates: polypropylene plates, ideal for long-term compound storage; can be purchased from Greiner Bio-One (Monroe, NC), Corning (Acton, MA), and other leader suppliers.

8. *Compound array solution*: 2% DMSO (v/v) and 10% glycerol (v/v).

2.2. Proteases and Histone Deacetylases Assays

1. Caspase reaction materials: Purified caspases, peptide substrates with MCA tag, and peptide inhibitors were purchased from BioMol (Plymouth Meeting, PA).

2. Caspase reaction buffer: 50 mM HEPES (pH 7.4), 100 mM NaCl, 1 mM EDTA, 0.1% CHAPS (v/v), 10 mM DTT, 10% glycerol (v/v), and 2% DMSO(v/v).

3. Matrix Metalloproteinases (MMPs) reaction materials: Enzymes were purchased from BioMol, and FRET substrates [MMP-2 substrate, Cy3B-PLGLAARK(Cy5Q)-NH, MMP-3 substrate, and Ac-RPK(Cy3)PVENvaWRK (Cy5Q)-NH] were from Amersham Biosciences (Piscataway, NJ).

4. HDAC8, HeLa Nuclear extract and substrates, Boc-(Ac) Lys-AMC (4-methylcouMarin-7-amide), developing buffer and control inhibitor, Trichostatin A (TSA), were from BioMol.

5. HDAC reaction buffer: 50 mM Tris–HCl, pH 8.0, 137 mM NaCl, 2.7 mM KCl, 1 mM $MgCl_2$, 1 mg/mL bovine serum albumin (BSA), 10% glycerol (v/v).

6. The Polylysine slides were from FullMoon Biosystems.

2.3. Kinase Assays

1. Kinases p60[c-Src] and B-Raf (V599E); substrate biotinylated poly (Glu4-Tyr), and inactive MEK1 were purchased from Upstate (Lake Placid, NY).

2. Protein slides to immobilize MEK1 were purchased from Full Moon Biosystems. Streptavidin-coated slides to immobilize biotinylated poly (Glu4-Tyr) were manufactured internally. Hybridization cover slips, LifterSlip, were from Erie Scientific Company (Portsmouth, NH).

3. ATP (Adenosine 5′-triphosphate disodium salt), BSA, staurosporine, and other general chemicals were purchased from Sigma. GW5074, KDR inhibitor ((Z)-3-[(2, 4-Dimethyl-3-(ethoxycarbonyl) pyrrol-5 yl)methylidenyl]indolin-2-one), was from Calbiochem (San Diego, CA).

4. Primary antibodies for phosphotyrosine, P-Tyr-100, were purchased from Cell Signaling (Beverly, MA); secondary antibodies, Alexa Fluor 555-labeled goat antirabbit and antimouse IgG, were from Molecular Probes (Eugene, OR).

5. Kinase Reaction Buffer: 50 mM Tris–HCl (pH 7.5), 10 mM $MgCl_2$, 1 mM EGTA, 0.01% Brij 35 (v/v), 10% glycerol (v/v), 2 mM DTT, 0.1 mM Na_3VO_4, and 0.02 mg/mL BSA.

6. TPBS: 0.05% Tween-20 (v/v) in 1× PBS diluted from 10× PBS (Sigma).

7. Stopping Buffer: TPBS/EDTA (50 mM EDTA in TPBS).

8. Blocking Solution: 10 mg/mL BSA in TPBS.

9. Antibody Diluting Solution: 5 mg/mL BSA in TPBS.

2.4. Phosphatase Assays

1. PTP1B is from Upstate, substrates are the phosphorylated biotin-poly-$(Glu_4\text{-}pTyr)_{10}$ peptide generated by using tyrosine Src as enzyme and biotin-poly-$(Glu_4\text{-}Tyr)_{10}$ peptide as substrate *(14)*. The general PTP inhibitors, potassium bisperoxo(1,10-phenanthroline)oxovanadate(V) [*bpV(phen)*] was from BioMol.

2. Streptavidin-coated slides and antibodies for detecting phosphorylated peptides are the same as in the kinases assay.

3. Phosphatase Reaction Buffer: 50 mM Tris–HCl, pH 7.5, 10 mM $MgCl_2$, 0.01% Brij 35 (v/v), 10% glycerol (v/v), 1 mM DTT, and 2% DMSO (v/v).

3. Methods

3.1. Proteases and HDAC Screening

1. Compound preparation and printing: Aliquot chemical compounds in pure DMSO were transferred into Compound Array Solution and related substrate in 384-well plates (5–10 μL per well). For a typical primary HTS, the final compound concentration ranges from 10 to 30 μM. For IC_{50} confirmation, the compounds are typically prepared in a ten-dose level assay, starting from 20 μM, and following with a 1:3 dilution scheme. Polylysine slides should be cleaned free of dust before secured on the slide holders on the deck of printer. Forty-eight stealth pins (ideal choice to produce multiple microarrays of same library) or solid pins (easier to clean) are used for fast printing. Arrayer is instructed to produce 1 nL drops (reaction center), featured at 120–150 μmin diameter with a 350-μm center-to-center distance. Pins are thoroughly washed before and after picking different chemicals. After printing, the slides can be stored at –80°C for long-term or 4°C for short-term storage, or used right away for biological assays.

2. Reaction activation: To activate the microarray reactions, the compound microarrays are placed on the deck of Nano-Sprayer™; an enzyme mixed with reaction buffer is deposited as a fine aerosol with an average droplet size of 18 μm in diameter and 2.2 pL in volume, by spraying evenly on top of the microarray slide. These nanoliter-sized enzyme droplets will interact with the compound inside the nanoliter-sized reaction centers. The setting for enzyme flow rate is 400 nL/s and carrier nitrogen gas flow rate is 3 L/min. The activated microarrays are incubated in a humidity chamber before detections (*see* **Note** 7).

3. The caspase assays are simply one-step reactions. Enzymes will digest the Rhodamine-110 conjugated substrates to yield fluorescence. The fluorescence substrate $(Z-DEVE)_2-R110$ (10 μM) (Biomol International, Plymouth Meeting, PA) was mixed with compounds and Caspase Reaction Buffer, then arrayed on the polylysine slides. The diluted Caspase-7 in ddH_2O (5 U/μL) (Biomol International, Plymouth Meeting, PA) was then sprayed on top of the slides after NanoSprayer™ was equilibrated by 0.1 mg/mL BSA in ddH_2O. After 1 h incubation at 30°C and 95% humidity chamber, the slides were scanned with fluorescence microscope-based scanner at excitation at 460 and emission at 520. An example result is shown in **Fig. 1**.

4. Fluorescent-based HDAC assay is a two-step reaction based on the special synthesized substrate, which comprises an acetylated lysine side-chain and the fluorophore-AMC. The first enzymatic reaction is that HDAC activity removes the acetyl group from the substrate, which is followed by the second enzymatic reaction, where trypsin digests the deacetylated substrate to release the fluorescent AMC group. The fluorescence substrate (100 μM) from BioMol's HDAC Fluorimetric Assay/Drug Discovery Kit was mixed with HDAC Reaction Buffer, and then arrayed on the polylysine slides. For TSA IC_{50} determination, a five-dose series dilution from 50 nM to 50 pM is prepared in a 384-well plate and then arrayed on the polylysine slides in the same fashion. Then the diluted HeLa Nuclear Extract was sprayed on top of the slides. After 1 h incubation at 30°C in the humidity chamber, the slides were developed by spraying the developer. After 30 min incubation, the slides were scanned with fluorescence microscope-based scanner at excitation at 360 and emission at 460. An example result is shown in **Fig. 2**.

5. Reaction detection: Fluorescence microscope equipped with auto scanning function is used for detecting reactions with AMC as fluorophore. The filter with excitation of 360 nm/460 nm excitation/emission filter is used and the acquisition time

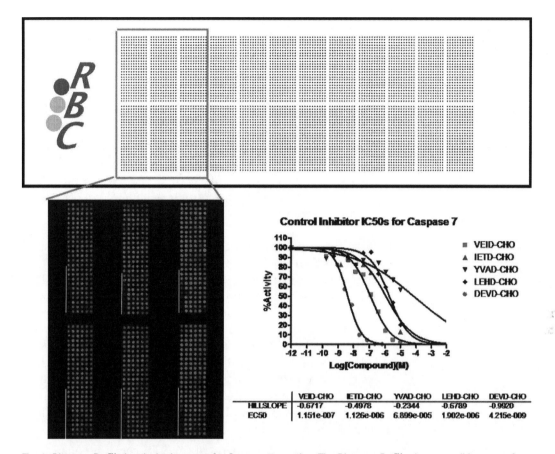

Fig. 1. DiscoveryDot™ chemical microarray for Caspase 7 reaction. The DiscoveryDot™ microarray slide can perform more than 6,000 reactions; a HTS run with peptide inhibitors against caspase 7 is demonstrated here. The Z′ > 0.6 [Z′ = $1-3 \times (SD_{signal} + SD_{background})$/abs $(M_{signal} - M_{background})$] is reached for this assay, which correlated to a signal to background ratio >30, and CV at 12%. The fluorescent detector was 50 µM fluorogenic substrate [p-toysl-DEVD]$_2$-R110.

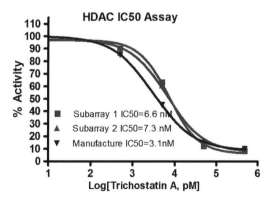

Fig. 2. DiscoveryDot™ chemical microarray for HDAC reaction. Nuclear extract from HeLa cells is used for this assay. The duplicated TSA inhibitions are performed in two separated subarrays. The IC$_{50}$ values are consistent in subarrays and also comparable to manufacture-suggested value.

is adjusted based on the reaction intensity. Either the fluorescence microscope or laser scanner equipped with blue laser can be used to detect Rhodamine-110 fluorophore, the 490 nm/520 nm excitation/emission filter or 10 µM resolution is selected. Rhodamine-110 is relatively less affected by photo bleaching; the reactions can be detected in a different time period to give out a time-dependent readout *(11)*.

6. Data analysis: The image of microarray is converted to signal by using DNA microarray software of Acuity (Axon Instruments) or ArrayPro (MediaCybernetics, Silver Spring, MD). Nonlinear regression and plotting to obtain IC_{50} and K_m were performed with Graphpad Prism (San Diego, CA).

3.2. Kinase Assays

1. Slide coating: Both natural protein substrate and biotinylated peptide substrate can be used to coat microarray surface. To coat MEK protein, apply 50 µL/slide of 1 µM MEK in PBS on the center of the protein slides (FullMoon) or Epoxy slides (TeleChem), and spread out by covering with a hybridization cover slip (LifterSlip), and incubate at room temperature (RT) for 2 h in a humidified chamber with >90% relative humidity (Rh); then wash the slides with PBS, rinse with double-distilled water, and spin-dry prior to microarraying. Coating biotinylated peptide substrate was performed in a similar manner as above on streptavidin-coated slides (*see* **Notes 8** and **10**).

2. Compound preparation: To perform HTS, chemical compounds are diluted in pure DMSO at a concentration of 500 µM, and then 1:50 diluted into Kinase Reaction Buffer with desired concentrations of ATP. The final compound concentration at 10 µM in 2% DMSO was printed on precoated substrate slides. For kinetic study and dose–response inhibition curves, the concentrations of ATP and inhibitors were adjusted accordingly before printing.

3. Reaction activation: After loading the compound slide on the NanoSprayer™, kinases, such as p60[c-Src] and B-Raf (V599E), are diluted into a solution containing 0.1 mg/mL BSA and sprayed on slides, which are incubated in a humidity/temperature-controlled incubator at 30°C/96% humidity for 2–3 h. An example result is shown in **Fig. 3**.

4. Reaction detection and data analysis: After incubation, reactions are stopped by a gentle but quick wash (*see* **Note 9**) in a five-slide holder with Stopping Buffer. The slides are then washed again with fresh stopping buffer with rotation for 10 min before placing into the Blocking Buffer, then incubated for 1 h at RT, and then washed in PBS three times at 10 min for each washing. The slides are then treated with the primay antibody,

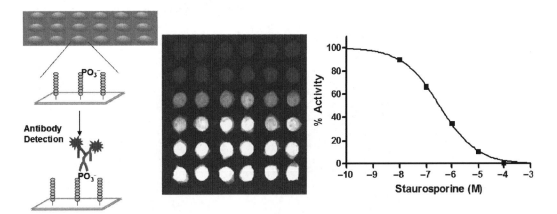

Fig. 3. DiscoveryDot™ microarray for tyrosine kinase assay. The Biotinylated poly-$(Glu_4\text{-}Tyr)_{10}$ peptide is coated on streptavidin slides. Staurosporine is diluted and arrayed on microarray with reaction buffer of 10 mM $MgCl_2$, 1 mM EGTA, 50 mM Tris–HCl, pH 7.5, 2 mM DTT, 0.02 mg/mL BSA, 0.01% Brij 35, and 10 ATP. Src (p60$^{c\text{-}Src}$) is sprayed on top of the slides, which is then incubated for 2 h and then blocked, immunostained, and imaged with a laser scanner. IC_{50} value is 322 nM at 10 μM ATP.

mouse monochronal P-Tyr-100 for phosphorylated poly Glu-Tyr detection or rabbit antiphospho-MEK1 (Ser 218/22) /MEK2 (Ser 222/226) antibody for B-Raf reaction. Antibody is diluted to 1:400 in Antibody Dilution Buffer and the incubation time is 1 h at RT. After washing three times with PBS for 10 min each, slides are then treated with Alexa Fluor 555-labeled secondary antibody for 1 h at RT in the dark. Slides are washed in the dilution buffer three times at 10 min for each washing, followed with a one-time washing with distilled water, and a spin dry (*see* **Note 10**). Fluorescence detection was done by using laser scanner at Ex = 532 nm with 575 nm cutoff filter for emission. The data analysis is performed in the same manner as in the protease-based assay.

3.3. Phosphatase Assays

1. Compound preparation and reaction activation: Chemical compounds (10 μM for HTS) and 2 μM of the phosphorylated biotin-peptide are mixed with Phosphatase Reaction Buffer and printed on streptavidin slides. For determining the IC_{50} value of inhibitor bpV (phen), the compound is serially diluted from 100 μM to 0.6 nM concentration and arrayed on a slide. An example result is shown in **Fig. 4a**.

2. Reaction activation and detection: The reaction started by spraying 3 nU/μL PTP1B on top of the microarray slides and followed with 2 h incubation in temperature-controlled humidity chamber. The reactions are then stopped and blocked in the same way described in the kinase assay. An example result is shown in **Fig. 4b**.

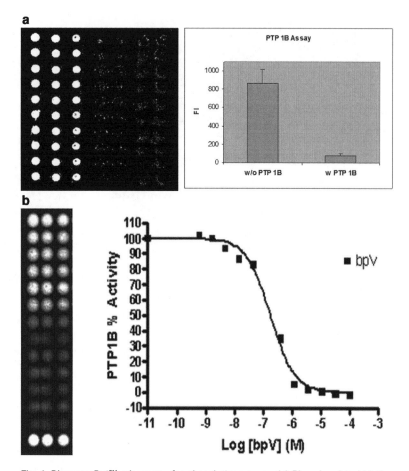

Fig. 4. DiscoveryDot™ microarray for phosphatase assay. (**a**) Phosphorylated biotin-poly-$(Glu_4$-$pTyr)_{10}$ peptide is arrayed on streptavidin slide, which was followed with activation by spraying PTP1B. The slide is then washed and stained with antiphos-photyrosine antibody **(a)**. The first three columns demonstrated the original signal from the phosphorylated peptide without PTP1B exposure. After PTP1B was applied (Col 4–7), the fluorescent signals dropped to a level just slightly above the background. The full dynamic range of the assay (without phosphatase/phosphatase) was 11-fold, demonstrating that this protocol is amenable to phosphatase HTS. (**b**) The same method is used to test the IC_{50} for a potent general PTP inhibitors, potassium bisperoxo(1,10-phenanthroline)oxovanadate (V) [*bpV(phen)*]. The 10-point dilution from 10 nM to 100 μM was obtained from 100× DMSO dilutions of compounds. bpV (phen) was serially diluted from 100 μM to 0.6 nM concentration and arrayed on slide with triplicate. No inhibitor control with 1× Reaction Buffer was arrayed on both ends of dilutions (*row 1* and *14*). Reaction condition: 2 μM Tyr2 Phospho-peptide, 3 nU/μL PTP1B; 50 mM Tris–HCl, pH 7.5, 10 mM $MgCl_2$, 0.01% Brij 35, 10% glycerol, 1 mM DTT, and 2% DMSO. Curve fits with Prism graph. The IC_{50} value of bpV (phen) for PTP1B was 180 nM.

4. Notes

1. Unless stated otherwise, all solutions should be prepared fresh for better results (unpublished observation).

2. The morphology of the printing dot is affected by many factors, such as: pin quality, glass material quality, slide coating evenness, cleanliness, etc. User should test a few brand slides to select the best one for a particular reaction, because manufactures may sell similar products made with totally different formulas and processes *(18, 19)*.

3. To have better printing, the slide should be clean free of dust by using gentle nitrogen stream and the slide dock will be closed during printing to avoid dust *(18)*.

4. Slides with certain functional groups can absorb moisture on the surface, which can affect the size of an array. Therefore, slides should be kept in consistent storage conditions for printing *(18)*.

5. Certain chemicals are very sticky, especially for stealth pin. To avoid carry over, DMSO can be used in a sonication bath (unpublished observation).

6. To create uniformed printing, the time of dipping and blotting should be pre-tested *(18, 19)*.

7. The NanoSprayer™ is cleaned with distilled water and reaction buffer before infusing enzyme-spraying solution. The sprayer will be cleaned with low-concentration ethanol, followed by distilled water after using.

8. Protein-coated microarray slides should only be used fresh, because natural protein has limited stability. Peptide-coated chips can be stored at low temperature for over 6 months.

9. When the peptide substrate is arrayed in the phosphatase reaction, excess peptide in the arrayed dots will bind on the area without reaction dots to create a comet tail, if the slide is washed slowly. Therefore, the reaction slide should be washed quickly with the reaction-stopping buffer.

10. During slide coating, blocking, and antibody detection, slide transfers among solutions should be performed quickly. The slides should be kept wet between washing, blocking, and staining to avoid uneven coating.

Acknowledgments

We would like to thank Ms. Debra Barninger for critical editing. This work is partially supported by NIH grant RO1HG003818, R44CA114995, and R44DE017485 to H.M.

References

1. Ma, H. and Horiuchi, K.Y. (2006) Chemical microarray: a new tool for drug screening and discovery. *Drug Discov. Today* **11**, 661–668

2. Sundberg, S.A. (2000) High-throughput and ultra-high-throughput screening: solution- and cell-based approaches. *Curr. Opin. Biotechnol.* **11**, 47–53

3. MacBeath, G., Koehler, A.N. and Schreiber, S.L. (1999) Printing small molecules as microarrays and detecting protein-ligand interactions en masse. *J. Am. Chem. Soc.* **121**, 7967–7968

4. Kuruvilla, F.G., Shamji, A.F., Sternson, S.M., Hergenrother, P.J. and Schreiber, S.L. (2002) Dissecting glucose signaling with diversity-oriented synthesis and small-molecule microarrays. *Nature* **416**, 653–657

5. Uttamchandani, M. (2005) Small molecule microarrays, recent advances and applications. *Curr. Opin. Chem. Biol.* **9**, 4–13

6. Uttamchandani, M., Wang, J. and Yao, S.Q. (2006) Protein and small molecule microarrays: powerful tools for high-throughput proteomics. *Mol. Biosyst.* **2**, 58–68

7. David, C.A., Middleton, T., Montgomery, D., Lim, H.B., Kati, W., Molla, A., Xuei, X., Warrior, U., Kofron, J.L. and Burns, D.J. (2002) Microarray compound screening (microARCS) to identify inhibitors of HIV integrase. *J. Biomol. Screen.* **7**, 259–266

8. Gopalakrishnan, S.M., Karvinen, J., Kofron, J.L., Burns, D.J. and Warrior, U. (2002) Application of Micro Arrayed Compound Screening (microARCS) to identify inhibitors of caspase-3. *J. Biomol. Screen.* **7**, 317–323

9. Anderson, S.N. (2004) Microarrayed Compound Screening (μARCS) to identify activators and inhibitors of AMP-activated protein kinase. *J. Biomol. Screen.* **9**, 112–121

10. Gosalia, D.N. and Diamond, S.L. (2003) Printing chemical libraries on microarrays for fluid phase nanoliter reactions. *Proc. Natl. Acad. Sci. U.S.A.* **100**, 8721–8726

11. Ma, H., Horiuchi, K.Y., Wang, Y., Kucharewicz, S.A. and Diamond, S.L. (2005) Nanoliter homogeneous ultra-high throughput screening microarray for lead discoveries and IC_{50} profiling. *Assay Drug Dev. Technol.* **3**, 177–187

12. Horiuchi, K.Y., Wang, Y. and Ma, H. (2006) Biochemical microarrays for studying chemical biology interaction: DiscoveryDot™ technology. *Chem. Biol. Drug Des.* **67**, 87–88

13. Ma, H., Wang, Y., Pomabo, A. and Tcai, C. (2006) A homogenous microarray for enzymatic functional assays. In: *Frontiers in Biochip Technology*. Springer, New York, pp. 3–18

14. Horiuchi, K.Y., Wang, Y., Diamond, S.L. and Ma, H. (2006) Microarrays for the functional analysis of the chemical-kinase interactome. *J. Biomol. Screen.* **11**, 48–56

15. Simmons, G., Gosalia, D.N., Rennekamp, A.J., Reeves, J.D., Diamond, S.L. and Bates, P. (2005) Inhibitors of cathepsin L prevent severe acute respiratory syndrome coronavirus entry. *Proc. Natl. Acad. Sci. U.S.A.* **102**, 11876–11881

16. Gosalia, D.N., Salisbury, C.M., Maly, D.J., Ellman, J.A. and Diamond, S.L. (2005) Profiling serine protease substrate specificity with solution phase fluorogenic peptide microarrays. *Proteomics* **5**, 1292–1298

17. Gosalia, D.N., Salisbury, C.M., Ellman, J.A. and Diamond, S.L. (2005) High throughput substrate specificity profiling of serine and cysteine proteases using solution-phase fluorogenic peptide microarrays. *Mol. Cell. Proteomics* **4**, 626–636

18. Zong, Y., Wang, Y., Shi, J., and Zhang, Z., (2006) The Application of Novel Multi-Functional Microarray Slides for Immobilization Biomolecules. In: *Frontiers in Biochip Technology*. Springer, New York, pp. 3–18

19. Mann, C.J., Stephens, S.K., and Burke, J.F. (2004) Production of Protein Microarries. In: *Protein Microarray Technology*. Wieley-VCH, Weinheim, pp. 165–194

Chapter 10

Fluorescence Polarization and Time-Resolved Fluorescence Resonance Energy Transfer Techniques for PI3K Assays

Kurumi Y. Horiuchi and Haiching Ma

Summary

Fluorescence-based biochemical assays are sensitive and convenient to use; therefore, they are widely employed for enzyme assays and molecular interaction studies. However, when this method is applied for screening of a compound library for drug discovery, high fluorescence compounds, which usually exist in large numbers in chemical libraries, are problematic. Fluorescence Polarization (FP) and Time-Resolved Fluorescence Resonance Energy Transfer (TR-FRET) assays are less affected by compound fluorescence and suitable for large-scale high-throughput screening (HTS). In this section, we describe homogenous FP and TR-FRET methods for PI3-kinase (PI3K), a family of lipid kinases that is "difficult-to-do-HTS" since traditional radioisotope assays are hard to apply to HTS format. The application of FP and TR-FRET techniques for PI3K HTS will be described and advantages and disadvantages of these assays will be discussed.

Key words: FP, FRET, TR-FRET, HTS, Kinase, Lipid kinase, PI3K

1. Introduction

Phosphoinositide 3-kinase (PI3K) regulates important cellular processes such as mitogenesis, apoptosis, and cytoskeletal functions, by phosphorylating phosphatidylinositol-4,5-bisphosphate (PIP_2) to phosphatidylinositol-3,4,5-trisphosphate (PIP_3) to activate the serine/threonine kinase Akt/mTOR pathway. Overexpression of PI3K is involved in many cancer forms (1) and the PI3K pathway becomes a drug target for various diseases (2–4). The traditional PI3K assay is performed by radioisotope assays with extraction of lipid products by organic solvent and/or separation using TLC plates (5, 6). Since PI3K is an important drug

Ana Cecília A. Roque (ed.), *Ligand-Macromolecular Interactions in Drug Discovery: Methods and Protocols,*
Methods in Molecular Biology, vol. 572,
DOI 10.1007/978-1-60761-244-5_10, © Humana Press, a part of Springer Science+Business Media, LLC 2010

target, assays applicable to high-throughput screening (HTS) have been long awaited. Assays using Fluorescence Polarization (FP) and Time-Resolved Fluorescence Resonance Energy Transfer (TR-FRET) are available, and we describe here the detailed assay methods. Both techniques require special instrumentation; however, once the conditions are set up, the results are stable and consistent.

1.1. Fluorescence Polarization

FP is a technique that monitors molecular movement and rotation by labeling the molecule with a fluorescent dye; the molecule's speed of rotation will be changed after it changes size. Since first introduced by Perrin in 1926, FP has been applied widely in the fields of biochemistry and drug discovery *(7, 8)*. FP assays are homogeneous reactions and easy to automate, with a single incubation of premixed tracer-receptor reagent, the reactions are very rapid, and take seconds to minutes to reach equilibrium. FP reagents are stable and easily prepared, with high reproducibility. In addition, since FP is independent of fluorescent intensity, it is less affected by the inner filter effect and fluorescent compounds. It is also relatively insensitive to instrument changes such as drift, gain settings, lamp changes, etc. Furthermore, fluorescence intensity is obtained in addition to polarization, if required.

FP (P) is defined by the following equation:

$$P = \frac{\left(I_{\parallel} - I_{\perp}\right)}{\left(I_{\parallel} + I_{\perp}\right)} \tag{1}$$

where $I_{\parallel}$ equals the emission intensity parallel to the excitation plane, and $I_{\perp}$ equals the emission intensity perpendicular to the excitation plane of a fluorophore when excited by polarized light. As can be seen from **Eq. 1**, P ("polarization unit") is a dimensionless entity and is not dependent on the intensity of the emitted light or on the concentration of the fluorophore. This is the fundamental power of FP. For convenience, the term "mP" (pronounced "millipee") is now in general use, where 1 mP equals one thousandth of a P. Many instruments contain built-in functions to calculate P. In the real world, when the fluorophore is free to rotate, the polarization is a smaller number. After the fluorophore binds to a larger molecule, which limits the rotation of molecule, the polarization becomes a larger number. Since the rotation of molecule can be affected by the viscosity of a buffer, the reaction buffer must be chosen carefully.

1.2. Time-Resolved Fluorescence Resonance Energy Transfer

Fluorescence resonance energy transfer (FRET) is a very powerful method for obtaining both structural and dynamic information about macromolecules and macromolecular complexes. Lakowicz *(9)* and Eftink *(10)* provide good introductions to the phenomenon of energy transfer. A more detailed treatment can be found in the chapter by Cheung in Vol. 2 of the Lakowicz series *(11)*. The energy transfer occurs when the donor fluorophore is in close proximity (typically less than 75Å) to the acceptor fluorophore after excitation of the donor, resulting in observation of acceptor emission. The efficiency of transfer is related to the inverse of the sixth power of the distance between the donor and acceptor molecules *(9–11)*. Other factors such as the overlap of the emission of the donor with the absorption of the acceptor and the relative orientations of the two chromophores also affect the energy transfer. Thus, the donor/acceptor pair needs to be chosen carefully, and many pairs at different wavelengths are available in Invitrogen's Website.

TR-FRET uses the principle of FRET, but the donor fluorophore has a long excited state lifetime; a millisecond or longer compared to usual fluorophores in the nanosecond range. Commonly, Terbium or Europium complex is used for a long lifetime donor. Because of long lifetime donor emission, short lifetime emissions from test compounds are eliminated. The only concern is compounds with strong absorbance at donor and acceptor emission wavelengths, which may cause a decrease of output fluorescence signal.

2. Materials

2.1. Fluorescence Polarization Assay for PI3K

1. Black 384-well low-volume plate (Low protein binding; Nalgen, NY).

2. PI(4,5)P$_2$ Substrate, 1 mM stock (Echelon Biosciences Inc., UT).

3. PI(3,4,5)P$_3$ FP Detector and diluent (Echelon Biosciences Inc., UT).

4. PI(3,4,5)P$_3$ Standard, 40 μM stock (Echelon Biosciences Inc., UT).

5. Fluorescence FP Probe (2.5 μM stock) and diluent (Echelon Biosciences Inc., UT).

6. Assay buffer: 20 mM HEPES, pH 7.5, 4 mM MgCl$_2$, 10 mM NaCl, 2 mM DTT (DTT must be added fresh each time).

7. ATP (Sigma-Aldrich, MO).

8. EnVision Fluorescence Plate Reader: Ex filter; BODIPY TMR FP (531 nm), Em filter; BODIPY TMR FP P-pol and BODIPY TMR FP S-pol (595 nm), Mirror; TMR FP Dual (D555fp/ D595) (Perkin-Elmer, MA). EnVision has a function of calculation for Polarization (mP). If using an instrument without

calculation function, Polarization is calculated using **Eq. 1** in the Introduction.

9. Echo acoustic liquid transfer instrument for compound transfer (Labcyte, CA).

2.2. Time-Resolved Fluorescence Resonance Energy Transfer

1. Black 384-well low-volume plate (Low protein binding; Nalgen, NY).

2. $PI(4,5)P_2$ Substrate (Echelon Biosciences Inc., UT).

3. $PI(3,4,5)P_3$ Standard (Echelon Biosciences Inc., UT).

4. Stop Solution (EDTA and Biotin-$PI(3,4,5)P_3$; mix solution A and B in 1:3, Millipore, CA).

5. HTRF Detection Mix (GST-pleckstrin homology (PH) domain, Europium-labeled anti-GST Ab, and Streptavidin-APC; mix detection solution A, B, and C in 18:1:1, Millipore, CA).

6. Assay buffer, add 5 mM DTT in fresh (Millipore, CA).

7. ATP (Sigma-Aldrich, MO).

8. EnVision setting: Flash; 100, Delay time; 50 µs, Integration time; 400 µs, Ex filter; 320 nm, Em1 filter; 615 nm, Em2 filter; 665 nm, Mirror; LANCE/DELFIA (D400) (Perkin-Elmer, MA).

9. Echo acoustic liquid transfer instrument for compound transfer (Labcyte, CA).

3. Methods

All methods require optimization of enzyme concentrations by titration since each method has a different sensitivity. All fluorescently labeled materials are light sensitive, so that they need to be protected from light during incubation by covering the plate with aluminum sealing tape.

3.1. Fluorescence Polarization Assay for PI3K

The detection of phosphorylation of substrate is a competitive FP assay. When there is no phosphorylation, the fluorescent PIP_3 FP Probe binds to the PIP_3 Detector protein and polarization of the probe is the maximum. After PI3K phosphorylates the substrate, the number of product (PIP_3) increases, which competes with the probe for binding to the Detector protein. As a result, polarization decreases. The kinase activity is monitored by the decrease of mP value, and the amount of PIP_3 product is estimated based on a standard curve.

3.1.1. Enzyme Titration
of PI3K by FP

1. Prepare 2× enzyme dilution in the assay buffer. Typically the enzyme is diluted threefold serial dilution starting at 100 nM (or 10 nM if active) including no enzyme control, and two sets are prepared for 2 time points. Deliver 2.5 μL per well (*see* **Note 1**).

2. Prepare 2× substrate mixture in the assay buffer; make 20 μM of PIP_2 substrate and 20 μM ATP for final 10 μM of each substrate. The reaction is initiated by the addition of 2.5 μL of 2× substrate mixture into the enzyme dilution above, and incubated at room temperature for 30 and 60 min (*see* **Note 2**).

3. During the incubation, standards are prepared in the reaction buffer for a standard curve. Make seven twofold serial dilutions starting at 4 μM including buffer alone for base line, and deliver 5 μL/well. Standard curve can be run duplicate or triplicate.

4. To stop the reaction at 30 min, 5 μL of 250 nM FP Detector diluted in Detector diluent is added into each reaction well of the first set. The detector solution quenches the kinase reaction. After the reaction of another set is stopped at 60 min as above, the same amount of Detector was added into standards.

5. Subsequently, 2.5 μL of 50 nM Fluorescence FP Probe is added into all wells. The plate is shaken on plate shaker and incubated for 10 min in the dark to equilibrate (*see* **Note 3**).

6. The plate is read in EnVision with setting mentioned in **Subheading 2**. EnVision has the function of calculation of Polarization as follows:

$$\text{Polarization (mP)} = 1,000 \times \frac{(S - GP)}{(S + GP)} \qquad (2)$$

where S and P represent the emission intensity parallel and perpendicular to the excitation polarization, respectively, and G is an instrument and wavelength-dependent correction factor to compensate for the polarization bias of the detection system. In this measurement, G-factor is 0.97. Typical standard curve is shown in **Fig. 1**.

3.1.2. Inhibitor Evaluation
of PI3K Activity by FP

1. Based on the result of enzyme titration, prepare 2× enzyme dilution in the assay buffer. Deliver 2.5 μL/well. Include no enzyme wells by delivering buffer instead of enzyme solution (*see* **Note 4**).

2. Add inhibitor in dose-response manner by Echo; 5–20 nL/well. DMSO (1–2%) can be added into the assay buffer to help compound solubility (*see* **Note 5**).

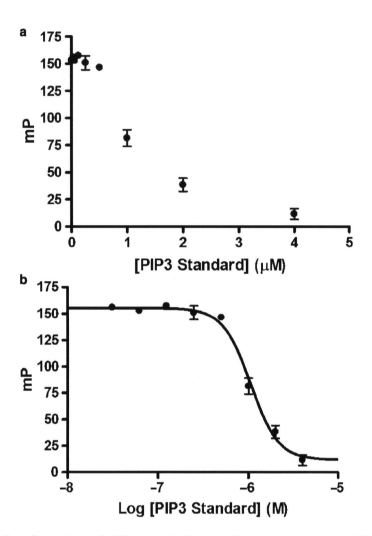

Fig. 1. Standard curve for PIP$_3$ product by FP assay. mP values are plotted against PIP3 concentration in micromolar (μM) (**a**), or in logarithm molar (Log M) (**b**). The curve fit of (**b**) was performed with Sigmoidal dose–response curve (variable slope) using **Eq. 5** (*see* **Note 7**). Parameters obtained from fit are 11.93, 155.4, −5.976, and −2.187, for Bottom, Top, LogEC50, and Hill Slope, respectively.

3. Add 2× substrate mixture in the assay buffer; make 20 μM of PIP$_2$ substrate and 20 μM ATP. The reaction is initiated by the addition of 2.5 μL of 2× substrate mixture into all wells, and incubated for 60 min at room temperature (*see* **Note 6**).

4. Add 5 μL of 250 nM FP Detector diluted in Detector diluent to stop the reaction. Also add into standards prepared as above.

5. Subsequently, 2.5 μL of 50 nM Fluorescence FP Probe is added into all wells. The plate is shaken on plate shaker and incubated for 10 min in the dark to equilibrate.

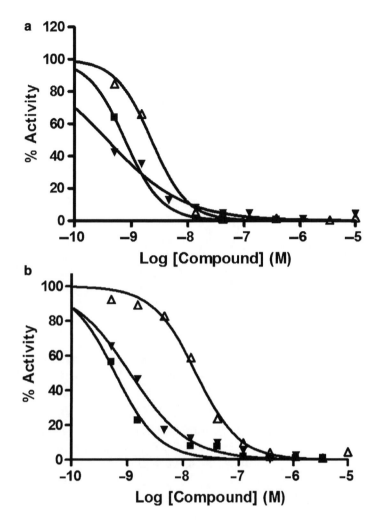

Fig. 2. Dose-response inhibition curves by FP assay for PI3Kα (a) and PI3Kδ (b). (**a**) IC50 values are 0.78, 2.3, and 0.4 nM for Wartmannin (*filled square*), PI3Kγ Inhibitor (*open up triangle*), and PI-103 (*filled down triangle*), respectively. (**b**) IC50 values are 0.6, 16.8, and 1.1 nM for Wartmannin (*filled square*), PI3Kγ Inhibitor (*open up triangle*), and PI-103 (*filled down triangle*), respectively.

6.   The plate is read in EnVision with setting mentioned in **Subheading 2**. Typical dose-response inhibition curves are shown in **Fig. 2**.

3.2. Time-Resolved Fluorescence Resonance Energy Transfer Assay for PI3K

The PI3K activity is also detected by a competitive assay. Biotinylated PIP_3 forms a complex with streptavidin-APC, which binds to GST-tagged PH domain, binding specifically to PIP_3 but not to PIP_2. Europium-labeled anti-GST antibody binds to GST-tagged PH domain complex with Biotinylated PIP_3, resulting in APC emission (665 nm) via energy transfer from Eu emission (620 nm) to APC when excited at 330 nm. Phosphorylated product, PIP_3, competes with biotin-PIP_3 binding to Detector protein, which breaks energy transfer complex, resulting in the decrease of the energy transfer signal. Taking a ratio of Eu emission (620 nm) and APC emission (665 nm), the kinase activity is estimated based on a standard curve.

3.2.1. Enzyme Titration of PI3K by TR-FRET

1. Prepare Stop solution and Detection mixture. These solutions need to incubate at least 2 h at room temperature before use. Protect Detection mixture from light.

2. Prepare 2× enzyme dilution by threefold serial dilution starting at 10 nM in constant 2× concentration (20 μM) of PIP$_2$. The enzyme/substrate mixture is delivered into the plate at 5 μL/well.

3. Prepare 2× ATP (20 μM) in the assay buffer. The reaction is initiated by the addition of 5 μL of 2× ATP into the enzyme dilution above, and incubated at room temperature for 30–60 min. For the first time, time course measurement of enzyme activity is recommended to determine enzyme stability (*see* **Note 2**).

4. To stop the reaction, 2.5 μL of Stop solution is added into each reaction well and standards. Standards are made by ten twofold serial dilutions starting at 4 μM including buffer alone for base line, and deliver 10 μL/well. Standard curve can be run duplicate or triplicate.

5. Subsequently, 2.5 μL of Detection mixture was added into all wells. Plate is shaken on plate shaker and incubated for 6 h to overnight at room temperature. Protect from light.

6. The plate is read in EnVision with setting mentioned in **Subheading 2**. Calculate the ratio of two emissions as follows:

$$\text{FRET Ratio} = \left(\frac{\text{Emission at } 665 \, \text{nm}}{\text{Emission at } 615 \, \text{nm}} \right) \times 10,000 \qquad (3)$$

7. Typical standard curve is shown in **Fig. 3**.

3.2.2. Inhibitor Evaluation of PI3K Activity by TR-FRET

1. Based on the result of enzyme titration, prepare 2× enzyme and 20 μM of PIP$_2$ Substrate in the assay buffer. Deliver 5 μL/well. Include substrate alone wells as no enzyme controls (*see* **Note 4**).

Fig. 3. Standard curve for PIP3 product by TR-FRET assay. Emission ratios obtained using **Eq. 3** are plotted against the PIP3 concentration in micromolar (μM) (**a**, **b**), or in logarithm molar (Log M) (**c**). (**b**) The same data but expanded scale of (**a**), and *solid* and *broken lines* are drawn from "No enzyme control" to two enzyme concentrations of "No inhibitor control." (**c**) The curve fit was performed with Sigmoidal dose–response curve (variable slope) using **Eq. 5** (*see* **Note 7**). Parameters obtained from fit are 2756, 77068, −6.731, and −0.804, for Bottom, Top, LogEC50, and Hill Slope, respectively.

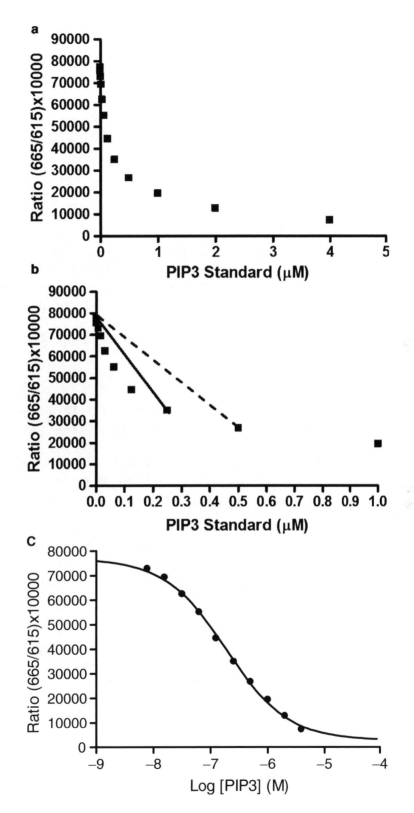

2. Add inhibitors in dose-response manner by Echo; 5–20 nL/well. DMSO (1–2%) can be added into the assay buffer to help compound solubility.

3. Prepare 2× ATP (20 μM) in the assay buffer. The reaction is initiated by the addition of 5 μL of 2× ATP into all wells, and incubated for 30–60 min at room temperature.

4. During the incubation, standards are prepared as above. Standards can be run duplicate or triplicate.

5. To stop the reaction, 2.5 μL of Stop solution is added into each reaction well. At the same time, the same amount of Stop solution is added into standards.

6. Subsequently, 2.5 μL of Detection mixture is added into all wells. Plate is shaken on plate shaker and incubated for 6 h to overnight at room temperature. Protect from light.

7. The plate is read in EnVision with setting mentioned in Subheading2 and FRET ratio are calculated using Eq. 3.

8. The inhibition of compounds are calculated as following equation, or based on standard curve (*see* **Subheading 3.3** and **Note 7**).

$$\% \text{ Enzyme Activity} = \left(\frac{\text{Sample} - \text{No Enzyme Control}}{\text{No Inhibitor Control} - \text{No Enzyme Control}} \right) \times 100 \quad (4)$$

3.3. Results and Discussion

3.3.1. Standard Curves and Activity Calculations

Both FP and TR-FRET assays described here are competition assays; the products of kinase reaction (PIP$_3$) compete with the probe for each method. Therefore, standard curves are not simple linear lines. Since each method has a different sensitivity, it is important to include a standard curve for every assay to confirm whether the control signal is in the sensitive range (*see* **Note 4**). However, since standard curves are not linear, it is not so simple to estimate the kinase activities from standard curves. There are three methods to estimate kinase activity and the inhibition profile of compounds: (1) Calculate activity using **Eq. 4**, (2) Directly read from a standard curve for product formation, and (3) Calculate activity from curve fit of a standard curve (*see* **Note 7** for detail).

Method 1 is the simplest way to estimate the inhibition activities of compounds. However, this method assumes a linear relationship between signal and product. If a line drawn between the emission ratio of "No enzyme control" and "No inhibitor control," in TR-FRET assay for example (**Fig.3b**, *solid line*), is close enough to the standard curve, this method is reasonable. However, when the activity of "No inhibitor control" is high (the product of enzyme activity for a period is 0.5 μM for example), the line is far from the standard curve (**Fig.3b**, *broken line*). In this case, the estimated inhibition may be off from real inhibition. One can use method 2 to directly read the enzyme activity from the standard curve by converting the obtained FRET ratio to μM product, and then calculate the % Enzyme Activity. However, it is time consuming and difficult to apply this processing to a large number of data. Method 3 is somewhat complicated but most accurate, and can process a large number of data once a template is made in Excel (*see* **Note** 7 for formula). The IC50 values obtained from method 1 (linear assumption, *broken lines*) and method 3 (curve fit, *solid lines*)·are compared in **Fig.4**. There is about a fivefold difference between these two methods.

3.3.2. Inhibitor Evaluation by FP and TR-FRET Assays

Although both assays here are competition assays, each method uses a different detection complex. FP assay uses fluorescence-labeled PIP_3 which binds to the PIP_3-specific binding protein, resulting in polarization change. Therefore, equilibrium is quick (10–15 min). Since it monitors the change in polarization signal from one kind of fluorophore, this assay is simple and easy to apply for HTS. On the other hand, TR-FRET assay involves two fluorophores and many protein bindings including antibodies: It needs to incubate pre-mixed reagents for more than 2 h before use in addition to the 6 h to overnight incubation for detection after stopping the reaction by adding all reagents. It is time consuming, yet signals are stable and results are consistent.

Using both assays, known PI3K inhibitors are tested against PI3-Kinase isoforms. Wortmannin is known as a potent and selective inhibitor, which inhibits PI3-Kinase irreversibly *(12)*. PI3-Kγ Inhibitor (5-Quinoxalin-6-ylmethylene-thiazolidine-2,4-dione) is a selective and ATP-competitive inhibitor *(13)*. PI-103 (3-(4-(4-Morpholinyl)pyrido[3′,2′:4,5]furo[3,2-d]pyrimidin-2-yl)phenol) is also a potent and ATP-competitive inhibitor for PI3-Kinase as well as DNK-PK and mTOR *(14)*. IC50 values obtained by both assays, FP and TR-FRET assays, are consistent with published values (**Figs.2** and **4**).

3.3.3. Compound Screening

A small library of natural products is screened against PI3-Kinase α using TR-FRET assay. Most compound libraries have compounds which have a fluorescence background. This fluorescence affects measurement signals, resulted in false-positives or false-negatives. TR-FRET assay is less affected by compound fluorescence: It uses long lifetime donor fluorescence from Europium, so that short lifetime fluorescence from compounds is eliminated. In addition, since the acceptor/donor emission ratio is taken, disturbance of excitation light by compounds does not affect the results. However, if compounds have strong absorbance at donor emission (615–635 nm) or acceptor emission (665–668 nm) range, the FRET ratio may be affected. In order to avoid false-positives or -negatives, a counter assay can be performed. In this assay, compounds are added into wells in a similar manner to the screening: Each well contains buffer

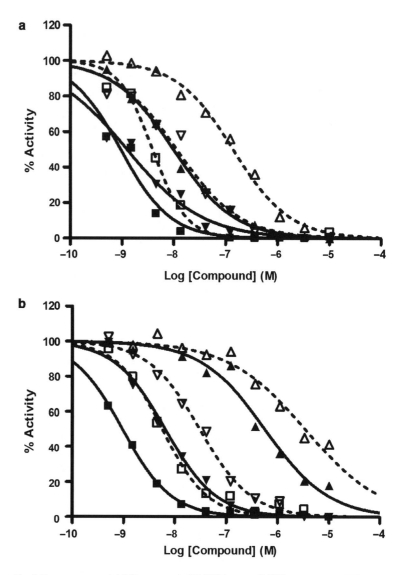

Fig. 4. Dose–response inhibition curves by TR-FRET assay. *Solid lines* represent % Enzyme Activity obtained by curve fit of a standard curve (method 3), and *broken lines* represent % Enzyme Activity obtained by **Eq. 4** (method 1) from the same data set. (**a**) IC50 values for PI3Kα obtained by method 3 (*solid line*) are 0.94, 9.8, and 1.3 nM for Wartmannin (*filled square*), PI3Kγ Inhibitor (*filled up triangle*), and PI-103 (*filled down triangle*), respectively, and IC50 values obtained by method 1 (*broken line*) are 3.9, 140, and 11.4 nM for Wartmannin (*open square*), PI3Kγ Inhibitor (*open up triangle*), and PI-103 (*open down triangle*), respectively. (**b**) IC50 values for PI3Kβ obtained by method 3 (*solid line*) are 0.95, 564, and 6.5 nM for Wartmannin (*filled square*), PI3Kγ Inhibitor (*filled up triangle*), and PI-103 (*filled down triangle*), respectively, and IC50 values obtained by method 1 (*broken line*) are 5.7, 3,400, and 30.5 nM for Wartmannin (*open square*), PI3Kγ Inhibitor (*open up triangle*), and PI-103 (*open down triangle*), respectively.

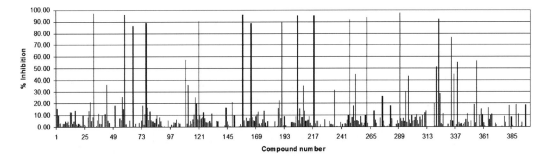

Fig. 5. Screening results of a natural product library (400 compounds) against PI3Kα by TR-FRET assay. Ordinate represents % Inhibition of control activity, and abscissa represents compound number. Hit rate (more than 50% inhibition) was 4.8%, and Z′-factor was 0.78.

and a certain concentration of PIP_3 (i.e., 0.5–0.8 µM) that mimics the PI3K control activity but without enzyme. Signals are detected as usual by adding the stop solution and the detection mixture. If there is no interference from compounds, all signals and FRET ratios will be the same. In this way, the compound effects are determined without enzyme reactions. The results of PI3Kα screening against 400 compounds from a natural product library are summarized in **Fig. 5**. Hit rate (percentage of number of compounds that showed more than 50% inhibition) was 4.8%, and Z′-factor *(15)* was 0.78.

4. Notes

1. For developing an assay the first time, the time course measurement with multitime points of enzyme activity is recommended for determination of enzyme stability and reaction linearity.

2. ATP concentrations can be varied for Km studies and ATP competition assays. Incubation can be done at 30°C for better enzyme activity.

3. The FP signal is stable for up to 6 h at room temperature.

4. Enzyme concentration must be chosen carefully since each assay format has a different sensitivity. For example, based on the product standard curve for FP assay in this study **(Fig.1)**, most sensitive range is 0.5–2 µM while 0.05–1 µM for TR-FRET assay **(Fig.3)**. Therefore, the optimal enzyme concentration for compound screening is to produce around 1.5–2 µM products for FP assay, in other words, product amount that give a signal range of 15% and 40% of the maximum signal (No Enzyme control signal) for FP assay **(Fig.1b)** and

0.5–1 μM products for TR-FRET assay (**Fig.3c**) for a period of incubation. If the enzyme concentration employed is higher than the optimal range mentioned above, the enzyme inhibition does not correspond to the signal change properly. Also, it needs to make sure the reaction is in the linear range within the time frame; the product is linearly increased for 1 h (or the incubation period).

5. If compound transfer needs intermediate dilution because of instrument limitations, care should be taken whether compounds are not crushed out in the intermediate dilution. This can be avoided by a larger volume of direct transfer; however, DMSO concentration in the final reaction is better kept lower than 2% since high concentrations of DMSO affect enzyme activity.

6. If the enzyme is unstable for 1 h, incubation can be performed for 30 min at room temperature or at 30°C.

7. First, plot mP values or FRET ratios versus PIP$_3$ molar concentrations in log scale, and perform curve fit with displacement equation below (or Sigmoidal dose–response curve (variable slope) in GraphPad Prism, **Figs.1b** and **3c**).

$$Y = \text{Bottom} + \frac{\text{Top} - \text{Bottom}}{1 + 10^{(\text{Log EC50} - X) \times \text{HillSlope}}} \qquad (5)$$

Or

$$\left[\begin{array}{l} Y = \text{Bottom} + (\text{Top} - \text{Bottom})/ \\ (1+10\,^\wedge((\text{Log EC50} - X)\text{HillSlope})) \end{array} \right]$$

where X is the logarithm of concentration, y is the response, Top and Bottom are the maximum and minimum signals, respectively, and EC50 in this case means the PIP$_3$ concentration which gives 50% signal of the maximum. After obtaining parameters such as Top, Bottom, LogEC50, and Hill Slope from fit, all signals are converted into μM product using the following equation:

$$x = 10^{\left(\text{LogEC50} - \frac{Log\left(\frac{\text{Top}-\text{Bottom}}{Y-\text{Bottom}}-1\right)}{\text{HillSlope}} \right)} * 1000000 \qquad (6)$$

Or

$$\left[\begin{array}{l} x=10\,^\wedge(\text{Log EC50} - \text{Log10}((\text{Top} - \text{Bottom})/ \\ (Y - \text{Bottom}) - 1) / \text{HillSlope}) \times 1,000,000 \end{array} \right]$$

where x is now unit of μM, and no more logarithm. In an Excel sheet, choose the cell where the calculation result goes, and type in the following:

$$=10 \char`^ (\text{"Log EC50"} - \text{Log} 10((\text{"TOP"} - \text{"BOTTOM"}) /$$
$$(\text{"Cell"} - \text{"Bottom"}) - 1) / \text{"HillSlope"}) \times 1,000,000 \qquad (7)$$

where "Cell" is the cell position of measured signal (or FRET ratio). By entering all values from fit such as Top, Bottom, LogEC50, and Hill Slope, product in μM is obtained. The % activity (or % Inhibition) then is calculated from product in μM.

Acknowledgments

This work is partially supported by NIH grants RO1HG003818 and R44CA114995 to H.M.

References

1. Engelman, J. A., Luo, J., and Cantley, L. C. (2006) The evolution of Phosphatidylinositol 3-kinases as regulators of growth and metabolism. *Nat Rev Genet* 7, 606–19

2. Martelli, A. M., Tazzari, P.L., Evangelisti, C., Chiarini, F., Blalock, W. L., Billi, A. M., Manzoli, L., McCubrey, J. A., and Cocco, L. (2007) Targeting the phosphatidylinositol 3-kinase/Akt/mammalian target of rapamycin module for acute myelogenous leukemia therapy: from bench to bedside. *Curr Med Chem* 14 (19), 2009–23 (review)

3. Maiese, K., Chong, Z. Z., and Shang, Y. C. (2007) Mechanistic insights into diabetes mellitus and oxidative stress. *Curr Med Chem* 14 (16), 1729–38 (review)

4. Rückle, T., Schwarz, M. K., and Rommel, C. (2006) PI3Kgamma inhibition: towards an 'aspirin of the 21st century'? *Nat Rev Drug Discov* 5 (11), 903–18 (review)

5. Kaplan, D. R., Whitman, M., Schaffhausen, B., Pallas, D. C., White, M., Cantley, L., and Roberts, T. M. (1987) Common elements in growth factor stimulation and oncogenic transformation: 85 kd phosphoprotein and phosphatidylinositol kinase activity. *Cell* 50, 1021–9

6. Leopoldt, D., Hanck, T., Exner, T., Maier, U., Wetzker, R., and Nurnberg, B. (1998) Gbetagamma stimulates phosphoinositide 3-kinase-gamma by direct interaction with two domains of the catalytic p110 subunit. *J Biol Chem* 273 (12), 7024–9

7. Albizu, L., Teppaz, G., Seyer, R., Bazin, H., Ansanay, H., Manning, M., Mouillac, B., and Durroux, T. (2007) Toward efficient drug screening by homogeneous assays based on the development of new fluorescent vasopressin and oxytocin receptor ligands. *J Med Chem* 50 (20), 4976–85

8. Mathias, U. and Jung, M. (2007) Determination of drug-serum protein interactions via fluorescence polarization measurements. *Anal Bioanal Chem* 388 (5–6), 1147–56

9. Lakowicz, J. (1983) *Principles of Fluorescence Spectroscopy*. Plenum Press, New York

10. Eftink, M. (1990) Fluorescence techniques for studying protein structure. In: Suelter, C. H. (ed.) *Protein Structure Determination: Methods of Biochemical Analysis*, 35. Wiley, New York

11. Cheung, H. C. (1991) Resonance energy transfer. In: Lakowicz, J. R. (ed.) *Topics in Fluorescence Spectroscopy*, 2. Plenum Press, New York

12. Cross, M. J., Stewart, A., Hodgkin, M. N., Kerr, D. J., and Wakelam, M. J. O. (1995) Wortmannin and its structural analogue demethoxyviridin inhibit stimulated phospholipase A2 activity in Swiss 3T3 cells. Wortmannin is not a specific inhibitor of phosphatidylinositol 3-kinase. *J Biol Chem* 270, 25352–5

13. Barber, D. F., Bartolomé, A., Hernandez, C., Flores, J. M., Redondo, C., Fernandez-Arias, C., et al. (2005) PI3Kgamma inhibition blocks glomerulonephritis and extends lifespan in a mouse model of systemic lupus. *Nat Med* 11, 933–5

14. Raynaud, F. I., Eccles, S., Clarke, P. A., Hayes, A., Nutley, B., Alix, S., et al. (2007) Pharmacologic characterization of a potent inhibitor of class I phosphatidylinositide 3-kinases. *Cancer Res* **67**, 5840–50

15. Zhang, J. H., Chung, T. D., and Oldenburg, K. R. (1999) A simple statistical parameter for use in evaluation and validation of high throughput screening assays. *J Biomol Screen* **4**, 67–73

Chapter 11

Small Molecule Protein Interaction Profiling with Functional Protein Microarrays

Lihao Meng, Dawn Mattoon, and Paul Predki

Summary

Small molecules possess the ability to interact with proteins and perturb their specific functions, a property that has been exploited for numerous research applications and to produce therapeutic agents in disease treatment. However, commonly utilized mass spectrometry-based approaches for identifying the target proteins for a small molecule have a number of limitations, particularly in terms of throughput and time and resource consumption. In addition, current technologies lack a mechanism to broadly assess the selectivity profile of the small molecule, which may be important for understanding off-target effects of the compound. Protein microarray technology has emerged as a powerful tool in the systems biology arsenal. Here, we describe how protein microarray technology can be applied to the study of small molecule protein interactions, with sensitivity sufficient to detect interactions with low µM affinity. These assays are highly reproducible, sensitive, and scalable, and provide an enabling technology for small molecule selectivity profiling in the context of drug development.

Key words: Protein small molecule interaction, Small molecule, Protein microarray, Staurosporine, Kinase inhibitor, Off-target effect, Selectivity profiling, Specificity profiling

1. Introduction

Functional protein microarrays, with immobilized, correctly folded, and functional proteins, are intended for the systematic study of protein bioactivities. Functional protein microarrays are an important new tool ideally suited to the mapping of biological pathways. Nearly all functional elements of these pathways, including both binary and multicomponent molecular interactions as well as enzymatic activity, have been recapitulated and studied on protein microarrays (1). Protein microarray technology has been

Ana Cecília A. Roque (ed.), *Ligand-Macromolecular Interactions in Drug Discovery: Methods and Protocols*,
Methods in Molecular Biology, vol. 572,
DOI 10.1007/978-1-60761-244-5_11, © Humana Press, a part of Springer Science+Business Media, LLC 2010

validated for use in profiling multiple types of post-translational events, including phosphorylation, ubiquitination, and methylation. Additionally, protein microarrays have been successfully employed to study a variety of binding events, including protein–protein interaction profiling, antibody specificity profiling, and immune response profiling. The recent generation of microarrays comprising thousands of proteins has enabled this technology to make significant contributions toward our understanding of the molecular etiology of disease.

The binding of small molecules to proteins is an integral facet of many biological pathways, including the intracellular binding of hormones to nuclear hormone receptors, as well as the extra-cellular binding of small molecule ligands to a variety of G protein-coupled receptors. In addition, numerous small molecule drugs exert their biological effects, both desired, and undesired, through interaction with protein-binding partners. Functional protein microarrays provide a powerful approach to the identification and characterization of protein–small molecule interactions.

The feasibility of detecting small molecule–protein interactions on functional protein arrays was first demonstrated through the binding of fluorescent molecules to the FK506-binding protein FKBP12 immobilized on chemically derivitized glass slides *(2)*. Molecules with affinities in the nM to µM range exhibited readily detectable protein-binding properties in the microarray format. Specific binding of small molecule ligands to G protein-coupled receptors has also been shown on arrays, indicating that the receptors in lipid bilayers can retain membrane-like properties when arrayed on a solid support *(3)*. Additionally, the thyroid hormone receptor has been shown to interact specifically with its ligand, triiodothyronine, when the receptor is immobilized in an array format *(4)*.

High-content protein microarrays comprising thousands of purified recombinant human proteins have established a new paradigm for studying interactions with proteins, and studies are now underway to use these high-content protein arrays to identify the protein targets for a variety of organic and synthetic small molecules. The binding properties of GTP to GTP-binding proteins on arrays have been shown to be comparable to those observed in solution, with disassociation constants determined from array-based experiments in good agreement with those derived from solution-based measurements *(5)*. High-content protein arrays have also been used to identify targets for small molecules that suppress a chemical-induced growth phenotype in the yeast *S. cerevisiae(6)*. These studies clearly demonstrate that this platform introduces a new mechanism for the rapid identification of binding targets for small molecules, with the potential to greatly accelerate the pace of drug discovery. The sensitivity and ease-of-use support the widespread adoption of protein microarray technology for profiling small molecule–protein interactions. High-density protein microarray technology has the potential to dramatically accelerate the development of therapeutic small molecule compounds with improved safety profiles. Here, we used a known kinase inhibitor, staurosporine, to profile Invitrogen's ProtoArray® Human Protein Microarray, which contains about 8,000 functional human proteins. We observed that staurosporine interacts with 83 protein kinases and 10 nonkinase proteins.

2. Materials

2.1. Reagents and Equipment

1. Protein Microarray Small Molecule Interaction Buffer (SMI Buffer): 50 mM Tris pH 7.5, 5 mM MgSO4, 0.1% Tween20, prepared fresh and stored at 4°C. A stock solution of 10% Tween20 should be used to make the buffer.

2. ProtoArray® Human Protein Microarrays (Invitrogen, Carlsbad, CA), stored at −20°C until use. See **Subheading 2.3** and http://www.invitrogen.com/protoarray for additional information on these protein array products.

3. Staurosporine conjugated to Alexa Fluor® 647 (Invitrogen, Carlsbad, CA) diluted to 100 nM in SMI Buffer. The compound is sensitive to light. Keep in dark at all times. Make the diluted staurosporine fresh just before use.

4. Unlabeled Staurosporine (EMD Chemicals, Gibbstown, NJ) at a stock concentration of 1 mM in DMSO (Sigma-Aldrich, St. Louis, MO). Recommended working concentration should be in 100-fold excess of labeled molecule. For use with 100 nM Alexa Fluor® 647-conjugated Staurosporine, dilute unlabeled Staurosporine to 10 μM.

5. 4-well quadriPERM incubation tray (Greiner, Monroe, NC).

6. Platform shaker (Lab-Line Instruments, Maharashtra, India).

7. LifterSlip™ cover slips (Thermo Scientific, Waltham, MA).

8. Polyacetal slide rack (RA Lamb, Durham, NC) and Eppendorf plate centrifuge (model 5,810, Fisher Scientific).

2.2. Scanning and Data Analysis

1. Genepix 4000B Fluorescent Microarray Scanner (Molecular Devices, Sunnyvale, CA) or any scanner that can detect Alexa Fluor® 647 dye can be used (excitation 647 m emission 666 nm). Wavelength is specified by excitation/emission spectrum of detection reagent or fluorophore attached to the small molecule.

2. Genepix Pro 6.1 (Molecular Devices, Sunnyvale, CA) suggested, or other microarray analysis software.

3. A "GAL" file or other file containing information about the location of proteins on the microarray (Gal files for Invitrogen products can be downloaded from http://www.invitrogen.com/protoarray).

4. ProtoArray® Prospector provides automated data analysis for Invitrogen microarray products (can be downloaded from http://www.invitrogen.com/protoarray).

2.3. Protein Microarray Manufacturing

Human clones used to produce proteins for ProtoArray® Human Protein Microarrays (Invitrogen) were obtained from Invitrogen's Ultimate™ ORF (open reading frame) collection or from a Gateway® collection of kinase clones developed by Protometrix. The nucleotide sequence of each clone was verified by full-length sequencing. All clones were transferred into a system for expressing recombinant proteins in insect cells via baculovirus infection. Using a proprietary high-throughput insect cell expression system, thousands of recombinant human proteins were produced in parallel. Each protein is tagged with Glutathione-S-Transferase (GST), which enables high-throughput affinity purification under conditions that retain activity. After purification, a sample of every purified protein is checked to ensure that the protein is present at the predicted molecular weight. ProtoArray® microarrays are manufactured using a contact-type printer equipped with 48 matched quill-type pins. Each protein is deposited along with a set of control proteins in duplicate spots on 1" × 3" glass slides that have been coated with a thin layer of nitrocellulose. Thin-film nitrocellulose slides are manufactured by GenTel® BioSciences, Inc. using a proprietary surface chemistry owned by Decision Biomarkers, Inc. Thin-film nitrocellulose slides are covered by US Patent 6,861,251; 7,297,497; and 7,384,742. The printing of these arrays is carried out in a cold room under dust-free conditions in order to preserve the integrity of both samples and printed microarrays. Before releasing protein microarrays for use, each lot of slides is subjected to a rigorous quality control (QC) procedure, including a gross visual inspection of all the printed slides to check for scratches, fibers, and smearing. Since each of the proteins on the array contains an N-terminal GST tag, a GST-directed antibody detects human proteins in a second QC assay. The procedure measures the variability in spot morphology, the number of missing spots, the presence of control spots, and the amount of protein deposited in each spot. The arrays are designed to accommodate 12,288 spots. For the ProtoArray® Human Protein Microarray, samples are printed in 150-μm spots arrayed in 48 subarrays (4,400-μm² each) and are equally spaced in vertical and horizontal directions with 20 columns and 20 rows per subarray. Spots are printed with a 220-μm spot-to-spot

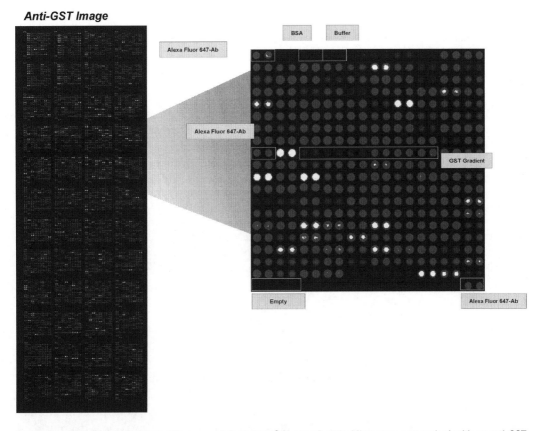

Anti-GST Image

Fig. 1. ProtoArray® Human Protein Microarray. A ProtoArray® Human Protein Microarray was probed with an anti-GST antibody conjugated to Alexa Fluor® 647. The array was dried and scanned at 635 nm on an Axon 4000B scanner. The ProtoArray® microarray is divided up into 48 individual subarrays, each comprised of an identical set of negative and positive control elements, and variable human protein content. An enlarged image of a single subarray is shown on the right, with fluorescent positional mapping markers and a subset of the control elements highlighted. Note the fluorescence pattern associated with the spotting of features as adjacent duplicates. © Copyright 2008 Invitrogen.

spacing. An extra 100-μm gap between adjacent subarrays allows quick identification of subarrays (**Fig. 1**). The proteins printed on the microarray retain functionality even after extensive storage at –20°C, as demonstrated by the auto-phosphorylation of spotted kinases following incubation with ^{33}P-ATP (data not shown).

3. Methods

As a versatile tool in proteomics, protein microarrays have been used to study enzyme–substrate modifications and various types of binding events including protein–protein interactions, antibody-target binding, and immunological profiling with serum or other antibody-containing biological fluids. Additionally, this

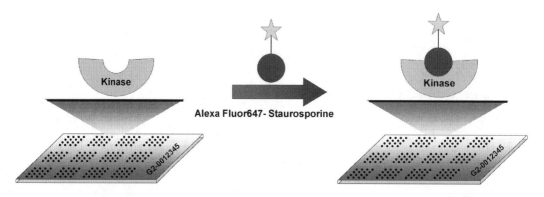

Fig. 2. Overview of Small Molecule–Protein Interaction Profiling Protocol on ProtoArray® Protein Microarrays. After blocking arrays, small molecules are added, allowed to bind, and then the arrays are washed to remove unbound small molecule. Subsequently, if the small molecule is conjugated to biotin, a detection reagent is added (i.e., streptavidin conjugated to Alexa Fluor® 647). Arrays are then washed and scanned with a fluorescent microarray scanner. If the small molecule is istopically labeled, the image is acquired with either X-ray film or a phosphoimager. © Copyright 2008 Invitrogen.

technology is well suited for rapidly profiling small molecule–protein interactions for identifying on- and off-target binding events. The ability to test thousands of proteins in an addressable format minimizes the experimental bias introduced by more focused approaches involving limited content. The method described here is based on the use of ProtoArray® Human Protein Microarrays (Invitrogen). The workflow for small molecule–protein interaction profiling using protein microarray technology is diagrammed in **Fig. 2**.

3.1. Small Molecule Interaction Profiling on Functional Human Protein Microarray

1. Prior to initiating the small molecule interaction assay, an appropriate number of protein microarrays for the experimental design are obtained (ProtoArray® Human Protein Microarrays from Invitrogen are recommended). The protein microarrays are stored at –20°C, and must be allowed to equilibrate to 4°C for 10 min prior to initiating the blocking step (*see* **Note 1**). To help preserve the protein activity, assays should be carried out at 4°C (*see* **Note 2**).

2. The Small Molecule Interaction (SMI) buffer should be prepared fresh and 5 mL added to each well of the quadriPERM tray (*see* **Note 3**). Protein microarrays are placed protein-side up in the tray, one array per well. If the protein microarrays are manufactured on barcoded slides, ensure that the barcode end of the slide is near the end of the tray with the indented numeral. The indent in the bottom of the tray will be used as the site of buffer exchange. The trays are gently rocked to ensure that each slide is completely immersed in the blocking buffer. Cover the trays and place onto platform shaker for 1 h. Rotate at slow speed (~50 rpm) at 4°C (*see* **Note 4**).

3. During the protein microarray blocking step, the small molecule probing solutions are prepared. Small molecules

are diluted to suitable concentrations (*see* **Note 5**) in SMI buffer. For probing protein microarrays with Alexa Fluor® 647-labeled staurosporine, two identical solutions containing 100 nM Alexa Fluor® 647-labeled staurosporine are prepared in a total volume of 100 μL SMI buffer (*see* **Note 6**). One tube is supplemented with 1 μL of 1 mM unlabeled staurosporine in DMSO for a final concentration of 10 μM; the other is supplement with 1 μL of DMSO only. In a competition assay, the ability to out-compete the binding of a labeled molecule with an unlabeled molecule provides some assurance that the observed binding is likely a bona fide interaction.

4. Following the 1-h blocking step, the array is removed from the SMI buffer and tapped gently on a Kim wipe to remove excess fluid. Arrays are then placed on a flat surface with the proteins facing up (*see* **Note 7**).

5. 100 μL of each dilute small molecule solution is pipetted on top of the array, taking care to avoid contacting the array surface directly with the pipette. A LifterSlip™ cover slip is then overlaid onto the array (taking care to avoid bubbles). This is most effectively achieved by resting one of the short ends of the LifterSlip™ on the slide near the barcode, and slowly lowering (*see* **Note 8**). Arrays are then placed in a new 4-well tray without disturbing the LifterSlip™, and incubated in the dark for 90 min at 4°C without agitation (*see* **Notes 9** and **10**).

6. Following incubation, the protein microarrays are washed three times for 5 min with 5 mL of SMI buffer. For the first wash, 5 mL of probing buffer is added to the each well of the tray. The LifterSlip™ should dislodge from the surface of the array and can be removed with forceps (*see* **Note 11**). For each wash, 5 mL of probing buffer is added to each well of the tray and allowed to incubate for 5 min with gentle agitation before removal by aspiration or pipetting and addition of the next wash (*see* **Note 12**).

7. Once the final wash has been completely aspirated, the protein microarrays are removed from the tray and placed in a polyacetal slide rack (RA Lamb) (*see* **Note 13**). To facilitate removal, forceps can be inserted into the indented numeral and used to gently pry the edge of the slide upward. The arrays are then transferred to a slide drying rack with a gloved hand, taking care to only touch the slide by its edges. The arrays are quickly dipped into distilled water to remove any salt residue and the arrays are then centrifuged at low speed (2,000 rpm, or 800 × *g*) for 1 min in a centrifuge equipped with a plate rotor.

8. Arrays are then scanned with a fluorescent microarray scanner (GenePix 4000B, Molecular Devices is recommended) at 635 nm with a PMT gain of 600, a laser power of 100%, and a focus point of 0 μm. Laser power and PMT should be adjusted such that the feature signals on the array are not saturated. This is

done to keep signals within the linear range. These images are saved in Tagged Image File Format (.tif files) and are used in extracting pixel intensity information (*see* **Note 14**).

3.2. Data Analysis and Hit Identification

1. The array list file (.gal file) is uploaded to the image analysis software (GenePix 6.1 from Molecular Devices is recommended). This text file describes the layout of the protein microarray and contains the details of the microarray content, including relevant control elements. The .gal file is used to map the location of each array feature, initially with a fixed feature size based on the diameter of the spotted protein microarray features. To maximize accuracy, a pixel-based segmentation algorithm is recommended for pixel intensity data extraction (Irregular Feature Finding setting, located under the *Alignment* tab in the *Options* menu of GenePix 6.1) (*see* **Note 15**). After aligning all features using fluorescent positional mapping markers, pixel intensities for each spot on the array are calculated by the software and saved to a text file formatted for use in GenePix as a GenePix Result file (.gpr filename extension). These files are subsequently opened in other text editing or spreadsheet programs for analysis.

2. Assuming ProtoArray® Human Protein Microarrays are used in the assays, quantitated spot files are processed using the ProtoArray® Prospector freeware to determine which proteins interact with the small molecule probes. The software incorporates background subtraction, Z-Factor and Z-Score calculations, and replicate spot coefficient of variation (CV) filtering.

3. Alternately, Microsoft Excel can be used to analyze the raw.gpr data (the.gpr file is a text file which can be open directly in Excel). The background-subtracted signal values (F635median-B635median) may be used to calculate Z-factors using the following formula:

$$Z\text{-Factor} = 1 - 3 \times (\text{stdev feature signal} + \text{stdev negative signal})//(\text{avg feature signal} - \text{avg negative signal})/$$

Interacting proteins are defined as those yielding a Z-factor ≥0.4 and a replicate spot CV < 0.5.

$$\text{Coefficient of variation (CV)} = \text{standard deviation}/\text{mean}$$

The percentage of competition is calculated as:

$$\%\text{ completion} = 1 - (\text{avg feature signal of completion assay}/\text{avg feature signal of binding assay})$$

The specific interaction between the small molecule and the target protein(s) can then be defined as interacting proteins exhibiting

a

	Signal-Background		Z-Factor		Replicate Spot CV	
Name	100nM AF647-Staurosporine	100nM AF647-Staurosporine + 10µM Staurosporine	100nM AF647-Staurosporine	100nM AF647-Staurosporine + 10µM Staurosporine	100nM AF647-Staurosporine	100nM AF647-Staurosporine + 10µM Staurosporine
NTRK1	39044	1375	0.95	0.86	1.4%	1.6%
CAMKK2	27275	867	0.99	0.77	0.4%	3.5%
DAPK2	23341	860	0.90	0.63	3.1%	6.7%

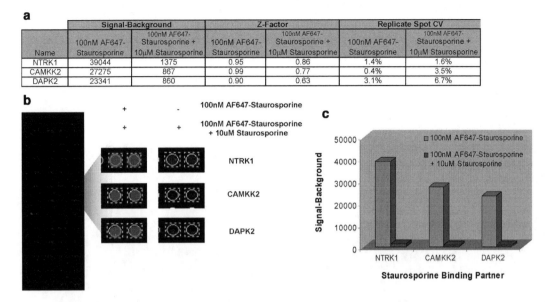

Fig. 3. Representative Alexa Fluor® 647-Staurosporine Interacting Proteins. Alexa Fluor® 647-staurosporine was probed on a ProtoArray® Human Protein Microarray at a concentration of 100 nM. Following completion of the assay protocol, the arrays were scanned on a fluorescent microarray scanner, and pixel intensity data was used to calculate background-subtracted signal values, Z-Factors, and Coefficients of Variation (CV) for adjacent duplicate spots. The >8,000 human proteins were evaluated for features giving rise to a Z-Factor >0.4 in the Alexa Fluor® 647-staurosporine assay, with a corresponding competition of ≥30% in the competition assay including 100-fold excess unlabeled staurosporine. Data from three representative kinases that met these threshold criteria is indicated (**a**). Duplicate spots from the Alexa Fluor® 647-staurosporine binding and competition assays are shown (**b**). Background-subtracted signal values are plotted (**c**). © Copyright 2008 Invitrogen.

≥30% competition in the presence of unlabeled compound. As an example, the specificity profile for Alexa Fluor® 647-staurosporine generated on a high-content protein microarray is shown in **Figs. 3** and **4**.

4. Notes

1. Spots may smear or merge if arrays are not equilibrated before use due to the formation of condensation on the array surface.

2. For certain small molecule–protein interactions, such as interactions between ligand and G-protein-coupled receptors, the favorable incubation temperature might be higher (e.g., room temperature).

3. It has been found that certain blocking reagents, such as 1% BSA or 1% casein can enhance the binding of some small

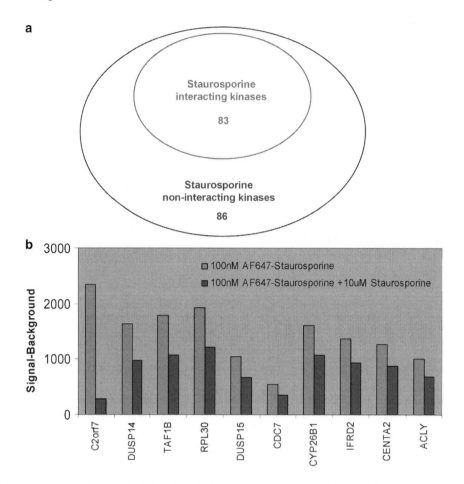

Fig. 4. Alexa Fluor® 647-Staurosporine interaction with kinases and nonkinase proteins. ProtoArray® v4.0 Human Protein Microarrays comprising more than 8,000 human proteins, including more than 150 kinases were probed with Alexa Fluor® 647-staurosporine in the presence and absence of excess unlabeled staurosporine. Eighty-three kinases exhibit specific binding to Alexa Fluor® 647-staurosporine as measured by proteins yielding a Z-Factor >0.4, with a corresponding competition >30% in the presence of excess unlabeled compound, consistent with its function as a broad spectrum kinase inhibitor (a). In addition, Alexa Fluor® 647-staurosporine also exhibited specific interaction with ten non-kinase proteins when the same threshold criteria were applied. Background-subtracted signal values for the non-kinase staurosporine interactors are plotted for both the binding and competition assays (b). © Copyright 2008 Invitrogen.

molecules to their protein targets. However, their use is not recommended as part of the standard protocol due to nonspecific binding between many small molecule compounds and these blocking agents.

4. Use a shaker that keeps the arrays in one plane during rotation. Nutating or rocking shakers are not to be used because of increased risk of cross-well contamination.

5. The concentration of the compound should be determined by the expected affinity for the protein target, and the labeling strategy employed. For example, fluorescently labeled small molecules with expected affinities for their protein targets in

the low nM range are commonly probed at concentrations ranging from 1.0 to 10 μM.

6. If the amount of compound is not limiting, compounds can be incubated on the microarray in a total volume of 5 mL. When probing using 5-mL volume, probe arrays for 90 min in 4-well quadriPERM incubation tray with gentle agitation, using a shaker that keeps the arrays in one plane during rotation.

7. Do not allow any part of the array surface to dry before adding the next solution as this can cause high and/or uneven background.

8. If air bubbles get trapped under the LifterSlip™, tap the slides to drive them out or lift one edge of the LifterSlip™ allowing bubbles to move to the fluid front, and then gently lower down again.

9. If probing with a biotinylated or radiolabeled small molecule, the assay can be carried out in ambient light.

10. If probing with a radiolabeled small molecule, appropriate positional mapping reagents must be included in the assay mixture to enable acquisition of the resultant pixel intensity data.

11. Do not remove the LifterSlip™ with the forceps if it is not dislodged from the array. Gently shake the array in the tray until the LifterSlip™ floats off.

12. Do not aspirate buffers from the surface of the slide in order to reduce the risk of surface scratches. The blocking buffer is aspirated by vacuum or by the use of a pipettor. The tip of the aspirator or pipettor is positioned in the indented numeral in order to remove as much of the liquid as possible. When each well is dry, the indented numeral end of the tray is lifted to facilitate removal of the liquid which pools at the base of the well.

13. If probing with a biotinylated small molecule, a second detection step must be performed. Detection with Alexa Fluor® 647-labeled streptavidin diluted to a concentration of 1.0 μg/mL in SMI buffer is recommended. The detection buffer should be applied in a total volume of 5 mL, and allowed to incubate for 90 min with gentle agitation on a planar shaking platform. Arrays should then be subjected to a series of washes as described above prior to drying and scanning.

14. If probing with a radiolabeled small molecule, the dried arrays should be exposed to a phosphoscreen or film for a suitable period of time, and high-resolution images acquired through phosphoimaging or film development and scanning.

15. In general, the use of pixel-based segmentation (irregular feature finding) results in more reproducible Signal Used values.

References

1. Predki, P. F. (2004) Functional protein microarrays: ripe for discovery. *Curr Opin Chem Biol* **8**, 8–13

2. MacBeath, G. and Schreiber, S. L. (2000) Printing proteins as microarrays for high-throughput function determination. *Science* **289**, 1760–3

3. Fang, Y., Frutos, A. G., and Lahiri, J. (2002) Membrane protein microarrays. *J Am Chem Soc* **124**, 2394–5

4. Ge, H. (2000) UPA, a universal protein array system for quantitative detection of protein-protein, protein-DNA, protein-RNA and protein-ligand interactions. *Nucleic Acids Res* **28**, e3

5. Schweitzer, B., Predki, P., and Snyder, M. (2003) Microarrays to characterize protein interactions on a whole-proteome scale. *Proteomics* **3**, 2190–9

6. Huang, J., Zhu, H., Haggarty, S. J., Spring, D. R., Hwang, H., Jin, F., Snyder, M., and Schreiber, S. L. (2004) Finding new components of the target of rapamycin (TOR) signaling network through chemical genetics and proteome chips. *Proc Natl Acad Sci U S A* **101**, 16594–9

Chapter 12

Capillary Electrophoresis in Drug Discovery

Milena Quaglia and Ersilia De Lorenzi

Summary

The several advantages that capillary electrophoresis (CE) offers in the study of protein folding, protein–ligand and protein–protein interactions, render this methodology appealing in several areas. In this chapter, a specific example is reported, where the use of affinity CE (ACE) in drug discovery is particularly advantageous over other separative and spectroscopic techniques. ACE is an analytical approach in which the migration patterns of interacting molecules in an electric field are recorded and used to identify specific binding and to estimate binding constants. A library of compounds has been tested, in free solution and with minimum sample consumption, for the affinity to two targets previously separated by CE, the native form and the partially structured intermediate of the folding of β_2-microglobulin (β_2-m) [Chiti et al. (J. Biol. Chem. 276:46714–46721, 2001), Quaglia et al. (Electrophoresis 26:4055–4063, 2005)]. β_2-m is an intrinsically amyloidogenic protein, and its tendency to misfold is responsible for dialysis-related amyloidosis, an unavoidable complication of chronic haemodialysed patients.

The criteria for choosing the compounds to be screened, the method conditions, and the possible data analysis strategies are detailed and discussed in this chapter.

Key words: Affinity capillary electrophoresis, Screening, Binding, Drug discovery, β_2-microglobulin

1. Introduction

A separative technique like capillary electrophoresis (CE) can be applied to drug discovery through its capability of gaining a precise and reliable quantitative description (binding constant) of the complexation process that takes place between a drug (ligand) and a macromolecule (e.g. proteins, carbohydrates, enzymes, lipids) *(1, 2)*. This can be achieved by using different set-ups and formats that in some cases can be also adapted to screening purposes. These modes can be ultimately collected under the term Affinity Capillary Electrophoresis (ACE) and regardless

Ana Cecília A. Roque (ed.), *Ligand-Macromolecular Interactions in Drug Discovery: Methods and Protocols,*
Methods in Molecular Biology, vol. 572,
DOI 10.1007/978-1-60761-244-5_12, © Humana Press, a part of Springer Science + Business Media, LLC 2010

of their variety, they all benefit by the intrinsic virtues of CE, namely low sample and reagent consumption, relatively short analysis times, and ease of automation. Importantly, there is no risk of altering the binding properties during the measurement, as ACE does not require immobilisation on a support of one of the interacting species, and physiologically relevant conditions can be easily applied.

The ACE formats can be divided into two categories: (1) the migration or mobility shift approaches, where interaction takes place *during* electrophoresis and weak-medium affinity complexations are described (2) the pre-incubation or pre-equilibrium approaches, where high affinity interactions take place *before* electrophoresis (*see* **Note 1**). This chapter will deal with the former approach and with its adaptation to drug screening, by presenting a specific example.

By applying the mobility shift mode that exploits the difference in mobility between the free and the complexed analyte, binding constants are calculated from a change in the electrophoretic mobility of the injected analyte, due to complexation with an interacting component which is added to the background electrolyte (BGE). In practice, a series of experiments with increasing concentrations of the interacting component in the electrophoretic buffer is carried out, and the binding constant is calculated by a binding isotherm, a plot where the change in the net electrophoretic mobility is reported as a function of increasing additive concentrations in the BGE. Accurate binding data can be obtained if one makes sure that the dissociation and association rates of the interaction are fast enough compared with the electrophoretic run time, to allow for a dynamic equilibrium to be established inside the capillary. In other words, to safely choose the migration shift mode, the complex dissociation half-time, expressed as $\ln 2 / k_{off}$ (k_{off} = dissociation constant), must be less than 1% of the peak appearance time. Consequently, as only very long separation times (i.e. a very long capillary) would enable strong interactions to be adequately studied by this approach, only weak-medium affinity complexations are generally considered. Assuming a 1:1 molecular association, migration data can generally be described by the rectangular hyperbolic form of the binding isotherm *(3)*. Non-linear and linear regressions lead to the binding constant and to the complex mobility values.

Given all these premises, when programming a set of ACE experiments in the migration shift mode to derive a reliable binding constant of, e.g. a drug–protein complexation, two set-ups are possible: either the protein is injected and the capillary is filled with the drug, or vice versa. In this chapter, a library of sulphonated molecules was evaluated for the affinity to β_2-microglobulin (β_2-m), using ACE in the migration shift mode (*see* **Note 2**).

β_2-m is a small protein responsible for the onset of a severe and unavoidable complication of long-term haemodialysis, named dialysis-related amyloidosis. Amyloidoses encompass a number of pathologies where certain proteins with a very diverse structure have the ability to change conformation, to partially unfold or misfold and to associate with each other into insoluble aggregates, named fibrils, that are localised at the extracellular level in various tissues and organs. In this context, the availability of a CE separation method for the native form (N) of β_2-m and its folding intermediate (I_2), which has a role in the pathology, was particularly important *(4)*.

A therapeutic approach for this amyloidosis could be based on the stabilisation of β_2-m, through the binding to a small molecule, to possibly inhibit protein misfolding and amyloid fibril formation. The search for a strong ligand of this protein is extremely challenging, as β_2-m does not have a specific binding site. Looking for a small compound of pharmaceutical interest that could somehow mimic the structures of the few known ligands to the protein, namely Congo Red, 8-anilino-1-naphthalene sulphonic acid and heparin *(4–7)*, we found that suramin, a bis-hexasulphonated naphthyl urea, had some affinity, albeit weak, for β_2-m *(8)*. The structure of suramin in turn inspired the selection of 208 sulphonated/suramin-like compounds, of which 193 have been tested for affinity to β_2-m, by applying ACE in the migration shift mode. The screening of the first 56 compounds of the library performed by ACE and ultrafiltration has been previously reported *(9)*. In this case, by selecting the set-up where the protein is injected and each library compound is added to the BGE, the primary extra advantage consists in the possibility of getting independent affinity results for the two separated forms of the protein, migrating as two independent peaks, namely the native form and the partially structured intermediate of the folding mentioned above. To get an on–off response on whether a given compound of the library would be endowed with a certain affinity for the protein, ACE experiments were carried out only at two representative compound concentration levels. The migration shift results were compared with those obtained by suramin, taken as a reference threshold to select interesting hits from the library. The binding of suramin with β_2-m has also been studied *(9)*, and an example of the calculation of the binding constants of the two forms of β_2-m for a pharmaceutical compound is also discussed.

2. Materials

2.1. Chemicals

1. Recombinant β_2-m was produced as inclusion body in *Escherichia coli* by following the procedure described by Esposito et al. *(10)*.

2. Suramin was purchased by Sigma Aldrich (St. Louis, MO, USA).

3. 208 sulphonated suramin-like compounds were purchased from Specs (Delft, The Netherlands) (*see* **Note 3**). Compounds have been provided as powder in two sets: aliquots of 500 µg and aliquots of 1.5 mg. Samples were stored at room temperature. Only compounds with predicted water solubility (Log S $\geq$ –4.7) were screened as potential ligands for β_2-m by ACE (*see* **Note 4**).

4. Deionised water was produced by using a Millipore Direct-Q™ device.

5. NaH_2PO_4 and Na_2HPO_4 were purchased by KGaA Merck (Darmstadt, Germany).

2.2. Instrumentation

1. A P/ACE MDQ capillary electrophoresis (Beckman Coulter, Fullerton, CA, USA) with a built-in diode array detector was employed to perform ACE (*see* **Note 5**).

2. 50-µm inner diameter fused silica capillaries were purchased from Micro Quartz (München, Germany). The total length of the capillary was 57.5 cm with a detection window at 47.5 cm. New capillaries were pre-treated with 1 M NaOH for 60 min, H_2O for 60 min, and 100 mM phosphate buffer pH 7.4 for 90 min, by applying a pressure of 14.5 psi.

3. Methods

3.1. Preparation of Samples

1. Aliquots of lyophilised β_2-m were solubilised in water to obtain a 25 µM solution. The solution was stored at +4°C and freshly prepared every other day.

2. 300-µM stock solutions of suramin were prepared in water and stored at +4°C. Solutions were prepared weekly.

3. Each sulphonated suramin-like compound was solubilised in water or water added with the lowest percentage of dimethyl sulphoxide (DMSO) to obtain 200-µM stock solutions, which were stored at +4°C.

3.2. Preparation of Background Electrolytes

1. The BGE for the evaluation of the mobility of β_2-m, suramin, and potential ligands was phosphate buffer at pH 7.4. 1.2 g of NaH_2PO_4 and 1.42 g of Na_2HPO_4 were solubilised in 100 mL of water to obtain a 100-mM solution (*see* **Note 6**). The two solutions were mixed to obtain a buffer at pH 7.4 (*see* **Note 7**).

2. The BGE for the evaluation of the binding properties of β_2-m for suramin were solutions containing increasing concentrations of suramin dissolved in phosphate buffer 100 mM. An appropriate volume of a 300-µM stock solution of suramin was diluted with an appropriate volume of water and 1 mL of

200-mM sodium phosphate buffer, pH 7.4, to obtain 2-mL solutions of suramin. Suramin concentrations in the BGE were 10, 20, 30, 50, 60, 80, 100, and 150 μM.

3. BGE for the screening of the sulphonated suramin-like compounds as potential ligands of β_2-m were 50-μM and 100-μM solutions of the individual compounds in 100-mM phosphate buffer. Aliquots of the 200-μM stock solution were diluted with an appropriate amount of water and 200-mM phosphate buffer to obtain 50 and 100-μM solutions of sulphonated/suramin-like compounds in phosphate buffer 100 mM.

3.3. Affinity Capillary Electrophoresis

ACE experiments can be successfully performed only if the effective mobility (μ_{eff}) of the injected sample is different from the effective mobility of the run buffer additive (*see* **Notes 8** and **9**). The set-up chosen for the ACE experiments in this work was based on the injection of β_2-m and on the addition of suramin or sulphonated/suramin-like compounds to the BGE (*see* **Note 2**).

The μ_{eff} of the N and I_2 peaks of β_2-m were therefore compared with the μ_{eff} of suramin (run buffer additive) *(9)*. The N μ_{eff} = $-4.699 \times 10^{-5} \pm 3.99 \times 10^{-7}$ cm^2/V/s ($n = 7$) and the I_2 μ_{eff} = $-5.524 \times 10^{-5} \pm 4.20 \times 10^{-7}$ cm^2/V/s ($n = 7$) were one order of magnitude different from suramin μ_{eff} = $-2.8 \times 10^{-4} \pm 2.510^{-6}$ cm^2/V/s ($n = 3$) as this, and all other sulphonated compounds belonging to the library, were migrating much slower than the protein under the operative conditions, consistently with their negative charge and with their mass much smaller than that of the protein.

ACE based on mobility shift experiments can be applied when the binding is characterised by fast interactions. The complex dissociation time has to be shorter than the peak appearance time. A complex dissociation half-time ($\ln 2/k_{off}$) equal or less than 1% of the peak appearance time is generally enough. k_{off} was calculated by using surface plasmon resonance and was $1.76 \times 10^{-1} \pm 6.0 \times 10^{-3}$ s^{-1}. The complex dissociation half-time is therefore 3.9 s and the protein peaks (N and I_2) appearance time is ≥660 s (*see* **Note 10**).

ACE based on migration shift experiments has been therefore considered appropriate for the measurement of the binding constant of β_2-m for suramin and for the screening of potential ligands of the protein.

3.4. Calculation of the Binding Constant of Suramin for the N and I_2 Isoforms of β_2-M

1. The capillary was thermostatted at 15°C and the autosampler at 25°C. The difference in temperature between the capillary and the autosampler has been considered crucial for the successful separation of the two β_2-m isoforms N and I_2 *(4)*.

2. Before each run, the capillary was rinsed with NaOH 0.1 M for 1.5 min (*see* **Note 11**) and H$_2$O for 1.5 min by applying a

pressure of 14.5 psi. The capillary was then conditioned with BGE for 3 min. The rinsing and conditioning steps have been optimised in order to obtain repeatable mobility measurements of β_2-m in presence of suramin. The composition of the BGE used for the evaluation of the binding constant of β_2-m for suramin is described in detail in **Subheading 3.2**.

3. The baseline was auto-zeroed before each injection (*see* **Note 12**).

4. 25 μM of β_2-m were hydrodynamically injected by applying a pressure of 0.7 psi for 8 s (*see* **Note 13**).

5. A voltage of 18 kV was applied and the electrophoretic run was performed for 20 min and recorded at 200 nm (*see* **Note 14**).

6. The electro-osmotic flow (EOF) has been measured as a perturbation of the baseline given by the water used to solubilise the protein sample at 200 nm. In our experience, it can be used as reliable non-interacting species. The average EOF migration time was 9.03 min (RSD%: 0.002).

7. The effective mobilities of N and I_2 were calculated.

8. The mobility shifts of N and I_2 injected in BGE with or without suramin were calculated as the difference between the effective mobility of N and I_2 injected in BGE with (μ_i) and without (μ_f) suramin. Mobility shift is therefore defined as $\mu_i - \mu_f$ (*see* **Note 15**).

9. Due to the negative charge of suramin and the interactions of the N and I_2 isoforms with suramin, an anionic shift of the N and I_2 peaks was observed when β_2-m was injected in BGE containing suramin (**Fig. 1**).

10. The binding isotherm was constructed by plotting the mobility shift versus the concentration of suramin in the BGE (**Fig. 2**) (*see* **Note 16**).

11. By assuming a stoichiometry of 1:1 (*see* **Note 17**), the complexation can be described by the general rectangular hyperbolic form of the binding isotherm: $y = dx/f + ex$, where y is the measured response of the system (mobility shift), x is the concentration of the complexation additive (suramin), d, e, and f are constants or parameters related to the properties of each interacting species and of the complex *(1, 11, 12)*. By referencing the measured electrophoretic mobilities, the equation can be written as: $\mu_i - \mu_f = (\mu_c - \mu_f) \cdot K[S] / 1 + K[S]$, where μ_c is the complex mobility, K is the association constant, and [S] is the concentration of the running buffer additive, in our case, suramin (*see* **Notes 18** and **19**).

12. Data regression using a non-linear least square fitting (Levember-Marquardt algorithm) was carried out with Igor professional 3.1 software (WaveMetrics Inc., Lake Oswego, OR, USA).

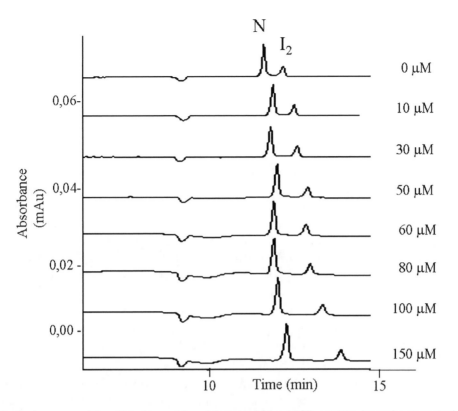

Fig. 1. Electropherograms of N and I_2 isoforms of β_2-m injected in BGE and BGE containing increasing concentrations of suramin (Reproduced from **ref. 9** with permission from Wiley Interscience.).

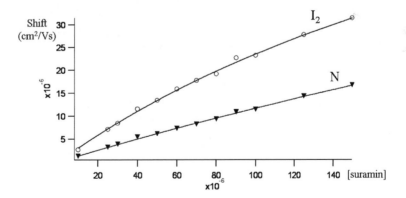

Fig. 2. Non-linear regression of the binding isotherm, used to calculate the binding constants of the complex β_2-m-suramin (Reproduced from **ref. 9** with permission from Wiley Interscience.).

13. The binding constants of suramin for the isoform N and for the isoform I_2 of β_2-m were $1.16 \times 10^3 \pm 2.20 \times 10^2$ M^{-1} and $3.32 \times 10^3 \pm 3.10 \times 10^2$ M^{-1}, respectively.

14. The complex mobility was $-1.39 \times 10^{-4} \pm 6.91 \times 10^{-6}$ cm^2/V/s for I_2 and $-1.58 \times 10^{-4} \pm 1.99 \times 10^{-5}$ cm^2/V/s for N.

3.5. Screening of Sulphonated/ Suramin-Like Compounds

1. In order to perform a screening based on ACE experiments by using mobility shift mode, the following requirements have to be fulfilled:
   - The compounds have to be soluble in water or in water with a percentage of DMSO < 5%, v/v (*see* **Note 3**).
   - The absorbance of the analyte (100 μM) in the running buffer should be low enough to allow the detection of the protein (*see* **Note 12**).
   - The effective mobility of each library compound injected in the running buffer has to be different from the effective mobility of β_2-m injected in the same buffer (*see* **Note 9**).

2. 193 out of 208 sulphonated suramin-like compounds were found to satisfy the requirements (161 were soluble in pure water, 32 in water with a percentage of DMSO < 5%) and their binding properties for β_2-m were screened.

3. The method applied for the determination of the binding constant of β_2-m for suramin was also applied for the screening of sulphonated/suramin-like compounds. Only two concentration levels (50 μM and 100 μM) of the compound added to the BGE were evaluated.

4. If no interaction occurred, the mobilities of N and I_2 in phosphate buffer and in phosphate buffer containing the potential ligand were comparable (*see* **Note 20**).

5. If interactions between β_2-m and the potential ligand occurred, a change in mobility of β_2-m in phosphate buffer and in phosphate buffer containing the ligand was observed.

6. Mobility shifts are indicative of the binding of the protein for the ligand. When the mobility shifts of N and I_2 injected in the same BGE were similar, a similar affinity of the ligand for the two isoforms was expected.

7. The mobility shift values for N and I_2 peaks, measured at the two concentration levels of suramin (50 μM and 100 μM), were taken as a reference threshold absolute value to judge a compound of the library as an interesting hit. When DMSO was present in the BGE, due to solubility issues of the potential ligand, the same percentage of DMSO was added to the reference BGE containing suramin.

8. Only one of the 193 compounds showed higher affinity for β_2-m than suramin (**Fig. 3**). This compound, named 573, contains a central core based on a quinoline structure flanked by two *p*-sulphonyl anilins. The binding constants of 573 for the isoform N and for the isoform I_2 of β_2-m were $7.66 \times 10^3 \pm 4.38 \times 10^2$ M^{-1} and $2.42 \times 10^3 \pm 3.27 \times 10^2$ M^{-1}, respectively, comparable to those of suramin (*13*).

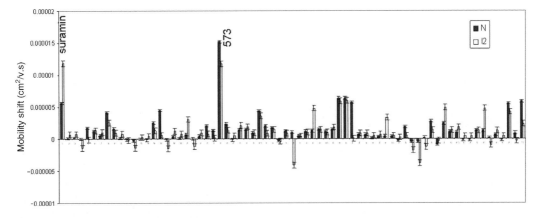

Fig. 3. Mobility shifts of N and I_2 obtained from the screening of 70 of the 209 sulphonated/suramin-like compounds. Each couple of histograms represents the shift induced on the two protein peaks by a single compound added to the BGE. The capillary is filled with 50-μM library compound dissolved in BGE buffer.

4. Notes

1. The plethora of different names found in the literature regarding the different ACE modes may well be a deterrent to the approach to the technique. Despite some confusion with old and new denominations, synonyms, classifications, different approaches within the same category etc., the reader should not be discouraged. To facilitate literature search and orientation, it is here reported a list of some of the published names and corresponding acronyms when available, together with the category to which they belong. The list does not mean to be exhaustive. Category 1 (migration or mobility shift approaches): Affinity Capillary Electrophoresis (ACE, now used as a general name that covers all modes), Dynamic Complexation Capillary Electrophoresis (DCCE), Kinetic Capillary Electrophoresis (KCE), Capillary Electrophoresis Mobility Shift Assay (CEMSA), On-line mixing CE, Partial Filling Affinity CE (PFACE), Flow Through Partial Filling ACE (FTPF-ACE). Category 2 (pre-incubation or pre-equilibrium approaches): Frontal Analysis Capillary Electrophoresis (FACE), Capillary Electrophoresis-Frontal Analysis (CE-FA), Frontal Analysis (FA), Frontal Analysis Continuous CE (FACCE), Preincubation ACE (PI-ACE), Non-Equilibrium CE of Equilibrium Mixtures (NECEEM), Pre-equilibrium CZE.

2. For the example presented in this chapter, the set-up where the protein is injected and the drug is added to the BGE is particularly advantageous, as with one series of experiments

the affinity for different folding states of the protein is gained. Other good reasons to choose this set-up would be: low protein availability; highly UV-absorbing protein; protein highly sticking to the capillary wall. The opposite set-up (drug injected, protein added to the BGE) would be advisable when: protein peak shape is bad (not adequate to obtain reproducible mobility measurements); complexation induces small protein mobility changes. In order to ensure good precision in the determination of the binding constant, the difference in the mobility between free and complexed analyte must be significantly larger than the error associated with the data points. This is, in general, easier by using this second set-up, as wider shifts are observed, according to the changes in both mass and charge of the complexed form. *See* also **Note 12**.

3. The 208 suramin-like compounds were selected through an accurate database search, based on the similarity of portions of the structure of suramin and Congo Red with the structure of the compounds contained in the Specs structure database, as updated in October 2003. The database search was carried out through the ISIS software from MDL. In the choice of the compounds, it is important to evaluate purity, solubility, and possible re-supply of the material. Most compounds were soluble in DMSO. However, DMSO can influence the degree of binding and/or denature the protein. In addition, DMSO has high absorbance at 200 nm (the operative wavelength chosen for the experiments) and causes instability of the current obtained by applying the voltage during the electrophoretic run. Solubility in water has been therefore considered as a discriminating factor. Impurities can also influence the binding with the protein and determine false positives. This is particularly important in ACE when the ligand is dissolved in the BGE and separation between the compound and possible impurities would not be detected. Purity of 95% was considered sufficient for this study. Not all companies that provide chemical libraries for screening in drug discovery can re-supply the compounds offered, especially when the synthesis processes are not performed in-house. In the selection of the 208 compounds, the possibility to obtain a re-supply has been considered fundamental. Appropriate packages also need to be chosen. Possible contaminations (especially with 96-well plates), final volume of stock solutions, and storing space are important issues to be considered.

4. Log S is a predicted value of water solubility, which does not always correspond to the real solubility of the sample in water and therefore needs to be experimentally verified. The cut-off value of log S $\geq$ –4.7 ($\pm$10% margin for error prediction) was set to limit the selection to compounds that would be water soluble within the operative concentration range of 20–100 μM.

5. P/ACE MDQ system was chosen for the ACE experiments as it offers an accurate thermostatting of the capillary at 15°C, through the use of liquid coolant.

6. Buffer solutions were prepared daily and filtered through 0.45-μm Millipore membrane filters (Millipore, Bedford, MA, USA).

7. pH 7.4 was chosen because it is near to the physiological pH.

8. The effective mobility (μ_{eff}) is defined as $\mu_{eff} = \mu_a - \mu_{EOF}$, where μ_a is the mobility of the injected species and μ_{EOF} is the mobility of the EOF. μ_a and μ_{EOF} are measured from the electropherogram by using the following equation: $\mu = L \times 1/V \times t$, where L = total capillary length (expressed in cm), l = capillary length to the window (expressed in cm), V = voltage (expressed in volts), and t = time (expressed in seconds).

9. In order to appreciate mobility shift, the effective mobility of the free and drug-bound protein (and for the opposite set-up, where the drug is injected, of the free and protein-bound drug) must be significantly different.

10. As in practice the k_{off} of an interaction is rarely known, an inspection of the peak at different additive concentration levels in the BGE would help to have a rough idea of the interaction kinetics. Peak tailing, broadening, splitting, or disappearance may mean that on- and off-rates are too slow and therefore the kinetics of interaction of the chosen system may not be adequate for the migration shift approach. Pre-equilibration formats should be investigated (in this case).

11. Phosphoric acid should be used in preference to NaOH when BGE characterised by low pH are employed. SDS should be preferred to NaOH when proteins are dissolved in the BGE *(14)*.

12. Particular attention has to be paid to the absorbance of the species (in our case suramin) dissolved in the BGE. A reduced signal was observed at increasing concentrations of suramin in the BGE and the maximum concentration allowed has been considered 150 μM. In case of low binding constant this would be a drawback, as, to properly describe the curvature of the binding isotherm, enough data points have to be collected over a wide range of additive concentration. *See* also **Note 16**. This would have been a good reason to change set-up, but the possibility of obtaining affinity data for both protein peaks was considered a priority.

13. To maintain equilibrium during the analyte migration, the amount of analyte injected should be much less than the amount of additive present in the entire capillary. This is because the concentration of additive in the analyte plug is unknown and in the data elaboration it is assumed to be

equal to the concentration in the capillary buffer. However, in the case of high affinity interactions the opposite is advisable, i.e. to keep the additive concentration low and avoid rapid saturation of the analyte plug.

14. The wavelength of 200 nm was chosen as the maximum absorbance of the UV spectrum of β_2-m.

15. In this context, special care has to be taken to ensure that any changes in the electrophoretic mobility of the injected species caused by changing additive concentration are only due to complexation and not to other factors. Monitoring the EOF is useful to identify problems such as adsorption of the running buffer additive to the capillary walls or change in viscosity of the BGE at high concentrations of the running buffer additive. In particular: between-run conditioning cycles should be optimised to prevent adsorption of the interacting species to the capillary wall (variability of coated capillaries discourages their use in ACE); to monitor the actual conditions of the capillary wall, measurements of the protein mobility in plain buffer should be repeated before carrying out a new additive concentration level and, if necessary, updated as a new reference; a correction factor for the viscosity (η/η_0) should be used if the additive affects the buffer viscosity at high concentrations. A marker that does not bind can be co-injected, when the peak representing the not-retained solubilisation solvent of the injected species is not well defined. If EOF, like in this case, is representative of the mobility of a non-interacting species, then the reproducibility of EOF mobility over the entire set of experiments has to be proved.

16. It is recommended that data should be collected over a large portion of the binding isotherm, so that the plot exhibits a definite curvature. Error will be minimised when responses are measured over the central portion of the binding isotherm (where the fraction of the complexed analyte ranges from 0.2 to 0.8, assuming a 1:1 binding) *(12)*. The optimal additive concentration range is thus related to the strength of the interaction (*see* also **Note 12**).

17. Assuming a 1:1 interaction stoichiometry is often an approximation and nevertheless it is used in the vast majority of the available literature. The assumption may be supported by linear rearrangement (double reciprocal, x-reciprocal or Scatchard plot, y-reciprocal) of the rectangular hyperbolic form of the binding isotherm. When multiple-site binding types are present, a more appropriate data elaboration has to be considered *(15)*.

18. The mathematical derivations to treat 1:1 binding isotherm for CE have been discussed by several groups *(1, 11, 12)*.

Numerous variations have been published and despite the different formalism, they are substantially equivalent.

19. A non-linear least square fitting model to solve the constants directly should be preferable over the more frequent linearisations, to minimise both the error and the bias in the estimation of the binding constants. Nevertheless, if data points are properly weighed for linear regression, then linear and non-linear regressions generally produce comparable results. One should be aware of the fact that sometimes linear plots can mask deviations from linearity *(11, 12, 15, 16)*.

20. Mobility of N and I_2 injected in BGE with and without ligand have been considered the same when the mobility values fall within the experimental error, equal to two times the standard deviation for $n = 7$.

Acknowledgments

The authors would like to thank Professor Vittorio Bellotti for providing recombinant beta$_2$-microglobulin and for continuous support and inspiration. Dr. Chiara Carazzone and Dr. Raffaella Colombo are gratefully acknowledged for their invaluable expertise in the experimental part. This work was financed by MIUR (FIRB RBNEO1529H and PRIN 2005051707).

References

1. Tanaka, K. and Terabe, S. (2002) Estimation of binding constants by capillary electrophoresis. *J. Chromatogr. B Analyt. Technol. Biomed. Life Sci.* **768**, 81–92

2. Heegaard, N.H.H. and Kennedy, R. (1999) Identification, quantitation and characterization of biomolecules by capillary electrophoretic analysis of binding interactions. *Electrophoresis* **20**, 3122–3133

3. Connors, K.A. (1987) Binding Constants, The Measurement of Molecular Complex Stability. New York: Wiley

4. Chiti, F., De Lorenzi, E., Grossi, S., Mangione, P., Giorgetti, S., Caccialanza, G., Dobson, C.M., Merlini, G., Ramponi, G. and Bellotti, V. (2001) A partially structured species of beta$_2$-microglobulin is significantly populated under physiological conditions and involved in fibrillogenesis. *J. Biol. Chem.* **276**, 46714–46721

5. Heegaard, N.H.H., Sen, J.W. and Nissen, M.H. (2000) Congophilicity (Congo red affinity) of different beta$_2$-microglobulin con-

formations characterized by dye affinity capillary electrophoresis. *J. Chromatogr. A.* **894**, 319–327

6. Heegaard, N.H.H., Sen, J.W., Kaarsholm, N.C. and Nissen, M.H. (2001) Conformational intermediate of the amyloidogenic protein beta$_2$-microglobulin at neutral pH. *J. Biol. Chem.* **276**, 32657–32662

7. Heegaard, N.H.H., Roepstorff, P., Melberg, S.G. and Nissen, M.H. (2002) Cleaved beta$_2$-microglobulin partially attains a conformation that has amyloidogenic features. *J. Biol. Chem.* **277**, 11184–11187

8. De Lorenzi, E., Grossi, S., Massolini, G., Giorgetti, S., Mangione, P., Andreola, A., Chiti, F., Bellotti, V. and Caccialanza, G. (2002) Capillary electrophoresis investigation of a partially unfolded conformation of beta$_2$-microglobulin. *Electrophoresis* **23**, 918–925

9. Quaglia, M., Carazzone, C., Sabella, S., Colombo, R., Giorgetti, S., Bellotti, V. and De Lorenzi, E. (2005) Search of ligands for

the amyloidogenic protein beta$_2$-microglobulin by capillary electrophoresis and other techniques. *Electrophoresis* **26**, 4055–4063

10. Esposito, G., Michelutti, R., Verdone, G. and Viglino, P. (2000) Removal of the N-terminal hexapeptide from human beta$_2$-microglobulin facilitates protein aggregation and fibril formation. *Protein Sci* . **9**, 831–845

11. Rundlett, K.L. and Armstrong, D.W. (2001) Methods for determination of binding constants by capillary electrophoresis. *Electrophoresis* **22**, 1419–1427

12. Bowser, M.T. and Chen, D.D.Y. (1998) Monte Carlo simulation of error propagation in the determination of binding constants from rectangular hyperbolae. 1. Ligand concentration range and binding constant. *J. Phys. Chem.* **102**, 8063–8071

13. Carazzone, C., Colombo, R., Quaglia, M., Mangione, P., Raimondi, S., Giorgetti, S., Caccialanza, G., Bellotti, V. and De Lorenzi, E. (2008) Sulfonated molecules that bind a partially structured species of β$_2$-microglobulin also influence refolding and fibrillogenesis. *Electrophoresis* **29**, 1502–1510

14. Lloyd, D.K. and Wätzig, H. (1995) Sodium dodecyl sulphate solution is an effective between-run rinse for capillary electrophoresis of samples in biological matrices. *J. Chromatogr. B Biomed. Appl.* **663**, 400–405

15. Galbusera, C., Thachuk, M., De Lorenzi, E. and Chen, D.D.Y. (2002) Affinity capillary electrophoresis using a low-concentration additive with the consideration of relative mobilities. *Anal. Chem.* **74**, 1903–1914

16. Bowser, M.T. and Chen, D.D.Y. (1999) Monte Carlo simulation of error propagation in the determination of binding constants from rectangular hyperbolae. 2. Effect of maximum-response range. *J. Phys. Chem.* **103**, 197–202

Chapter 13

SPR in Drug Discovery: Searching Bioactive Compounds in Plant Extracts

Maria Minunni and Anna Rita Bilia

Summary

Biosensors represent an interesting tool in the search of bioactive compounds. In particular, optical sensors based on Surface Plasmon Resonance transduction (SPR) allow monitoring of biomolecular interaction in real time and without any labelling of the interactants. The biosensor analysis can be applied to both pure compounds or to complex mixtures (e.g. plant extract). The SPR detection principle is here presented and the application to the analysis of plant extracts (i.e. of *Chelidonium majus* L.) as a paradigmatic example for the search of bioactive compounds able to interact with DNA, is also discussed.

Key words: Plant extract, DNA, Biomolecule immobilization, Affinity sensor

1. Introduction

Currently, the bottleneck of drug discovery has shifted from the generation of compound libraries to the identification of biologically active lead structures *(1, 2)*. Identification of biologically active molecules involves the use of bioassays intended not only at measuring their extent of interaction with specific receptors or binding proteins *(3)*, but also at evaluating their bioactivity *(4)*, bioavailability *(5)*, and pharmacokinetic behaviour *(6)*.

Behind conventional approaches, biosensor technology could play an important role in selecting compound exhibiting biological activity towards selected target receptors.

Biosensors represent new analytical devices which appear to be an analyst's dream: they are able to give rapid analysis responses; to operate directly on complex matrices, in many

Ana Cecília A. Roque (ed.), *Ligand-Macromolecular Interactions in Drug Discovery: Methods and Protocols,*
Methods in Molecular Biology, vol. 572,
DOI 10.1007/978-1-60761-244-5_13, © Humana Press, a part of Springer Science+Business Media, LLC 2010

cases; to be selective and sensitive enough for the required application; to be portable and sometimes also disposable; and, to have fast analysis times *(1)*. Biosensors have mainly been applied for analytical purposes in environmental chemistry, clinical practice, and analysis of food, but recently several examples dealing with natural compounds have also been reported in the literature.

In this chapter, SPR-based biosensor technology is described and the protocol to immobilize proteins on a chip is given. In particular, using streptavidin as immobilized protein and how to further immobilize biotinylated nucleic acid sequences for investigating the ability of selected plant extracts to interact with the immobilized nucleic acid double strand is also reported. The application is suitable for evaluating the ability of the extracted compound to interact with DNA and thus with potential interest as antitumour agents (i.e. natural alkaloids).

1.1. Biosensors

As defined by IUPAC in 1999, a biosensor is an integrated device able to give qualitative and quantitative or semiquantitative specific information through the use of a biological element of recognition in close spatial contact with a transducer. The biological element is responsible for the biological recognition of the target analyte and thus for the sensor specificity. The biomolecule is immobilized on a physical transducer that translates the biorecognition event into a useful electrical signal (**Fig. 1**). Biosensors can be divided mainly into two categories: catalytic and affinity biosensors *(7)*.

Catalytic biosensors involve a catalytic event in which a substrate is converted into a product.

Catalysis occurs at the transducer interface and substrate depletion or product formation is measured by a transducer. The well-known enzyme-based sensor for glucose, widely used in clinical practice for glycaemia measurements and marketed by many different companies belongs to this category.

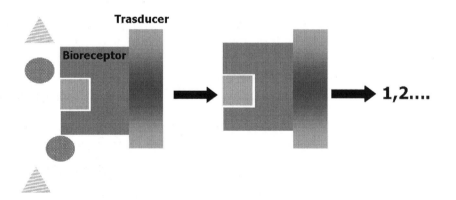

Fig. 1. Biosensor scheme.

In affinity biosensors, recognition of the analyte in solution by the immobilized biological element is based on an affinity reaction i.e. an antigen–antibody binding, nucleic acid hybridization, or receptor–ligand interaction. On the basis of these different interactions, the affinity biosensors can be divided into immunosensors, DNA sensors, aptasensors *(8)*, etc.

The transduction principles can be various and the choice is based on different considerations, such as the electrochemical or optical behaviour of the analyte, the molecular weight, the binding properties, etc.

Transduction can require the use of labels or can eventually be label-free. Moreover, the sensor can be single or multi-use, with this meaning that the chip is disposable or it can be used for more than one analysis. Portable or bench instrumentation is also available.

Relatively to drug discovery, different transduction principles have been reported *(9, 10)*. However, the most applied one is optical detection and in particular Surface Plasmon Resonance (SPR) *(11)*.

1.2. Optical Detection: Surface Plasmon Resonance-Based Sensing

SPR is a widely used optical method for biospecific interaction analysis since it is possible to monitor the affinity interactions in real time. Surface Plasmon is a charge density oscillation that may occur at the interface between two media with dielectric constants of opposite sign, such as a metal and a dielectric. Metal used are mainly gold (Au) and silver (Ag).

Surface plasmon can be excited by a *p*-polarized optical wave. Light is directed through a transparent optical substrate and the intensity of the resulting reflected light is measured with a detector. At certain incident angle of the light, part of the energy will couple into a surface plasmon wave travelling at the interface Au/sample. This angle will be called resonant angle. This coupling (resonance phenomenon) is observed as a sharp attenuation in reflectivity (in total reflection conditions) **(Fig. 2)**.

There are different parameters influencing the resonance angle at which the coupling occurs: among others, the refractive index of the media involved in the phenomenon (air, glass, and solution). So a change in the refractive index of the media involved (i.e. the solution flowing on the metal layer/glass) would be recorded as a change in the resonant angle at which the minimum in reflectivity occurs. It is possible thus to follow interface phenomena (i.e. changing the refractive index in the solution flowing on the glass/metal surface) by monitoring the shift in the resonant angle (as also the shift in the reflectivity minimum). This optical principle can be easily used for biosensor development. As mentioned above, a biosensor relies on the immobilization of a biological molecule on the transduction surface; in the case of SPR transduction, the molecule can be

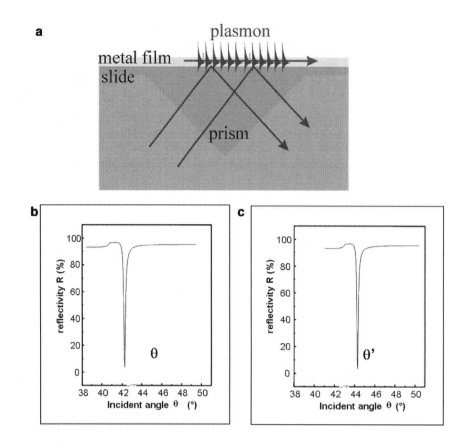

Fig. 2. Surface Plasmon Resonance (SPR) optical transduction: (**a**) SPR theory. At certain incident angle of the light, part of the energy will couple into a surface plasmon wave travelling at the interface Au/sample. This coupling is observed in (**b**) as a sharp attenuation in reflectivity; here a minimum in the reflectivity recorded when SPR occurs at the metal interface (the minimum is observed at the resonant angle) is shown at the resonant angle. When an interface phenomenon occurs (i.e. binding between a receptor and an analyte at the surface), a shift in the angle at which the minimum occurs (from θ to θ) can be observed in (**c**).

immobilized directly on the metal surface by physical adsorption or by optimized immobilization chemistries. The immobilization step as well as the subsequent interaction of the immobilized molecule with an analyte/interactant in solution can be followed in real time without the use of any label. In fact, the interface changes induced by the flowing solution induce a shift in the resonant angle. This event can be displayed in function of time generating what is called a sensorgram. A sensorgram is thus a way to represent interface phenomena in function of time. In particular, on SPR sensorgrams, changes over time in arbitrary units, such as Resonant Units (RU) is reported. Alternatively, the signal is expressed as Reflectivity %.

From the Sensorgram important information relative to interaction between the immobilized molecule and its ligand, added in solution, can be obtained.

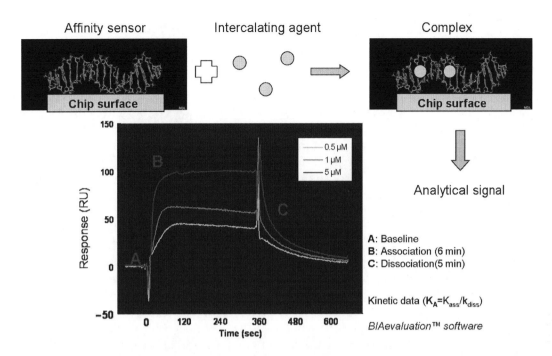

Fig. 3. Sensorgram reports the association and dissociation phases of the interaction between a binder agent (i.e. sanguinarine) and the DNA immobilized on the surface. Three different concentrations of the solution are tested.

In the Sensorgram reported in **Fig. 3**, different steps can be identified. We could consider a system with the immobilized analyte and the buffer solution flowing on the surface. As base line, the signal corresponding to the presence of buffer only is taken. After the addition of the interacting analyte, a complex is formed at the transducer surface, which is revealed by the shift in the recorded signal. The first part of the signal corresponds to the association phase and could eventually reach a plateau. When the sample containing the analyte is replaced by buffer at the end of the injection, dissociation is observed. The rates of association and dissociation are important parameters in the evaluation of the interaction. In other words, we can see how fast the complex is formed and how stable it is. Quantitative information about the analyte present in solution is also obtained by this technology: how much compound/analyte is in the sample? The concentration level can be in the range of submicromolar-nanomolar level, depending on the molecular mass of the ligand.

These findings are strategic when focused on drug development. In other words, one can evaluate if a molecule under study has the ability to bind a target receptor and eventually how fast is this interaction, how strong by calculating its affinity constant (K_A), how specific by testing chemically related molecules as well as negative control to test any unspecific binding.

Thinking more generally, one could have answers to the following questions: Does anything bind from this crude extract?

Is the interaction specific? Is sample activity reduced during purification? How much analyte/product is in the sample? How does the activity of a preparation compare with the last batch?

In terms dealing with drug discovery, this could be referred to a plant extract in natural compound screening. For example, this approach can be applied for the analysis of bioactive constituents with application in drug and herbal drug screening *(12)* or as new analytical tool in the Quality Control of Herbal Drugs, Herbal Drug Preparations, and Herbal Medicinal Products. In particular here is reported as paradigmatic example the case of the protocol applied using SPR sensing for herbal extract screening of some fractions obtained from an ethanolic extract of the fresh plant material of *Chelidonium majus* L. (great celandine). This plant contains benzo[*c*]phenanthridinium alkaloids, having multiple biological and pharmaceutical effects such as anti-inflammatory and anti-microbial activity, inhibition of SH-enzymes and microtubule assembly *(13, 14)* and, in particular, these alkaloids possess the ability to interact with DNA *(15)*.

For SPR-based sensing, commercially available bench-top instruments are perfectly suitable for laboratory use. Some have high degree of automation, contributing significantly to very good analytical performance. More recently, portable devices based on SPR have also been marketed demonstrating that the technology is also suitable for in-field measurements. In the last few years, the SPR transduction principle has been commercialized in an array format, which push a step forward this technology as interesting tool for screening purposes since it is possible to monitor simultaneously hundreds of interactions *(16)*.

2. Materials

2.1. Immobilization Step

1. 1-mM ethanolic solution of 11-mercaptoundecanol (Aldrich, 2 mg of the thiol in 10 mL of ethanol) is freshly prepared before use.

2. 600-mM epichlorohydrin in a 1:1 mixture of 400-mM NaOH.

3. Activating solution: 50 mM N-hydroxysuccinimide (NHS).

4. 200 mM N-(3-dimethylaminopropyl)-N′-ethylcarbodiimide hydrochloride (EDAC) in water.

5. Basic dextran solution: Add 3 g of dextran T500 (Amersham Biosciences) in 10 mL of 100 mM NaOH.

6. 1-M bromoacetic acid (Sigma) in 2-M NaOH.

7. Blocking solution: 1-M ethanolamine hydrochloride in water, pH 8.6.

8. Oligonucleotides (Amersham Biosciences-Uppsala, Sweden) are received, lyophilized, and then diluted with H_2O in stock solutions, divided into aliquots, at a concentration of 100 μM. The stock solutions are stored at –20°C. Freshly prepared diluted solution (concentration 1 μM) in hybridization buffer are used for the immobilization on the chip.

   The base sequences of the 5′-biotinylated probe (25-mer) and of the complementary oligonucleotide (25-mer) are as follows:

   Probe: 5′-biotin-GGCCATCGTTGAAGATGCCTCTGCC-3′;

   Complementary sequence: 5′-GCAGAGGCATCTTCAACG-ATGGCC-3′

   Prepare freshly diluted solution of complementary target (concentration 1 μM) in hybridization buffer (concentration 1 μM) and inject into the instrument, after the probe immobilization, to obtain a double helix to further expose to the plant extracts.

9. Streptavidin (protein from *Staphylococcus aureus*) (Sigma); the protein is diluted in 10-mM acetate buffer, pH 5.0 (concentration 1,000 μg/mL), divided into aliquots, and stored at –20 °C. Before immobilization, it is diluted down to 200 μg/mL streptavidin in the same buffer.

10. Immobilization buffer: 0.01 M N-2-hydroxyethylpiperazine-N′-2-ethanesulfonic acid (HEPES), 0.15 M NaCl, 3 mM EDTA, and 0.005% polyoxyethylene sorbitan monolaurate (Tween 20), pH 7.4.

11. Binding buffer: 0.01 M HEPES, 0.15 M NaCl, 3 mM EDTA, 0.005% Tween 20, and 0.1% BSA, pH 7.4.

12. Regeneration solution: HCl 1 mM, 30 s.

2.2. Samples for Testing

2.2.1. Pure Compounds

1. Doxorubicin (Doxorubicin hydrochloride – Sigma, Aldrich) and sanguinarine (sanguinarine nitrate – City Chemical Corporation, New York, NY), are well-known intercalator (17–19) and are used here as standard compound. This means that before testing plant extract with unknown DNA binding activity, the system must be checked with compounds of well-documented activity, to demonstrate its reliability avoiding, possibly, false negative results.

2. Doxorubicin and sanguinarine 1 mg/mL in DMSO.

2.2.2. Plant Extract and Fractions

1. *C. majus* L. (great celandine) grows in summer and can be collected in fields.

2. Ethanol (Merck) at analytical grade is used for extraction.

3. 150 g of whole-fresh plant material of great celandine is used to obtain a crude EtOH extract by percolation. Deposit the plant in an ordinary paper filter, into a glass fennel in contact with a flask. Then add solvent (1,000 mL) to extract the compounds.

4. The extract is concentrated on rotavapour under reduced pressure. Evaporate a portion of the extract to dryness to obtain a dry residue. Re-suspend 2 g of it in $CHCl_3$ (5 mL). Then filter the solution, using a paper filter, to remove the insoluble part. Place the chloroform solution on the top of a silica column and fractionated by gravity column chromatography (Sigel CC fractionation). Elute using mixtures of $CHCl_3$–MeOH (from 100:1 to 1:100), flow for elution: flow, 3 mL/min, to obtain 100 fractions of 20 mL. Fractions 1–35 are obtained from $CHCl_3$ 100%; fractions 36–40 are obtained by $CHCl_3$–MeOH 96:4; fractions 41–60 from $CHCl_3$–MeOH 9:1; fractions 61–73 from $CHCl_3$–MeOH 4:1; fractions 74–84 from $CHCl_3$–MeOH 70:30; fractions 85–96 from the mixture of $CHCl_3$–MeOH 1:1, and the last three fractions from 100% methanol. Fractions are examined by thin layer chromatography (TLC) to select and collect those containing the compound of interest i.e. alkaloids. Eight main fractions *(1–8)* can be obtained (**Fig. 4**).

5. These eight main fractions *(1–8)* are evaporated to dryness on rotavapour under reduced pressure to be analysed by biosensor analysis.

2.2.3. Thin Layer Chromatography of Purified Fractions from Great Celandine

1. Prepare Silica gel 60 F254 aluminium sheets (20 × 20 cm) (Merck).

2. Deposit a drop of the crude extract and fractions on the sheet.

3. Elute with $CHCl_3$–MeOH (9:1 and 1:1) using a light at 365 nm as detector.

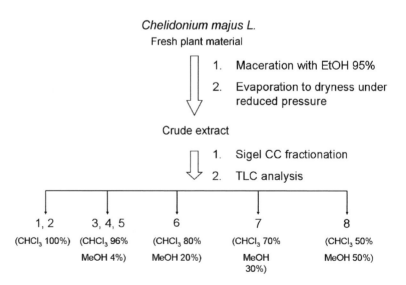

Fig. 4. Plant extracts fractionation by solid phase extraction on silica gel column.

2.3. Instrumentation

SPR measurements are performed using the BIACORE X ™ instrument (Biacore AB, Sweden) and carboxylated dextran-coated chips (CM5 chip, Biacore AB). The SPR signal is expressed in RU.

3. Methods

3.1. Immobilization of Biotinylated DNA Probe onto Streptavidin-Modified Chip Surfaces

To assure a strong surface anchoring of the DNA probe, the interaction occurring between streptavidin and biotin $K_A = 10^{15}$ M^{-1} is used. Thus, on commercial thiol/dextran-modified chip purchased by Biacore AB (Chip CM5), streptavidin is first immobilized to form a layer for biotin attachment.

1. Immobilization of streptavidin onto SPR Chips: The instrumentation operates in flow conditions and under a fine temperature control. The flow and the temperature can be varied. The immobilization procedure is performed at a constant flow rate of 5 µL/min and a temperature of 25°C. Immobilization buffer is used as running buffer. For the immobilization of the protein streptavidin, the dextran surface of the CM5 chip is further modified with 35 µL of streptavidin solution after treatment with 35 µL of activating solution. The remaining carboxylated sites on dextran are blocked with 35 µL of 1 M ethanolamine. After the blockings step, the DNA immobilization is performed.

2. Immobilization of the DNA probe: The biotinylated probe (1 µM) is injected (35 µL at a flow rate of 5 µL/min corresponds to 7 min of contact time between the solution and the surface). After the injection, the excess of unbound probe is removed by the running buffer and the relative signal in RU recorded. For a probe of 20 bases in length, the immobilization is considered successful when a signal of at least 500 RU is recorded.

   After immobilizing the probe, a dsDNA stand, to be used as receptor, is obtained by adding to the surface the probe's corresponding complementary sequence.

3. The complementary sequence in hybridization buffer is thus injected (50 µL) and the formation of the double helix (dsDNA) can be followed in real time. The amount of hybrid formed is given by the corresponding RU recorded after the end of injection and washing off of the excess of the unbound oligo by the running buffer.

   The instrument provides two flow cells: the probe is immobilized only on one of the two. In the other flow cell, where the probe is not immobilized, there is only the streptavidin attached to the chip. This second flow cell is referred as control cell, meaning that this latter is used for checking the specificity of the interaction between the DNA and the testing compound/ sample. If the interaction with the surface in flow cell one (where the DNA is immobilized) is specific, then no signal should be recorded on the control cell.

3.2. Binding Measurements

1. After the immobilization of small oligonucleotides onto the SPR chip, the interaction between the immobilized receptor and the samples in binding buffer is monitored with an association time of 6 min followed by 5 min of dissociation. Solutions of each compound are injected in the system and association and dissociation phase of the interaction were recorded (**Fig. 5**). The flow cell without DNA is used as control cell to verify the absence of non-specific adsorption of the compounds on the sensor chip.

2. Binding interactions are monitored at a constant flow rate of 5 μL/min at a temperature of 25°C. The different compounds can be monitored with both in the association and dissociation phase of the interaction with the DNA. In particular, here the used association time is 6 min while the dissociation phase is monitored for 5 min.

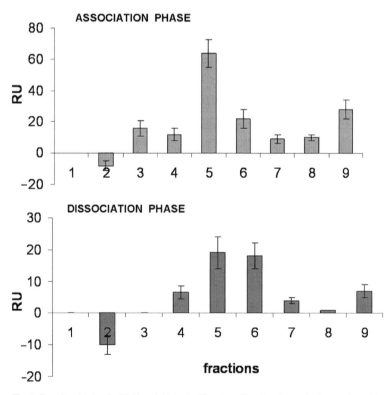

Fig. 5. Results obtained with the eight tested fraction; the signal reported are referred to the association and dissociation phase of the interaction.

3. The surface of the chip can be regenerated to remove the sample and obtain the receptor; however, the solution will also destroy the dsDNA, leaving only ssDNA probe on the surface. To restore the dsDNA, an injection of the complementary target should be done (*see* **Subheading 2.1**, **item 12**) and then the surface is ready for another binding measurement.

4. Negative controls can be tested to prove the specificity of the interaction, for example, compound that are well known that do not interact with DNA, i.e. emodine, chrysophanol.

5. The response of the sensor to blank solutions has to be tested to make sure that the signal is generated by the interaction between the compounds under study and the receptor (i.e. dsDNA on the surface) and not by a matrix effect. The blank solution should contain the same concentration of BSA that is present in the binding buffer.

6. The affinity constants (K_A) of the interaction between the immobilized DNA and pure compounds are estimated by the use of the dedicated BIAevaluation software™ (Biacore AB, Uppsala, Sweden). A variety of methods is available for measuring affinity of biomolecular interactions. SPR-detection can be used more generally than fluorescence or absorbance-based methods since it is sensitive to changes in mass (influencing the refractive index of the media) and no labelling of the reagents is required. The theoretical introduction (*see* **Note 8**) is limited to the description of one to one interactions between independent interaction sites, used for calculating the constant in this application. However, the software available with the instrumentation offers different possibilities for data fitting.

4. Notes

1. The NHS and EDAC solution must be prepared immediately before the use to avoid loss of activity.

2. The regenerating solution removes both the interactant and the target DNA strand, restores initial conditions (*see* **Subheading 1.2**, **step 3**) before performing a new cycle.

3. Pure sample solutions diluted in DMSO cannot be directly applied to the system, but should be diluted to have maximum 5–10% presence of organic solvent when tested with the biosensor (*20*).

4. For the dried crude extract (*Chelidonium*) the estimated detection limit (DL) is below 5 ppm (5 µg/mL) (*12*).

5. Affinity constant for the interaction with dsDNA for Doxorubicin (Molecular Weight 579.99) and sanguinarine (Molecular Weight 394.22) results respectively in $9 \pm 4 \times 10^5$ and $9 \pm 3 \times 10^5$ (M^{-1}) (*12*).

6. The association and dissociation phases of the interaction can be clearly distinguished, as shown in the sensorgram reported in **Fig. 3**. In the association phase, the ligand interacts with the DNA forming a complex; in the dissociation phase, the complex resulting from the interaction starts to dissociate, while the chip is washed with flowing buffer (phase C). The results are reported as resonance shift ΔRU, which is the difference between the final value of resonance units at the end of the association or dissociation phase and the initial value with buffer.

7. In the dissociation phase, some compounds in the fractions can be washed off very quickly by the flowing buffer while others remain bound to DNA. This is indicative somehow of the "strength" of the binding between the active compounds and immobilized DNA.

8. General consideration on measuring affinity and rate constants (*20*).

   In the present application, one should refer to the fact that molecular interactions are characterized in part by the extent of complex formation, described by an equilibrium constant. If two interacting molecular species A and B that can form a complex are mixed in solutions, complex formation will proceed until an equilibrium is eventually reached:

$$A + B \leftrightarrow AB.$$

According to the law of Mass Action, the position of the equilibrium is dependent on the concentration of A, B, and AB, respectively, but can always be described by the equilibrium constant K. Depending on the "direction" in which the reaction is written, K is either expressed as an association equilibrium constant K_A (unit M^{-1}) or a dissociation equilibrium constant K_D (unit M). These constants are given as the inverse of each other:

$$\frac{[AB]}{[A][B]} = K_A \text{ and } \frac{[A][B]}{[AB]} = K_D$$

Affinity constants do not characterize the reaction completely: rates of association and dissociation are equally important. The equilibrium constant does not contain any information on the time span that will elapse before reaching the equilibrium. The rates of association and dissociations are described by association and dissociation rate constants respectively:

Association

$$A + B \rightarrow AB$$

The increase in concentration of AB complexes over time can be written as:

$$\frac{d[B]}{dt} = k_{ass}[A][B],$$

where k_{ass} is the association rate constant (unit M^{-1} s^{-1}).

Dissociation

$$AB \rightarrow A + B$$

The decrease in concentration of AB complexes over time can be written as:

$$\frac{d[AB]}{dt} = -k_{diss}[AB],$$

where k_{diss} is the dissociation rate constant (unit s^{-1}).

The dissociation rate constant k_{diss} indicates the fraction of AB complexes that dissociate per second. Values for k_{diss} typically range from 10^{-5} to 10^{-2} s^{-1}, corresponding to half-lives for biological complexes from a few minutes to several hours.

The observed net rate of formation or dissociation of AB complexes when approaching equilibrium from either excess of A, B, or AB is the sum of the two rate expressions

$$\frac{d[AB]}{dt} = k_{ass}[A][B] - k_{diss}[AB]$$

At equilibrium, there is a balance between association and dissociation and the net rate of complex formation will be zero:

$$\frac{d[AB]}{dt} = 0, \quad \text{i.e.} k_{ass}[A][B] = k_{diss}[AB]$$

Rearranging this expression shows that the equilibrium constant K is the ration of the association and dissociation rate constants:

$$\frac{k_{ass}}{k_{diss}} = \frac{[AB]}{[A][B]} = K_A$$

$$\frac{k_{diss}}{k_{ass}} = \frac{[A][B]}{[AB]} = K_D$$

The approach on which is based the work illustrated in this chapter is based on the Biacore developed theory.

To apply the general rate equation to the formation of AB complexes, the progress of the reaction must be monitored in some fashion. This can be done by a series of time-resolved determinations of the concentration either of AB complexes or of reactant A or B.

Generally, using subscripts 0 and t to denote time: $[B]_t = [B]_0 - [AB]_t$ and substituting in the general rate equation for formation of AB gives:

$$\frac{d[AB]}{dt} = k_{ass}[A]_t \left([B]_0 - [AB]_t \right) - k_{diss}[AB]_t$$

In real-time analysis, such as the one provided by Biacore instrumentation in the case study reported here, complex formation is monitored directly as the change in the response over time. The concentration of free analyte is kept constant through a continuous flow of fresh analyte solution passed on the sensor surface. Denoting the analyte as A, (concentration C) and the ligand as B gives,

$$\frac{dR}{dt} = k_{ass}C\left(R_{max} - R_t \right) - k_{diss}R_t,$$

where C is the concentration of the analyte free in solution; R_{max} is the total amount of binding sites of the immobilized ligand B, expressed as SPR response in RU; $R_{max} - R_t$ is the amount of remaining free binding sites at time t, expressed as SPR response in RU; dR/dt is the rate of formation of surface complexes expressed as SPR response in RU/s, i.e. the derivative of the response curve.

In accordance with the general rate equation, the observed response curve will be the sum of the association term and the dissociation term. During sample injection, complexes will continue forming until eventually, when the two terms balance each other, equilibrium is reached. Note that only the association term $k_{ass}C$ $(R_{max} - R_t)$ is dependent on the concentration of the free analyte.

Dissociation is observed at a rate depending on the amount of complex formed when the sample is replaced by buffer at the end of the injection.

Rearranging the rate equation shows more clearly that the derivative of the binding curve is linearly related to the response. In principles, therefore, rate constant can be derived from a plot of dR/dt against R.

$$\frac{dR}{dt} = k_{ass}CR_{max} - \left(k_{ass}C + k_{diss} \right)R_t$$

Sometimes it could be difficult to estimate R_{max}, i.e. biological interaction can take several minutes or even hours to reach equilibrium, but this can be circumvented. From the rearranged rate equation, it can be seen that the slope ($k_{ass} C + k_{diss}$) of the line obtained by plotting dR/dt against R is itself linearly related to the concentration of the free analyte concentration C.

By plotting the slopes of the dR/dt vs. R lines as a function of analyte concentration C, a new line is obtained, with slope k_{ass} and intercept of the abscissa k_{diss}.

$$\text{Slope}\left(dR / dt \text{ vs.} R\right) = k_{ass}C + k_{diss} \quad \text{in other words:} \, y = k_{ass}x + q.$$

Relatively to the dissociation phase, we know that when the sample pulsed has passed on the sensor surface and is replaced by buffer, the concentration of the free analyte drops suddenly to zero. The derivative of the response curve then reflects the dissociation rate:

$$\frac{dR}{dt} = -k_{diss}R_t.$$

This is of course valid only under the assumption that re-association of released analyte is negligible, i.e. analyte is efficiently removed by the buffer flow. Integrating with respect to time gives:

$$\ln\frac{R_{t_1}}{R_{t_n}} = k_{diss}\left(t_n - t_1\right),$$

where R_{t_n} is the response at time n and R_{t_1} is the response at an arbitrary starting time 1. The dissociation rate constant k_{diss} can thus be obtained by plotting the log of the drop in response against time interval.

An important point to be considered is that association and dissociation reaction in surface-bound system are ultimately governed by mass transport of the analyte between surface and bulk solution. In order to have real and not apparent constant values it is important that the interaction does not occur in mass transport limited conditions.

9. The Biacore X™ instrument has two cells. The two cells can be modified with different receptors to have a working and a reference cell. Otherwise, one of the cells can be kept with only carboxylated dextran to check the effect of a "blank" to eliminate the signal due to eventual non-specific adsorption of the target onto the medication layers.

References

1. Dove, A. (1999) Drug screening - beyond the bottleneck. *Nat. Biotechnol.* **17**, 859–863

2. Rademann, J. and Jung, G. (2000) Techview: drug discovery: integrating combinatorial synthesis and bioassays. *Science* **287**, 1947–1948

3. Lee, T. R. and Lawrence, D. S. (2000) SH$_2$-directed ligands of the Lck tyrosine kinase. *J. Med. Chem.* **43**, 1173–1179

4. Campanella, L., Favero, G., Persi, L. and Tomassetti, M. (2000) New biosensor for superoxide radical used to evidence molecules of biomedical and pharmaceutical interest having radical scavenging properties. *J. Pharm. Biomed. Anal.* **23**, 69–76

5. Danelian, E., Karlén, A., Karlsson, R., Winiwarter, S., Hansson, A., Löfas, S., Lennernäs, H. and Hämäläinen, M. (2000) SPR biosensor studies of the direct interaction between 27 drugs and a liposome surface: correlation with fraction absorbed in humans. *J. Med. Chem.* **3**, 2083–2086

6. White, R. E. (2000) High-throughput screening in drug metabolism and pharmacokinetic support of drug discovery. *Annu. Rev. Pharmacol. Toxicol.* **40**, 133–157

7. Turner, A. P. F., Karube, I. and Wilson G. S. (eds.) (1987) *Biosensor – Fundamentals and Applications*, 1st edn., Oxford University Press, UK

8. Tombelli, S., Minunni, M. and Mascini, M. (2007) Aptamers for diagnostics, environmental and food quality applications. *Biomol. Eng.* **24**(2), 191–200

9. Minunni, M. and Bilia, A. (2007) Biosensing approach in natural products research. In: Colegate, S. M. and Molyneaux, R. J. (eds.) *Bioactive Natural Products. Detection, Isolation and Structural Determination*, 2nd edn, Chapter 11. CRC Press, Inc, Boca Raton, FL

10. Donghui, Y., Blankert, B., Vire, J. C. and Kauffmann, J. M. (2005) Biosensors in drug discovery and drug analysis. *Anal. Lett.* **38**, 1687–1701

11. Cooper, M. A. (2002) Optical biosensors in drug discovery. *Nat. Rev. Drug. Discov.* **1**, 515–528

12. Minunni, M., Tombelli, S., Mascini, M., Bilia, A. R., Bergonzi, M. C. and Vincieri, F. F. (2005) An optical DNA-based biosensor for the analysis of bioactive constituents with application in drug and herbal drug screening. *Talanta* **65**, 578–585

13. Colombo, M. L. and Bosisio, E. (1996) Pharmacological activities of Chelidonium majus L. (Papaveraceae). *Pharmacol. Res.* **33**, 127–134

14. Kokoska, L., Polesny, Z., Rada, V., Nepovim, A. and Vanek, T. (2002) Screening of some Siberian medicinal plants for antimicrobial activity. *J. Ethnopharmacol.* **82**, 51–53

15. Kim, D. J., Ahn, B., Han, B. S. and Tsuda, H. (1997) Potential preventive effects of Chelidonium majis L. (Papaveraceae) herb extract on glandular stomach tumor development in rats treated with N-methyl-*N'*-nitro-*N* nitrosoguanidine (MNNG) and hypertonic sodium chloride. *Cancer Lett.* **112**, 203–208

16. Fang, S., Lee, H. J., Wark, A. W. and Corn, R. (2006) Attomole Microarray detection of microRNAs by nanoparticle-amplified SPR Imaging measurements of surface polyadenylation reactions. *J. Am. Chem. Soc.* **128**, 14044–14046

17. Kong, X. B., Rubin, L., Chen, L. I., Cuszewska, G., Watanabe K. A., Tong, W. P., Sirotnak, F. M. and Chou, T. C. (1992) Topoisomerase II-mediated DNA cleavage activity and irreversibility of cleavable complex formation induced by DNA intercalator with alkylating capability. *Mol. Pharmacol.* **41**, 237–244

18. Das, S., Kumar, G. S. and Maiti, M. (1999) Conversions of the left-handed form and the protonated form of DNA back to the bound right-handed form by sanguinarine and ethidium: a comparative study. *Biophys. Chem.* **76**, 199–218

19. Sen, A., Ray, A. and Maiti, M., (1996) Thermodynamics of the interactions of sanguinarine with DNA: influence of ionic strength and base composition. *Biophys. Chem.* **59**, 155–170

20. Biacore, T. M. (1991) System manual, Uppsala, Sweden, *Parmacia Biosensor AB*

Chapter 14

Application of Frontal Affinity Chromatography with Mass Spectrometry (FAC–MS) for Stereospecific Ligand–Macromolecule Interaction, Detection and Screening

Jacek J. Slon-Usakiewicz and Peter Redden

Summary

Using frontal affinity chromatography coupled to mass spectrometry (FAC–MS) we have established a general stereoselective detection and screening method of intact racemates which can generate binding affinity information about the individual enantiomers that is also applicable to other ligand isomeric mixtures. FAC–MS has been shown to be a versatile technology utilizing direct binding in screening assays and extending its application toward chiral drug development, especially in the early discovery stages as well as its utility in secondary Structure-activity relationship (SAR) studies allow this platform to make a significant step toward facilitating the demand for pure enantiomeric drugs. Using renin, which is as an important drug target, we show that for detection and screening purposes there is no need to first use timely and costly methods of separating racemates in order to get precise information about the binding affinities of the composite enantiomers.

Key words: FAC–MS, Frontal affinity chromatography, Mass spectrometry, Chiral drugs, Affinity screening, Racemates, Enantiomers

1. Introduction

The biological action of the majority of drugs typically starts with the interaction or binding of the drug to its requisite target, either a receptor, enzyme, or other protein. At the early drug discovery stage the ongoing quest for the identification of novel and better drugs has been markedly enhanced with the use of combinatorial chemistry that has allowed for the generation of very large numbers

Ana Cecília A. Roque (ed.), *Ligand-Macromolecular Interactions in Drug Discovery: Methods and Protocols,*
Methods in Molecular Biology, vol. 572,
DOI 10.1007/978-1-60761-244-5_14, © Humana Press, a part of Springer Science+Business Media, LLC 2010

of novel compound libraries. These libraries typically contain compounds that range from being achiral to racemic in nature (and may also contain even more complicated compounds with multiple stereocenters). Screening these synthetic racemic compounds for their enantiomeric composition is very important due to differences typically seen in the biological activities of the enantiomers as well as in other pharmacological properties.

Currently, however, the typical screening paradigm involves first screening racemates followed by the physical separation of the individual enantiomers of only one or a limited number of racemates that demonstrate the greatest affinity or activity for a particular target. Moreover, significant resources *(1)* can, however, be wasted separating enantiomers that ultimately turn out to have the same or similar affinity. That being said and although the current state of the art for analytical as well as preparative enantioseparations is very high *(2)* it would still be very desirable to be able to elucidate the biological activity of the individual enantiomers in an intact racemate before undertaking the task of physically separating the enantiomers or designing complicated methods of stereoselective synthesis *(3)* and then conducting the biological testing.

Here we describe the use of frontal affinity chromatography coupled to mass spectrometry detection (FAC–MS) to determine the binding affinity of individual enantiomers to biological targets using intact racemates. FAC–MS and its ability to see individual m/z values for compounds certainly has the capability not only to assay mixtures, but also to discriminate between enantiomers if their binding affinities to a biological target are different.

The combination of FAC–MS detection *(4–6)*, although still a relatively new technique, is turning out to be a viable screening tool that has been successfully applied to a wide range of biological targets *(7–12)*. In the last 5 years FAC–MS has been used successfully to screen mixtures of compounds against a variety of immobilized targets *(13–15)*.

Basically, FAC–MS takes advantage of the ongoing equilibrium between binding ligands flowing through a column containing an immobilized protein target. As ligands flow through the column and interact (bind) with the target, individual ligands are retained in the column by their interaction with the protein target. This causes an increase in each ligand's specific "breakthrough volume" which is the effluent volume passing through the column that allows the output ligand concentration to equal the input ligand concentration. The breakthrough volume, characterized as a sigmoidal front (*see* **Fig.1**), can readily be detected by mass spectrometry based on the ligands m/z value and corresponds directly to the time that the front is observed to pass through the column (*see* **Note 1**). Hence tighter binding compounds elute later whereas compounds that do not interact

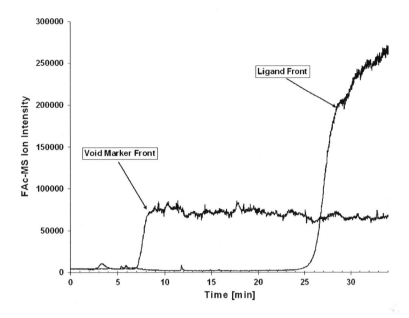

Fig. 1. Typical FAC-MS chromatogram for void marker and single compound.

with the target elute with the void volume of the column. The order of compounds eluting parallels their affinity meaning that FAC–MS offers a convenient method for measuring and ranking the relative binding affinities of ligands in a mixture against an immobilized protein target. FAC–MS can also be used in a more rigorous mode for determining thermodynamic binding constants, Dissociation constant (K_d), of individual compounds.

2. Materials

1. Controlled Pore Glass (CPG) beads, type CBX1000C (Millipore, Billerica, MA).
2. α Man$(1 \rightarrow 3)[\alpha$ Man$(1 \rightarrow 6)]\beta$ ManO-octyl (M3) (EMD, San Diego, CA) is dissolved in water at 1 mM stock solution. Store at 4°C.
3. 1-(3-dimethylaminopropyl)-3-ethyl carbodiimide (EDC) (Sigma) is dissolved fresh at 100 mg/mL in 1× activation buffer.
4. N-hydroxysuccinamide (NHS) (Sigma) is dissolved fresh at 200 mg/mL in 1× activation buffer.
5. Recombinant human renin (Proteos, Kalamozoo, MI) 100 µg in 100 mM sodium acetate of pH 4.2. Store at –20°C.

6. The peptides His-Pro-Phe-His-(D,L-Leu)-Leu-Val-Tyr, His-Pro-Phe-His-(L-Leu)-Leu-Val-Tyr and His-Pro-Phe-His-(D-Leu)-Leu-Val-Tyr (AnaSpec, San Jose, CA) were dissolved in water at 1 mM stock solution. Store at –20°C.

7. Bovine Serum Albumin (BSA) (Sigma, Oakville, Ontario, Canada) is dissolved fresh in activation buffer.

8. Activation buffer (5×): 0.1 M 2-(N-morpholine)-ethane sulfonic acid (MES) (Sigma, Oakville, Ontario, Canada), 0.5 M NaCl, pH 6.4. Store at room temperature.

9. Running buffer: 12.5 mM NH_4HCO_3 containing 1% DMSO. Store at room temperature.

10. Makeup buffer: 90% methanol containing 0.1% acetic acid in water. Store at room temperature.

11. FAC–MS equipment. An API 3000 triple-quadrupole mass spectrometer equipped with an ESI interface (Applied Biosystems/Sciex, Canada). Syringe pumps (PicoPlus) were obtained from Harvard Biosciences (Holliston, MA). The HPLC micropump (1100 series) was obtained from Agilent (Santa Clara, CA), and the autosampler (Famos) and the column switcher (Switchos) were obtained from LC Packings/Dionex (Bannockburn, IL). FAC–MS capillary columns loaded with protein beads were held in a modified custom column holder (obtained from Upstate) with frits included to hold the beads in place.

12. FAC–MS hardware. The general set-up of the FAC–MS hardware has been described before *(4, 7, 16)*; however, in order to automate the system an autosampler and column switcher was placed between the running buffer syringe and the MS. Also an HPLC micropump was used to deliver the makeup buffer. This micropump serves two purposes: firstly, syringe pumps do not have the capacity to run overnight unattended, and secondly the micropump delivers the makeup buffer at a consistent steady rate that markedly improves the MS signal-to-noise ratio.

13. The syringe pump containing the running buffer was connected in series to the autosampler and then to the column switcher with one or more columns and then connected to an AB/Sciex API 3000 triple-quadrupole mass spectrometer (*see* **Note 2**). The makeup buffer was then teed into the system postcolumn. Typically in automated mode the selected column with immobilized protein was washed with the running buffer for a preselected period of time for automated injections. The data were analyzed using proprietary software to determine the breakthrough times of the void marker, M3, and the various ligands (*see* **Note 3**).

3. Methods

3.1. Protein Immobilization

1. The CPG CBX1000C beads (5 mg) suspend in 900 μL activation buffer.

2. To the suspension add 50 μL freshly prepared EDC (100 mg/mL) and 50 μL NHS (200 mg/mL) and incubate the mixture for 45 min at room temperature with 360° vertical rotation to keep beads in suspension.

3. Centrifuge the suspension and remove supernatant and resuspend the beads in 500 μL activation buffer containing renin (200 μg); incubate at room temperature for 2 h with 360° vertical rotation; then incubate overnight at 4°C.

4. Add BSA (10 μL of 100 mg/mL fresh solution) and the beads; store at 4°C.

5. Use this procedure to prepare column with blank beads for nonspecific interaction detection, by immobilizing 200 μg of BSA onto CPG CBX1000C beads (*see* **Note 4**).

3.2. Determination of FAC-MS Binding and Nonspecific Interaction

1. After loading the beads with immobilized protein by syringe infusion, the FAC–MS capillary columns (250 μm id × 2.5 cm) (*see* **Notes 5** and **6**) are washed with 50 μL of phosphate-buffered saline (PBS) buffer followed by 50 μL of the running buffer at a flow rate of 200 μL/h.

2. Tune the void marker (M3) and compound (racemic peptide) using a 1 μM working solution in running buffer to optimize the mass spectrometry settings.

3. Prepare three analyte solutions to contain: (a) The ligand (racemic peptide His-Pro-Phe-His-(D,L-Leu)-Leu-Val-Tyr, 1 μM) and void marker M3 (1 μM) in running buffer. (b) The ligand (L-Leu5 peptide His-Pro-Phe-His-(L-Leu)-Leu-Val-Tyr, 1 μM) and void marker M3 (1 μM) in running buffer. (c) The ligand (D-Leu5 peptide His-Pro-Phe-His-(D-Leu)-Leu-Val-Tyr, 1 μM) and void marker M3 (1 μM) in running buffer.

4. Connect a blank column to the mass spectrometer and syringe pump (precolumn) (*see* **Note 7**) and set flow rate to 100 μL/h (*see* **Note 8**).

5. Connect the mass spectrometer to the makeup buffer (post-column) and set flow rate to 800 μL/h.

6. Allow the blank column to equilibrate with the running buffer until the ligand [M + H]⁺ signal is stable, and then start data acquisition with Analyst v1.4 software (*see* **Note 9**).

7. After 1 min, switch the system to analyte solution 1 and continue data collection until the lined signal had maximized for at least 10 min.

8. Wash the blank column with running buffer until the ligand and void marker signals had reduced to their background levels to regenerate the column (*see* **Note 10**).

9. Analyze the collected data to determine the breakthrough times of both the ligand and the void marker (*see* **Note 3**).

10. Replace the blank column with a protein (renin) column and follow steps 4–9 to analyze the first analyte. After completing this run again replace the used column with a new protein (renin) column (*see* **Note 11**) and analyze analytes 2, 3, and so on (*see* **Notes 12** and **13**).

4. Notes

1. At the early hit discovery stage in the drug discovery process, FAC–MS can be used in one of two modes, either "indicator" or "Q1 scan" mode. The indicator mode has been well documented and basically encompasses determining the extent (or percentage) in which a compound(s) shifts an indicator (a ligand with a known affinity to a known binding site) for a particular target *(4, 5, 13)*. With this method the FAC–MS readout (% shift of the indicator) can be used to rank the binding of ligands, that is the greater the % shifts, the greater the degree of competition for the indicator. As shown pictorially in **Fig. 2**, the % shift can be determined from the equation:

$$\% \, \text{Shift} = \frac{\left(t_I - t\right)}{\left(t_I - t_{NSB}\right)} \times 100\%$$

where t is the breakthrough time difference, measured at the inflection point, of the sigmoidal fronts between the indicator and void marker (a compound that does not interact with the target and gives the void volume of the column) in the presence of any competing ligand(s), t_{NSB} is the nonspecific binding breakthrough time difference in the absence of immobilized protein (and is a constant for the indicator used), and t_I is the breakthrough time difference in the absence of any competing ligands. In this manner the FAC–MS % shifts can be used to rank the binding affinity of compounds and only those mixtures where a significant displacement (or shift) of the indicator merit further interest and require deconvolution. Moreover we have recently demonstrated that the relative displacement of an "indicator" molecule binding at a specific binding site of an immobilized protein target by an individual member of a compound mixture correlates well with both the K_d and the IC_{50} value of the individual compound *(13)*. The second FAC–MS screening mode has also been described

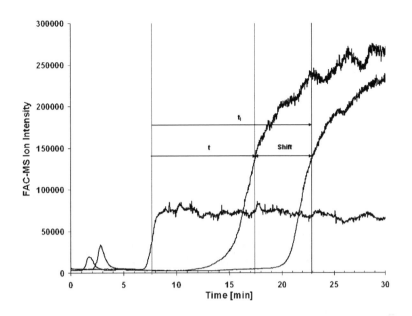

Fig. 2. Typical FAC-MS chromatogram seen using "indicator" screen mode. The % shift is determined from the equation % Shift = $(t_i - t)/(t_i - t_{NSB}) \times 100\%$, where t is the breakthrough time difference, measured at the inflection point, of the sigmoidal fronts between the indicator and void marker in the presence of any competing ligand(s), t_{NSB} is the nonspecific binding breakthrough time difference in the absence of immobilized protein (and is a constant for the indicator used), and t_i is the breakthrough time difference in the absence of any competing ligands.

in detail, the so-called Q1 scan, where multiple compounds are analyzed simultaneously *(16)*.

2. Matrix-assisted laser desorption/ionization (MALDI) mass spectrometry could be used as well as electrospray ionization in FAC–MS experiments *(17 and literature cited)*.

3. The breakthrough front times for the FAC–MS chromatogram could be determined manually. Since, however what is important is the inflection point (or the point where the first derivative is zero) for each m/z versus time for each compound, proprietary Excel-based macro was written that calculates this automatically from the recorded chromatogram.

4. It is important to choose the right blank column for the nonspecific interaction test and depends on the beads used for target immobilization. For CPG CBX1000C beads and direct target immobilization, beads that have been immobilized with a neutral protein (like BSA) typically will be the best blank reference. For targets immobilized via the biotin–streptavidin linkage streptavidin-coated CPG beads should be used that have been saturated with free D-biotin.

5. Sizes of FAC–MS columns could be different depending on the results of immobilization, ligand affinities, and breakthrough times.

6. The typical amount of protein immobilized within listed parameters is in range of 30–100 pmol. A relatively high density of immobilized protein ensures that unwanted nonspecific binding sites are minimized *(17)*.

7. An electric actuator could be used to connect the FAC–MS column system with syringe pumps for semiautomatic changes of fluid directed to the column (running buffer and analyte).

8. At low flow rates, variations in the flow rates could be significant and generated binding data will have relative high error; thus, other micro- or nanopump systems could be considered *(16)*.

9. Any type of mass spectrometry software with capability of recording ion intensity versus time could be utilized.

10. Blank columns used to test nonspecific interaction if properly maintained, washed (indicator and void marker signals have to be reduced to their background level), and stored (store in running buffer at 4°C) could be reused many times.

11. In contrast to blank columns it is highly recommended to use freshly prepared protein columns for every individual run. Protein columns could be regenerated only if low affinity ligands ($K_d > 1$ µM) were analyzed. For ligands of high affinity (as well as unknown) regeneration is always not complete which can introduce significant error to investigated samples.

12. Using previously described FAC–MS methodology to demonstrate the stereospecific interaction between ligands and macromolecule we chose a peptide containing D,L,-leucine at the fifth position as well as the individual D- and L-Leu peptides with known *(18)* and differing binding affinities to renin. The racemic leucine containing peptide His-Pro-Phe-His-(D,L-Leu)-Leu-Val-Tyr was infused at a concentration of 1 µM in 12.5 mM ammonium carbonate buffer through the renin column generating the double plateau FAC–MS chromatogram shown in **Fig. 3**. This double plateau arises as a result of the two D- and L-Leu peptides having different binding affinities to renin with presumably the weaker L-Leu peptide's breakthrough front of 9 min over the breakthrough time of the void marker (M3), and eluting earlier than the stronger binding D-Leu peptide's front of 23.5 min over the breakthrough time of M3. To confirm this we infused the D- and L-Leu containing peptides separately through the renin column again each at a concentration of 1 µM in 12.5 mM ammonium carbonate. Since there is always the possibility that compounds can bind to the FAC–MS hardware components (capillary lines, columns, beads, etc.) this nonspecific binding of compounds is always determined in

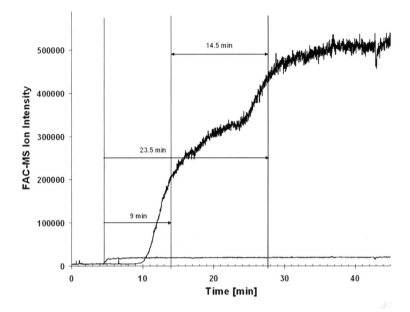

Fig. 3. FAC-MS chromatograms using immobilized renin. Racemic leucine containing peptide, His-Pro-Phe-His-D,L-Leu-Leu-Val-Tyr, was infused at a concentration of 1 μM in 12.5 mM ammonium carbonate buffer through the immobilized renin column generating a double plateau. This double plateau arises as a result of the two D- and L-Leu peptides having different binding affinities. The first breakthrough front at 9 min over and above the breakthrough front of the void maker, M3, corresponds to the weaker L-Leu peptide with the second breakthrough front of 23.5 min over the void marker corresponding to the stronger binding D-Leu peptide.

the absence of any immobilized protein or with a neutral immobilized protein such as BSA. Hence for these renin peptides, the less potent His-Pro-Phe-His-L-Leu-Leu-Val-Tyr peptide with a K_i = 39 μM generated a breakthrough binding time 1.6 min more than its corresponding nonspecific breakthrough time (**Fig. 4a**). In contrast to this the more potent His-Pro-Phe-His-D-Leu-Leu-Val-Tyr peptide with a K_i = 3 μM generated a breakthrough binding time of 19 min over the nonspecific breakthrough time (**Fig. 4b**).

13. Evaluating ligands whether achiral, racemic, or isomerically pure over a blank column (either with no protein present or with an immobilized natural protein like BSA) is an integral part of the FAC–MS screening experiment, and as shown in **Fig. 4c** the racemic peptide generates only one plateau sigmoidal-shaped breakthrough front. Similarly, if for a racemic ligand only one plateau breakthrough front (with a particular breakthrough time and will depend on the affinity) is generated with an immobilized protein target then this indicates that both enantiomers comprising the racemic mixture have the same affinity toward the target. Moreover, FAC–MS chromatograms not only generate information

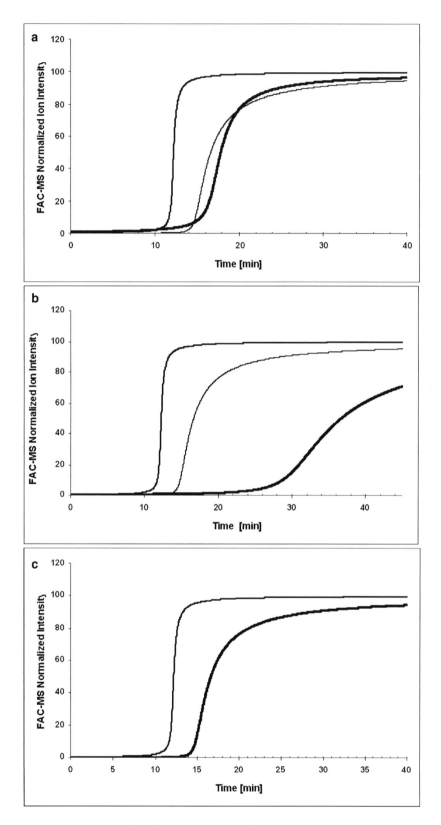

Fig. 4. Normalized, superimposed FAC-MS chromatograms with immobilized proteins and nonspecific binding. (a) The less potent His-Pro-Phe-His-L-Leu-Leu-Val-Tyr peptide with a $K_i = 39$ μM was infused through the immobilized renin column and generated 1.6 min of binding (*thick line*) over the nonspecific binding (*thin line*) to the FAC-MS components (*capillary lines*, columns, beads, etc.) in the neutral immobilized protein – BSA. (b) The more potent His-Pro-Phe-His-D-Leu-Leu-Val-Tyr peptide with a $K_i = 3$ μM generated a breakthrough binding time of 18 min (*thick line*) over the nonspecific breakthrough time (*thin line*). (c) Nonspecific interaction chromatogram for racemic peptide.

about binding potency but also differences in the enantiomer affinities. In most cases only those racemates with relatively high differences between the affinities of the enantiomers need be considered for further analytical development of separation processes leading to significant time and cost savings. It is also worthwhile noting that in addition to providing enantiomeric-binding affinities, the presence of a double plateau in FAC–MS screening of multiple compound mixtures (including natural product extracts) could also indicate the presence of two compounds that have the same m/z values but with different binding affinities toward the immobilized target.

Although the biological world is innately chiral, there has not been extensive use of chirality during the initial drug discovery process. The reasons are straightforward in that the synthesis of enantiomerically pure compounds is very challenging. Clearly, however, libraries of compounds with three-dimensionally well-defined structures should be expected to improve the odds of finding hits with increased binding affinities for important drug targets. With the use of FAC–MS screening of intact racemic compounds there is no need to shy away from using diverse or focused libraries containing racemic compounds nor should drug companies or library vendors shy away from producing racemic containing libraries. The expensive and complicated process of enantioselective synthesis or physically separating chiral compounds only needs to be reserved for the most potent chiral compounds where significant differences exist between enantiomers.

References

1. Zhao, Y., Woo, G., Thomas, S., Semin, D., and Sandra, P. (2003) Rapid method development for chiral separation in drug discovery using sample pooling and supercritical fluid chromatography – mass spectrometry. *J. Chromatogr. A* **1003**, 157–166

2. Chankvetadze, B., Yamamoto, C., and Okamoto, Y. (2000) Enantioseparations using cellulose tris(3,5-dichlorophenylcarbamate) during high-performance liquid chromatography with analytical and capillary columns potential for screening of chiral compounds. *Comb. Chem. High Throughput Screen.* **3**, 497–508

3. Federsel, H.-J. (2005) Asymmetry on large scale: the roadmap to stereoselective processes. *Nat. Rev. Drug. Discov.* **4**, 685–697

4. Schreimer, D.C. and Hindsgaul, O. (1997) Microscale frontal affinity chromatography combined on-line with mass spectrometric detection: a new method for the screening of compound libraries. *Angew. Chem. Int. Ed.* **37**, 3383–3387

5. Slon-Usakiewicz, J.J., Ng, W., Dai, J.R., Pasternak, A., and Redden, P.R. (2005) Frontal affinity chromatography with MS detection (FAC–MS) in drug discovery. *Drug Discov. Today* **10**, 409–416

6. Deng, G. and Sanyal, G. (2006) Applications of mass spectroscopy in early stages of target based drug discovery. *J. Pharm. Biomed. Anal.* **40**, 528–538

7. Zhang, B., Palcic, M.M., Mo, H., Goldstein, I.J., and Hindsgaul, O. (2001) Rapid determination of the binding affinity and specificity of the mushroom *Polyporus squamosus* lectin using frontal affinity chromatography coupled to electrospray mass spectrometry. *Glycobiology* **11**, 141–147

8. Chan, N.W.C., Lewis, D.F., Hewko, S., Hindsgaul, O., and Schriemer, D.C. (2002) Frontal affinity chromatography for the screening of mixtures. *Comb. Chem. High Throughput Screen.* **5**, 395–406

9. Wang, J., Zhang, B., Fang, J., Sujino, K., Li, H., Otter, A., Hindsgaul, O., Palcic, M.M., and Wang, P.G. (2003) Frontal affinity chromatography coupled to mass spectrometry: an effective method for K_d determination and screening of α-gal derivatives binding to anti-gal antibodies (IgG). *J. Carbohydr. Chem.* **22**, 347–376

10. Zhu, L., Chen, L., Luo, H., and Xu, X. (2003) Frontal affinity chromatography combined on-line with mass spectrometry: a tool for the binding study of different epidermal growth factor receptor inhibitors. *Anal. Chem.* **75**, 6388–6393

11. Luo, H., Chem, L., Li, Z., Ding, Z., and Xu, X. (2003) Frontal immunoaffinity chromatography with mass spectrometric detection: a method for finding active compounds from traditional Chinese herbs. *Anal. Chem.* **75**, 3994–3998

12. Chan, N.W.C., Lewis, D.F., Rosner, P.J., Kelly, M.A., and Schriemer, D.C. (2003) Frontal affinity chromatography-mass spectrometry assay technology for multiple stages of drug discovery: applications of a chromatographic biosensor. *Anal. Biochem.* **319**, 1–12

13. Slon-Usakiewicz, J.J., Ng, W., Foster, J.E., Dai, J.R., Deretey, E., Toledo-Sherman, L., Redden, P.R., Pasternak, A., and Reid, N. (2004) Frontal affinity chromatography with MS detection (FAC–MS) of EphB2 tyrosine kinase receptor. I. comparison with ELISA. *J. Med. Chem.* **47**, 5094–5100

14. Toledo-Sherman, L., Deretey, E., Slon-Usakiewicz, J.J., Ng, W., Dai, J.R., Foster, J.E., Redden, P.R., Uger, M.D., Liao, L.C., Pasternak, A., and Reid, N. (2005) Frontal affinity chromatography with MS detection of EphB2 tyrosine kinase receptor. 2. Identification of small-molecule inhibitors via coupling with virtual screening. *J. Med Chem.* **48**, 3221–3230

15. Slon-Usakiewicz, J.J., Dai, J.R., Ng, W., Foster, J.E., Deretey, E., Toledo-Sherman, L., Redden, P.R., Pasternak, A., and Reid, N. (2005) Global kinase screening. Applications of frontal affinity chromatography coupled to mass spectrometry in drug discovery. *Anal. Chem.* **77**, 1268–1274

16. Ng, W., Dai, J.R., Slon-Usakiewicz, J.J., Redden, P.R., Pasternak, A., and Reid, N. (2007) Automated multiple ligand screening by frontal affinity chromatography-mass spectrometry (FAC–MS). *J. Biomol. Screen.* **12**, 167–174

17. Chan, N., Lewis, D., Kelly, M., Ng, E.S.M., and Schriemer, D.C. (2007) Frontal affinity chromatography – mass spectrometry for ligand discovery and characterization, in *Mass Spectrometry in Medicinal Chemistry* (Wenner, K., Hofner, G., eds), Wiley-VCH Verlag GmbH & Co. KGaA, Weinheim, pp. 217–246

18. Poulsen, K., Burton, J., and Haber, E. (1973) Competitive inhibitors of renin. Inhibitors effective at physiological pH. *Biochemistry* **12**, 3877–3882

Chapter 15

GPC Spin Column HPLC–ESI-MS Methods for Screening Drugs Noncovalently Bound to Proteins

Marshall M. Siegel

Summary

Secondary drug screening methods are described for determining the relative degree of non-covalent binding between drug candidates and a protein of therapeutic interest by gel centrifugation chromatography using GPC spin columns for isolating the protein–drug complexes, under native conditions, and reversed-phase HPLC coupled with ESI-MS for highly resolved and sensitive detection of the drug in the complex, under denaturing conditions. The necessary control samples and limitations of this work are fully described. The GPC spin column HPLC–ESI-MS methodology for screening of drugs non-covalently bound to proteins is illustrated for the non-covalent binding of geldanamycin with Hsp90cat protein.

Key words: Gel permeation chromatography, Gel filtration chromatography, Size exclusion chromatography, Gel centrifugation chromatography, Spin columns, Non-covalent protein–ligand complexes, Reversed-phase HPLC, Electrospray ionization-mass spectrometry, Hsp90, Geldanamycin

1. Introduction

High-throughput screening (HTS) has been the technique of choice in the pharmaceutical industry to discover exploratory drug leads from corporate compound libraries. A variety of methods are used in HTS to identify small molecules that inhibit or activate the cellular function of a target protein *(1,2)*. Generally, HTS methods take considerable time to develop and are unique for each biological system of interest. On the other hand, structure-based screening methods have been developed which rely on the direct or indirect observation of non-covalent binding

Ana Cecília A. Roque (ed.), *Ligand-Macromolecular Interactions in Drug Discovery: Methods and Protocols,*
Methods in Molecular Biology, vol. 572,
DOI 10.1007/978-1-60761-244-5_15, © Humana Press, a part of Springer Science+Business Media, LLC 2010

of a ligand (drug candidate, peptide, oligosaccharide, protein) to a biopolymer (protein, DNA, polysaccharide) of therapeutic interest. Usually, these methods are quite general and require little to no additional development, e.g., NMR methods *(3)*, surface plasmon resonance (Biacore) *(4)*, microcalorimetry *(5)*, ultrafiltration *(6)*, Frontal Affinity Chromatography *(7)*, and X-ray crystallography *(8)*. These structure-based screening methods often require high levels of biopolymers, especially in the case of proteins, which may be challenging to obtain or limit the application. A method that requires generally low levels of protein and is useful for a wide range of binding affinities is the use of gel permeation chromatography (GPC) in the spin column mode with reversed-phase HPLC and electrospray ionization-mass spectrometry detection (ESI-MS) *(9)*. This methodology has been demonstrated to require little development time, and to be a reliable structural screening technique that can be performed at high speed and high sensitivity with large numbers of compounds, especially when analyzed as mixtures. This structure-based screening method relies on the ability of gel filtration in the spin column mode to resolve non-covalently bound protein-ligand complexes from free ligands in the condensed phase under native conditions (*see* **Fig.1a, b**), the HPLC chromatographic separation of the dissociated protein-ligand complex in the condensed phase under denaturing conditions, and the atmospheric pressure ionization-mass spectral analysis of the freed ligand in the gas phase using ESI-MS (*see* **Fig.1c**). Previous reports describe *primary* screening of large corporate compound libraries using the GPC spin column HPLC–ESI–MS methodology *(10–12)*. This report will focus on the use of GPC in the spin column mode for *secondary* screening of small compound libraries to confirm non-covalent binding between a protein and drug candidates using HPLC-ESI-MS as the highly specific and sensitive detector while optimizing each of the steps in the methodology.

1.1. Principles of Gel Filtration

The principles of gel filtration are now briefly compared when performed under equilibrium conditions for column chromatography and when performed under nonequilibrium conditions for spin column separations, often referred to as gel centrifugation chromatography.

1.1.1. Gel Filtration Under Equilibrium Conditions (13–17)

GPC (gel filtration chromatography, size exclusion chromatography) separates molecules by size as they pass through a column consisting of a bed of porous particles equilibrated in a buffer solution of carrier solvent. The size of the pores is selected so that molecules of a given dimension [molecular weight (MW)] penetrate the pores. As the molecules flow through the pores of the gel column, they are retarded and resolved. Larger pores retain both large and small molecules while smaller pores retain

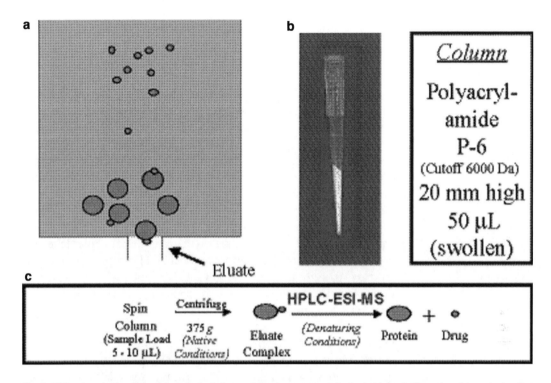

Fig. 1. GPC spin column methodology for isolating protein–drug non-covalent complexes in the spin column eluate after centrifugation, and the steps taken to detect the ligands present in the spin column eluate by HPLC–ESI-MS after denaturing the protein–drug non-covalent complexes. (a) GPC spin column image, (b) Photo of a miniature Glygen GPC spin column with slit orifice, (c) Schematic of GPC spin column HPLC–ESI-MS procedure.

lower MW molecules, so that higher MW compounds elute before lower MW compounds in the fractionation range for the gel. Lower MW molecules experience an effectively longer path relative to higher MW compounds as they travel through the column. Molecules larger than the pore sizes will pass through the void volume of the column unretarded. In many cases, the void volume is about 30% of the total gel volume. Very small low MW molecules penetrate deeply into the pores and their flow is retarded maximally, exiting from the column with a carrier solvent volume nearly equivalent to the total volume of the gel in the column.

1.1.2. Gel Filtration Under Nonequilibrium Conditions (17–20)

A GPC spin column is a short column packed with gel media that has been hydrated with buffer and gently centrifuged. The gentle gel centrifugation process removes nearly all the buffer solution present in the void volume while nearly retaining the pores fully hydrated with buffer. In many cases, the buffer volume in the pores constitutes about 50% of the total gel volume. A sample, consisting of a protein–drug complex, free protein and free drug, is loaded on to the column. The volume of the sample should be less than 50% of the gel volume, and 10–25% of the gel volume is generally used. The gel is chosen based on the pore size relative to the MW of the protein. The pores should exclude the protein but permit the penetration of the small molecules. Upon gentle centrifugation, the protein–drug complex and free protein pass rapidly through the void volume of the gel and are collected, while the free drug passes through the pores of the gel and is trapped, because the sample volume is less than the hydrated gel volume. Since this is a nonequilibrium process, the optimum conditions for the experimental parameters such as gel type, sample volume and g-forces are best optimized experimentally. The most common soft polymer gels used are the polyacrylamides *(21)* and the sephadexes *(22)*, and are listed in **Table 1** with their fractionation ranges for globular proteins. Either gel type can be used for hydrophilic proteins while the sephadexes are preferable for more hydrophobic proteins. Hydrophobic proteins tend to strongly bind to the gel beads reducing their transmission efficiency through the gel. Perhaps the two best methods to overcome this limitation are (i) to use the largest possible sample volume (without small molecule breakthrough) so that the maximum amount of protein would be eluted, and (ii) to coat the protein binding sites of the gel with protein by prewashing the column with protein prior to sample analysis. The optimal centrifugal force applied to the soft gels occurs when the maximum volume of excluded liquid is removed without collapsing the gel. Generally, g-forces between 735 and $1,000 \times g$ are used for the polyacrylamides and sephadexes G10–G50 and $200 \times g$ are used for sephadexes G75–G150.

Table 1
GPC soft gel types and fractionation ranges
(for globular proteins)

Polyacrylamide (Bio-Rad)		Sephadex (Pharmacia)	
Type	Fractionation range (Da)	Type	Fractionation range (Da)
P2	100–1,800	G10	<700
P4	800–4,000	G15	<1,500
P6	1,000–6,000	G25	1,000–5,000
P10	1,500–20,000	G50	1,500–30,000
P30	2,500–40,000	G75	3,000–80,000
P60	3,000–60,000	G100	4,000–150,000
P100	5,000–100,000	G150	5,000–300,000

2. Materials

Materials for secondary screening are listed as follows for the study of protein-drug complexes where individual complexes, drugs, and proteins are analyzed by GPC spin column HPLC–ESI-MS.

2.1. Spin Column Hardware

1. Soft polymer gel resins are listed in **Table 1**. Polyacrylamide resins are available from Bio-Rad Laboratories, Hercules, CA 94547 *(21)*. Sephadex gel resins are available from GE Health-care Bio-Sciences Corp. (formerly GE Amersham Pharmacia), Piscataway, NJ 08855 *(22)*.

2. Pipette-sized polypropylene GPC spin columns of different volumes are available empty or packed with gel material (micro: 50 µL gel, mini: 100 µL gel) either with a pinhole orifice or a slotted tip orifice that opens upon centrifugation (Glygen Corp, Columbia, MD 21045), or with a fritted orifice (Harvard Apparatus, Inc., Holliston, MA 01746) (*see* **Note 1**).

3. Reaction cups are equipped with a slotted tip that opens only under the *g*-forces used for gentle centrifugation of spin columns to deliver protein/drug reagents to the top of the spin column (Glygen Corp, Columbia, MD 21045, Part TT1EMTS, Empty Short Tip) (*see* **Note 2**).

4. Collection vials for spin column eluates. 100 µL polypropylene inserts with self-centering top for HPLC autosampler vials (Alltech Associates, Deerfield, IL 60015, Part 92102).

5. Eppendorf tube or equivalent used as a holder for the spin column collection vial [VWR, West Chester, PA 19380, Polypropylene Microcentrifuge Tube 1.7 mL Part 20170 or Harvard Apparatus, Inc., Holliston, MA 01745, Empty Sample Vial for Spin Column Eluate (2 mL)].

6. White Delrin adapter for mounting the spin column in the Eppendorf tube holder (Wyeth Bio-Engineering Department, Pearl River, NY 10965) (*see* **Fig. 2**) or adapted by tandem use of a Glygen Corp., Columbia, MD 21045, Part: Adapter for Centrifugation together with an autosampler Snap-On cap without the rubber seal (National Scientific Co., Rockwood, TN 37854, Part: C4011–51R Red Snap-It Seal).

7. Centrifuge (Eppendorf AG, Hamburg, Germany 22331, Model 5415C or equivalent). For this Eppendorf centrifuge, a relative centrifugal force (RCF) of $735 \times g$ corresponds to 3,000 rpm and $1,000 \times g$ corresponds to 3,500 rpm (*see* **Note 3**).

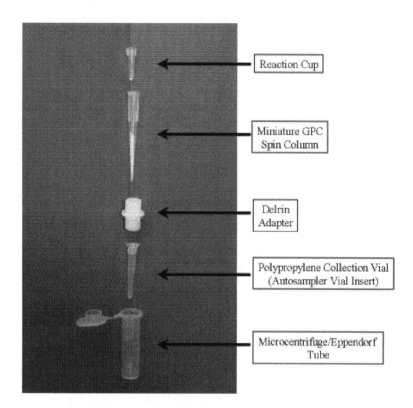

Fig. 2. GPC spin column hardware assembly: The reaction cup (loaded with sample) sits at the top of the freshly prepared GPC spin column pipette that fits into a Delrin adapter. The pipette tip loosely fits into an autosampler vial insert. The whole assembly is mounted into a 2.0-mL empty Microcentrifuge (Eppendorf) tube. Upon centrifugation of the assembly, the sample in the reaction cup passes through the GPC spin column. The spin column eluate is collected in the collection vial with no sample transfer losses.

2.2. Protein and Drug Samples

1. Protein solution in appropriate buffer 1–30 µM (lower concentration when seeking strong binders, higher concentrations for weaker binders).

2. Drug candidates 1–10 times protein concentration (lower concentration when seeking strong binders, higher concentrations for weaker binders).

3. Generally, DMSO concentrations, if necessary for solubilizing the drug candidates, should be less than 2% of the protein–drug mixture solution so as to maintain the protein in its native state.

2.3. HPLC Conditions

1. Atlantis T3 C18 HPLC column, 2.1 × 50 mm (w × l), 3 µm particle size (Waters, Milford, MA 01757, Part 186003717), [Aquasil C18 HPLC column, 1 × 50 mm (w × l), 3 µm particle size (Thermo Electron North America LLC, West Palm Beach, FL 33407, Part 77503–051030) or equivalent].

2. Agilent 1100 HPLC system (Agilent Technologies, Wilmington, DE 19808) or equivalent, Flow rate 300 μL/min, Gradient 98%/2% water/acetonitrile (0.1% formic acid) to 5%/95% water/acetonitrile (0.1% formic acid) in 12 min with 5 min re-equilibration at 98%/2% water/acetonitrile (0.1% formic acid). Peak widths at peak base of about 6 s are achieved with the Atlantis T3 C18 column.

3. Autosampler vial (Analytical Sales & Services, Pompton Plains, NJ 07444, Part 11211); Snap-On Cap with septum of Teflon-coated red rubber for the autosampler vial (National Scientific Co., Rockwood, TN 37854, Part: C4011–51R Red Snap-It Seal), 100 μL polypropylene inserts with self-centering top for HPLC autosampler vials (Alltech Associates, Deerfield, IL 60015, Part 92102).

2.4. ESI Mass Spectrometry

1. Waters Micromass LCT Premier XE Time-of-Flight Mass Spectrometer (Milford, MA 01757) or equivalent, equipped with a solvent divert valve. Very high sensitivity, better than 0.1 pmol for small molecules with resolution greater than 6,000, capable of identifying small molecules from their accurate masses.

3. Methods

The following steps should be taken to fully characterize the GPC spin column system for secondary screening for non-covalent protein–drug complexes: (1) the miniature GPC spin columns have to be properly prepared, (2) control samples have to be prepared and assayed to fully characterize the GPC spin column system, including (a) transmission efficiency of the protein through the spin column, (b) the transmission response of the drug candidate upon passing through the spin column without protein present (breakthrough), and (3) the MS response factors for the drug candidates have to be determined directly, without passing through the spin column, for quantitating the level of drug passing through the GPC spin column. Also, these GPC spin column non-covalent binding studies can be performed at subambient temperatures (e.g., 4°C) to extend the range of studies to even weaker binders by effectively lowering the off-rates of the dissociation reaction for the protein-ligand complex *(17)*.

3.1. Preparation of Miniature GPC Spin Columns

1. The volume of the protein-drug mixture to be analyzed should be 10–20% of the total volume of the gel in the column. If a sample volume of 10–20 μL is to be analyzed, the gel column volume should be about 100 μL when swollen.

If a sample volume of 5–10 μL is to be analyzed, the gel column volume should be about 50 μL when swollen. The larger the sample volume for a given gel column, the greater the transmission of the protein, but there is also a greater risk for the unbound free drug to pass unretained through the column.

2. Add dry powder of gel GPC beads to the column and saturate the beads with protein buffer solvent, preferably overnight. The gel will swell and contaminants will dissolve. If necessary, use a shaker to produce a uniform column of gel without air bubbles or channels and flick the column to remove any remaining air bubbles.

3. Mount the column with an adapter into an Eppendorf tube and centrifuge the column at 735–1,000 × *g* for 3–4 min to remove excess buffer and contaminants. Repeat this washing procedure and thereby produce a uniform miniature GPC column free of gel contaminants. The appearance of the gel should be uniformly white-powderly looking without traces of wetness. If the column appears wet, generally at the column tip, repeat the centrifugation process until all traces of excess water are removed.

3.2. GPC Spin Column Screening of Protein–Drug Non-covalent Complex Sample (With Suggested Sample Volumes and Concentrations)

1. Mix the drug and the protein and permit them to react from 30 min to 2 h to produce the expected protein–drug non-covalent complex. Suggested conditions for a weak binding complex using a 50-μL gel column: Mix 8 μL of 20 μM protein with 2 μL of 1,000 μM drug for 30 min in a reaction cup with a slotted tip. As dictated by the Law of Mass Action, the highest possible protein and drug concentrations should be chosen so as to produce the highest possible concentration of the weak protein-drug complex.

2. Assemble the following hardware (as illustrated in **Fig. 2**): Insert the reaction cup with sample into the top of the freshly prepared GPC spin column (50 μL gel P6 or G25). Place the column pipette tip through the homemade Delrin adapter and place the pipette tip into a 100-μL autosampler vial insert. Mount the whole assembly into a 2.0- mL empty Eppendorf tube.

3. Centrifuge the whole assembly at 735–1,000 × *g* (3,000–3,500 rpm for the Eppendorf centrifuge Model 5415C) for 3–4 min (*see* **Note 4**).

4. After centrifugation, the eluate, consisting of the complexed drug and free protein, is collected in the autosampler vial insert. The volume of eluate is the same volume as the original sample (10 μL) (*see* **Note 5**).

5. The eluate in the autosample vial insert is diluted with an equal or greater volume of 1:1 water/acetonitrile in 0.1%

formic acid to denature the sample and increase the volume to be analyzed for accurate injection into the HPLC for analysis by ESI-MS of the free drug.

3.3. Control Samples are Prepared to Fully Characterize the GPC Spin Column System

1. *Transmission Study of the Protein Through a Spin Column*: The highest possible transmission of protein through the spin column is key for obtaining the most reliable results by GPC spin column HPLC–ESI-MS drug screening (*see* **Note 6**). A protein control sample is used to determine the transmission efficiency of the protein through the spin column.

    (a) For a fixed amount of protein, the intensities of the protein peaks are computed from the HPLC–ESI mass spectra obtained after and before passing the protein through the GPC spin column. The average ratio of these intensities corresponds to the transmission efficiency of the protein through the spin column.

    (b) The quantity of protein passing through the GPC spin column should be identical to the quantity used to study a compound library. For the example described in **Subheading 3.2, step 1**, the quantity of protein used should be 8 μL of 20 μM protein with 2 μL buffer.

    (c) Protein transmission measurements should be made using a variety of gels (*see* **Table 1**) for determining the highest transmission gel. Often with hydrophobic proteins, the transmission is only ~30%. Generally, the sephadex gels (e.g., G 25) are preferred over polyacrylamide gels (e.g., P6) since they are less hydrophobic.

    (d) It is also advisable to study the response of a fixed amount of a given protein (e.g., 15 pmol) at different total volumes to determine the optimal detection volume for the protein, but not exceeding a volume corresponding to about 25% of the total gel volume (*see* **Note 7**).

    (e) If protein transmission values are still lower than desired, it would be advisable to prewash the spin column with the protein of interest so as to block the active sites on the gel responsible for the non-covalent binding of the protein to active sites on the gel.

2. *Transmission Study of Drug Candidates Through a Spin Column*: Ideally, the drug candidates should not pass through the spin column but should be trapped in the pores of the hydrated gel column, since the volume of the sample is less than the occluded volume of the buffer permeating the gel. This should be verified. If a considerable amount of low MW drug is transmitted through the spin column, then gels with a higher fractionation range cutoff should be used.

(a) The drug control samples used should have the same volume and concentration of drug (in the protein buffer) as prepared in the GPC spin column drug-protein screening experiment. For the example described in **Subheading 3.2**, **step 1**, the quantity of drug used should be 2 μL of 1,000 μM drug with 8 μL buffer, mixed in the reaction cup.

(b) The integrated area observed for this drug control sample should be subtracted from the observed area for the GPC spin column drug–protein screening sample.

3. *MS Response of Pure Drug Candidates*: The mass spectral response factors for the pure drug candidates should be determined by HPLC–ESI-MS. The concentration studied should be moderately above the limit of detection. Generally, single-point response factors should be determined for each of the samples, using from between 10 fmol and 1 pmol for each drug candidate, depending upon the sensitivity of the sample and MS instrument used. In this drug response study, GPC spin columns are not utilized.

3.4. HPLC–ESI-MS Data Analysis

1. The HPLC and ESI-MS instruments are set up for optimum sensitivity and resolution as recommended by the instrument manufacturers.

2. The solvent divert valve can be set to divert the low retention time (RT) carrier solvent containing salts and buffers from the samples, thereby minimizing contamination of the MS instrument that often suppresses sample ionization.

3. The suggested order of analysis of the samples is as follows: Drug alone through the GPC spin column, protein–drug mixture through the spin column, and drug control not through the spin column. This order of analysis goes from predicted lowest to highest levels of drug, minimizing possibilities of cross-contamination in the assays.

4. Generation of Ion Chromatograms and Determination of Relative Binding Affinities of Drugs to Protein:

(a) For each sample generate a Total Ion Chromatogram, and a Selected Ion Chromatogram, generally, over the nominal mass range of the parent ion of the drug candidate, viz., $[M + H]^{1+}$ and/or the $[M - H]^{1-}$. Combine the mass spectra over the retention time region of the drug candidate. Since the mass spectra were obtained under relatively higher resolution conditions (~6,000 resolution), generate a narrow mass range Selected Parent Ion Mass Chromatogram for the desired parent ion.

(b) Integrate the RT region of the narrow mass range Selected Parent Ion Chromatogram while properly setting the baseline for the integration. The integrated area of this narrow mass range Selected Parent Ion Chromatogram represents the abundance of the drug present in each experiment, viz., the quantity of drug passing through the GPC spin column complexed to the protein (Area I), the amount of drug alone passing through the GPC spin column (Area II), and the mass spectral quantitative response for a known quantity of the drug, viz., (Area III) per pmol of drug.

(c) The relative responses for each of the different drug candidates can be calculated as (Area I – Area II)/ Area III, and this quantitative number corresponds to the number of pmol of drug transmitted through the spin column for each drug candidate. Since the same amount of protein and drug was used for each drug candidate, a Table of (Area I – Area II)/Area III for each drug candidate effectively ranks the drugs by the strength of the non-covalent interaction between the drug candidates and the protein of interest, the higher the value the greater the degree of binding (*see* **Note 8**).

5. *HPLC–ESI-MS Data Analysis Example*: A model protein– drug non-covalent binding system studied by GPC spin column HPLC–ESI-MS is now illustrated for heat shock protein 90 catalytic domain (Hsp90cat, MW 25,925 Da) with geldan-amycin (MW 566), an anticancer inhibitor, having a dissociation constant in the low micromolar range *(23–25)*. These studies utilized 100 µL P2 gel columns with pinhole orifices, a Waters Atlantis T3 C18 reversed-phase column, an Agilent 1100 HPLC System, and a Waters-Micromass LCT Premier XE Time-of-Flight mass spectrometer.

(a) The transmission efficiency of the Hsp90cat protein was determined, using HPLC–ESI-MS, by measuring the signal intensities of Hsp90cat (RT: 7.2–7.3 min, base peak width: 6 s) with the same quantity of protein used in each screening study of weak non-covalent binders (249.6 pmol: 8 µL 31.2 µM Hsp90cat and 2 µL buffer). When using a P2 GPC spin column, the protein transmission efficiency was found to be 38.7% by computing the average ratio of the protein signals after and before the spin column treatment. A higher Hsp90cat protein transmission efficiency would have been desirable, but 38.7% is acceptable for secondary screening of a small library of compounds with moderately weak dissociation constants.

(b) **Figure 3** illustrates the (**a**) Narrow mass range Selected Parent Ion Mass Chromatograms (m/z 561.2–561.5), and (**b**) ESI mass spectra for the $[M + H]^{1+}$ ion for

242   Siegel

Selected Parent Ion Mass Chromatograms   **ESI Mass Spectra**

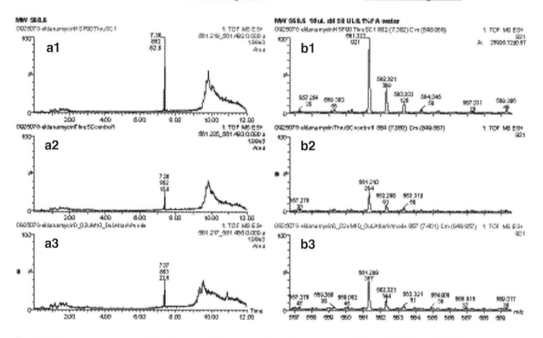

Fig. 3. (**a**) Narrow mass range Selected parent ion mass chromatograms (*m/z* 561.2–561.5), and (**b**) ESI mass spectra for the [M + H]¹⁺ ion for geldanamycin *m/z* 561.3 (predicted *m/z* 561.281) for the non-covalent binding studies of Hsp90cat and geldanamycin. (**a1, b1**) Detected HPLC-ESI-MS signal for the GPC spin column eluate for the reaction of geldanamycin (2 μL 1,000 μM) with Hsp90cat (8 μL 31.2 μM), 5× diluted. (**a2, b2**) Detected HPLC–ESI-MS signal for the GPC spin column eluate of geldanamycin alone (2 μL 1,000 μM geldanamycin mixed with 8 μL buffer), 5× diluted, and (**a3, b3**) Detected HPLC–ESI-MS signal for pure geldanamycin (0.2 pmol: 10 μL 0.02 μM). The selected parent ion mass chromatogram peaks (in **a1–a3**) are labeled (*top to bottom*) with retention time (7.36 min), scan number (852), and integrated area [for (**a1**) Area I, (**a2**) Area II, (**a3**), Area III (see text)]. The peak base width is 6.0-s wide. The ESI mass spectral peaks (in **b1–b3**) are labeled with the observed mass (*m/z* 561.31) and peak abundance (combined signals) in ion counts. The resolution of the peaks at half-height is ~6,000 resolution.

geldanamycin *m/z* 561.3 (predicted *m/z* 561.281) for the non-covalent binding studies of Hsp90cat and geldanamycin. The three basic measurements for these studies are (**Fig. 3a1, b1**) Detected HPLC–ESI-MS signal for the GPC spin column eluate for the reaction of geldanamycin (2 μL 1,000 μM) with Hsp90cat (8 μL 31.2 μM), 5× diluted, Area I = 62.8 counts-second, (**Fig. 3a2, b2**) Detected HPLC-ESI-MS signal for the GPC spin column eluate of geldanamycin alone (2 μL 1,000 μM geldanamycin mixed with 8 μL buffer), 5× diluted, Area II = 15.8 counts-second, and (**Fig. 3a3, b3**) Detected HPLC–ESI-MS signal for pure geldanamycin (0.2 pmol: 10 μL 0.02 μM), Area III = 22.5/0.2 counts-second/pmol = 112.5 counts-second/pmol. Area II corresponds to the breakthrough of geldanamycin through the spin column and serves as a correction for Area I. The value for Area II ideally should be close to zero but is observed here since a P2 gel has one of the lowest fractionation ranges.

(c) The ratio (Area I − Area II)/Area III when multiplied by a factor of 5, since the analyzed eluate was diluted by a factor of 5, corresponds to the number of pmol of geldanamycin that passes through the spin column non-covalently bound to Hsp90cat or [(62.8 − 15.8)/112.5] × 5 = 2.09 pmol of geldanamycin. The amount of Hsp90cat used in the experiment (8 μL 31.2 μM) corresponds to 249.6 pmol, but since only 38.7% of Hsp90cat was trans-mitted through the spin column, the amount of Hsp90cat passing through the spin column corresponds to 96.6 pmol. Therefore, under the experimental conditions used, the number of detected geldanamycin molecules non-covalently bound to Hsp90cat corresponds to about 1 geldanamycin molecule per 46 of Hsp90cat (see **Note 9**).

4. Notes

1. Larger empty spin column centrifuge tubes with polyethylene filters are available [Bio-Rad Laboratories, Hercules, CA 94547 (0.8 and 1.2 mL, Catalog #s: 732–6008 and 732–6204, respectively, and Pierce Scientific, Rockford, IL 61105 (0.8, 2, 5, and 10 mL, Catalog #s: 89868, 89896, 89897, and 89898, respectively)].

2. When preparing samples using reaction cups, all the drug-protein mixtures are transferred uniformly upon centrifugation to the GPC spin column and uncontrolled dissociation of the protein–drug complex is prevented when compared to manually loading the drug directly on to the top of the hydrated gel column.

3. The rpm for a centrifuge is calculated from RPM = 1,000 × [RCF/(1.12 × r)]$^{1/2}$ where r is the rotor radius measured from the spindle center to the bucket bottom.

4. This procedure causes the protein–drug complex and free protein to *rapidly* pass through the excluded volume of the spin column, with minimal loss of complexed drug due to its off-rate during this time period, while the free drug in the sample is trapped in the pores of the hydrated gel.

5. This procedure captures the eluate directly into the insert for the autosampler vial without losses due to sample transfers.

6. The higher the protein transmission, the greater the quantity of bound drug that is transmitted, thereby permitting the assay of weaker binders. If high protein transmission is not achievable, larger quantities of precious protein have to be used.

7. At about 25% of the total gel volume, the drug candidate of interest will pass through the pores of the column and instead of being excluded will now be detected.

8. The underlying assumption is that the off-rates for each of the complexes, as they rapidly pass through the void volume of the gel, are the same.

9. These experiments and calculations can be applied similarly to small libraries of compounds with a variety of proteins for secondary screening of non-covalently bound protein–drug complexes.

Acknowledgments

The author greatly appreciates the assistance provided by his colleagues Ashok K. Shukla, Chris Petucci, Jennifer Walter, Joseph Marini, Xidong Feng, Robert Powers, WeiDong Ding, Girija Krishnamurthy, Bart Zoltan, and Lynne Miller in improving the methodology described in this manuscript.

References

1. Hueser, J. (ed.) (2006) *High-Throughput Screening in Drug Discovery*, Wiley-VCH Verlag Gmbh & Co. KgaA, Weinheim, Germany

2. Seethala, R. and Fernandes, P. B. (eds.) (2001) *Handbook of Drug Screening*, Marcel Dekker, Inc., New York, NY

3. Powers, R. and Siegel, M. M. (2006) Applications of NMR and MS to anticancer drug discovery, (Chapter 5) in *Novel Anti-Cancer Agents: Strategies for Discovery and Clinical Testing*, (Adjei, A. A. and Buolamwini, J. K., eds.), Elsevier Science, New York, NY, pp. 107–190

4. Neumann, T., Junker, H-D., Schmidt, K., and Sekul, R. (2007) SPR-based fragment screening: advantages and applications. *Curr. Top. Med. Chem.* 7, 1630–1642

5. Beezer, A. E., Mitchell, J. C., Colegate, R. M., Scally, D. J., Twyman, L. J., and Willson, R. J. (1995) Microcalorimetry in the screening of discovery compounds and in the investigation of novel drug delivery systems. *Thermochim. Acta* **250**, 277–283

6. Cloutier, T. E. and Comess, K. M. (2007) Library screening using ultrafiltration and mass spectrometry, (Chapter 4) in *Mass Spectrometry in Medicinal Chemistry* (Wanner, K.T. and Höfner, G., eds.), Wiley-VCH Verlag Gmbh & Co. KgaA, Weinheim, Germany, pp. 157–183

7. Schriemer, D. C., Bundle, D. R., Li, L., and Hindsgaul, O. (1998) Micro-scale frontal affinity chromatography with mass spectrometric detection: a new method for the screening of compound libraries. *Angew. Chem. Int. Ed. Engl.* **37**, 3383–3387

8. Blundell, T. L. and Patel, S. (2004) High-throughput X-ray crystallography for drug discovery. *Curr. Opin. Pharmacol.* **4**, 490–496

9. Siegel, M. M. (2007) Drug screening using gel permeation chromatography spin columns coupled with electrospray ionization mass spectrometry, (Chapter 2) in *Mass Spectrometry in Medicinal Chemistry* (Wanner, K.T. and Höfner, G., eds.), Wiley-VCH Verlag Gmbh & Co. KgaA, Weinheim, Germany, pp. 65–120

10. Moy, F. J., Haraki, K., Mobilio, D., Walker, G., Powers, R., Tabei, K., Tong, H., and Siegel, M. M. (2001) MS/NMR: a structure-based approach for discovering protein ligands and for drug design by coupling size exclusion chromatography, mass spectrometry, and nuclear magnetic resonance spectroscopy. *Anal. Chem.* **73**, 571–581

11. Muckenschnabel, I., Falchetto, R., Mayr, L. M., and Filipuzzi, I. (2004) SpeedScreen: label-free liquid chromatography – mass spectrometry-based high-throughput screening for the discovery of orphan protein ligands. *Anal. Biochem.* **324**, 241–249

12. Schnier, P. D., Deblanc, R., Woo, G., Gigante, W., and Cheetham, J. (2003) Ultra-high throughput affinity mass spectrometry for screening protein receptors. *Proceedings of the 51ˢᵗ ASMS Conference on Mass Spectrometry and Allied Topics*, Montreal, Canada, June 8–12, 2003, Poster WPC 049

13. Mori, S. and Barth, H. G. (1999) *Size Exclusion Chromatography.* Springer, Berlin, Germany

14. Yau, W. W., Kirkland, J. J., and Blu, D. D. (1979) *Modern Size-Exclusion Liquid Chromatography: Practice of Gel Permeation and Gel Filtration Chromatography.* John Wiley & Sons, New York, NY

15. Ackers, G. K. (1973) Studies of protein ligand binding by gel permeation techniques, *Meth. Enzymol.* **27** (Enzyme Struct., Pt. D), 441–455

16. Andreu, J. M. (1985) Measurement of protein–ligand interactions by gel chromatography, *Meth. Enzymol.* **117** (Enzyme Struct., Pt. J), 346–354

17. Hulme, E. C. (1992) Gel-filtration assays for solubilized receptors, (Chapter 9) in *Receptor–Ligand Interactions; A Practical Approach* (Hulme, E. C., ed.), IRL Press at Oxford University Press, Oxford, England, pp. 255–263

18. Andersen, K. B. and Vaughan, M. H. (1982) Gel centrifugation chromatography for macromolecular separations. *J. Chromatogr.* **240**, 1–8

19. Nath, S., Brahma, A., and Bhattacharyya, D. (2003) Extended application of gel-permeation chromatography by spin column. *Anal. Biochem.* **320**, 199–206

20. GE Healthcare Product Booklet, illustra™ MicroSpin Columns, Part 27512001 PL Rev. D 2006, Buckinghamshire, UK

21. Bio-Rad Laboratories Literature, Bio-Gel P Polyacrylamide Gel Instruction Manual, LIT174 Rev B, Hercules, CA

22. Amersham Pharmacia Literature, Gel Filtration: Principles and Methods, A1 and 8th editions, Part 18–1022–18

23. Roe, S. M., Prodromou, C., O'Brien, R., Ladbury, J. E., Piper, P. W., and Pearl, L. H. (1999) Structural basis for inhibition of the Hsp90 molecular chaperone by the antitumor antibiotics radicicol and geldanamycin. J. Med. Chem. 42, 260–266

24. Panaretou, B., Prodromou, C., Roe, S. M., O'Brien, R., Ladbury, J. E., Piper, P. W., and Pearl, L. H. (1998) ATP binding and hydrolysis are essential to the function of the Hsp90 molecular chaperone in vivo. EMBO J. 17, 4829–4836

25. Chiosis, G., Rosen, N., and Sepp-Lorenzino, L. (2001) LY294002-geldanamycin heterodimers as selective inhibitors of the PI3K and PI3K-related family. Bioorg. Med. Chem. Lett. 11, 909–913

Chapter 16

A Scintillation Proximity Assay for Fatty Acid Amide Hydrolase Compatible with Inhibitor Screening

Yuren Wang and Philip Jones

Summary

Scintillation proximity assay (SPA) is a homogenous and versatile technology for the simple and sensitive detection of the interaction of protein targets with their ligands. Herein, we described a SPA assay developed to identify compounds that bind to human fatty acid amide hydrolase (FAAH). This SPA assay utilizes the specific binding of $[^3H]$-$R(+)$-methanandamide (^3H-MAEA), a competitive nonhydrolyzed FAAH inhibitor, to FAAH expressing microsomes and evaluates its displacement by FAAH inhibitors. In contrast to the classical SPA radioligand binding assay which detects bound ligand, in our assay the released radiolabel is detected through its interaction with the SPA beads. This novel SPA assay has been validated and demonstrated to be simple, sensitive, and amenable to high-throughput screening.

Key words: Scintillation proximity assay, Fatty acid amide hydrolase, Enzyme binding assay, Methanandamide, Enzyme inhibitors

1. Introduction

Fatty acid amide hydrolase (FAAH) is a member of the serine hydrolase superfamily whose substrates include several physiologically important endogenous fatty acid amides such as anandamide (arachidonyl ethanolamide, AEA), N-palmitoylethanolamine (PEA), and N-oleoylethanolamine (OEA) *(1–3)*. These endogenous fatty acid amides are involved in a variety of physiological activities such as pain sensation, mood modulation, sleep induction, and appetite suppression *(4–7)*. Several lines of evidence strongly suggest that FAAH plays a key role in regulating the physiological

Ana Cecília A. Roque (ed.), *Ligand-Macromolecular Interactions in Drug Discovery: Methods and Protocols*,
Methods in Molecular Biology, vol. 572,
DOI 10.1007/978-1-60761-244-5_16, © Humana Press, a part of Springer Science+Business Media, LLC 2010

tones and activities of these endogenous fatty acid amides in vivo. Thus FAAH has been regarded as an attractive therapeutic target for the treatment of pain, anxiety, inflammation, and other disorders (8–13). Tremendous efforts are underway to discover FAAH inhibitors with drug-like properties (14–19). However, current efforts to develop assays to identify FAAH inhibitors are predominantly focused on functional assays which monitor FAAH activity using either radiolabeled or fluorogenic substrates (20–26).

The Scintillation Proximity Assay (SPA) (27–29) is a homogenous and versatile technology for the simple and sensitive detection of the interactions between proteins and small-molecule ligands, protein–protein interaction, and protein–DNA interactions. SPA is readily amenable to high-throughput screening (HTS), including high-capacity robotic automation. One of the most common applications of SPA technology is the protein-ligand binding assay. The principle of classical SPAs involves radiolabeled ligand binding to the target protein immobilized on the surface of a SPA bead. The bound radioligand is held in close proximity to the bead to stimulate the scintillant within the bead to emit light. Unbound radioligand, including that displaced from its binding site by an inhibitor, is too distant from the bead to transfer energy and therefore goes undetected.

We described a method and experimental procedures for a modified application of the SPA technology, a homogenous and versatile assay for identification of FAAH inhibitors (30). A schematic representation of a SPA for FAAH is illustrated in **Fig. 1**. Three key components are employed in the assay, the FAAH microsomes, the wheat germ agglutinin (WGA) SPA beads, and [³H]-MAEA, which possesses a strong and specific binding affinity for FAAH. [³H]-MAEA is also able to bind to the SPA beads, albeit a relatively weak, nonspecific interaction. When the three components are incubated, [³H]-MAEA binds predominantly to the FAAH microsomes. Since only a small portion of radiolabel remains free and available to bind to the beads, low SPA signals are observed. However, in the presence of a FAAH inhibitor which competes with [³H]-MAEA for the FAAH binding site, the radiolabel is displaced and released into the assay solution where it interacts with the SPA bead, thus producing higher SPA signals. Therefore, unlike traditional binding assays that detect the radiolabel remaining bound to its target, this assay detects the increase in the free radiolabel concentration that occurs in the presence of an inhibitor, such as URB597 (see **Fig. 2**). The assay is specific for FAAH as microsomes prepared from cells expressing the vector alone had no significant ability to bind [³H]-MAEA (see **Fig. 3**).

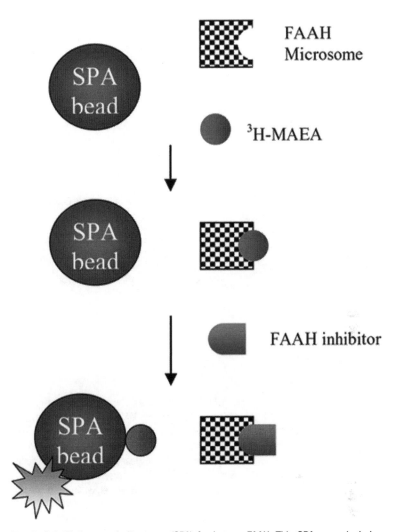

Fig. 1. Scintillation proximity assay (SPA) for human FAAH. This SPA assay includes three key components: FAAH microsomes, FAAH ligand [³H]-MAEA, and WGA SPA beads. When the three are incubated together, [³H]-MAEA binds predominately to the FAAH microsomes. In the presence of a FAAH inhibitor, the [³H]-MAEA is displaced from the enzyme's active site. The released [³H]-MAEA will interact nonspecifically with WGA SPA beads to stimulate a SPA signal (Reproduced from **ref. *30*** with permission from Elsevier.).

It is important to note that this unique SPA assay is different from classical receptor-ligand SPA assays in several aspects. First, FAAH microsomes do not bind to the SPA beads, whereas in the classical SPA, the target protein is tethered to the SPA beads by physical affinity or chemical linkages; for example, streptavidin-coated beads capture biotinylated proteins, antibody-coated beads capture the cognate antigen, and WGA-coated beads immobilize

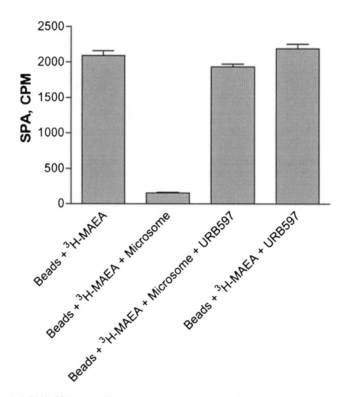

Fig. 2. A FAAH SPA assay. The reactions were carried out by incubating 0.5 mg WGA beads with 100 nM [³H]-MAEA (*1st column*), or 0.5 mg WGA beads with 100 nM [³H]-MAEA and 5 µg FAAH microsomes (*2nd column*), with or without 1 µM URB-597 (*3rd* and *4th columns*). Results are the means ±SD (*n* = 3). The data show that the incubation of the WGA SPA beads with [³H]-MAEA produced a robust SPA signal (~2,100 counts per minute), which represents the binding of [³H]-MAEA to the SPA beads in the absence of microsomes. However, when 5 µg of FAAH microsomes were included, the SPA counts decreased, indicating that ~92% of total [³H]-MAEA was now bound to FAAH microsomes and nonspecific binding to WGA beads was ~8%. URB-597, a selective FAAH inhibitor, blocked the binding of [³H]-MAEA to the FAAH microsomes, as shown by the increase in SPA signal. In contrast, the same concentration of URB-597 had no effect on the interaction of [³H]-MAEA and the WGA beads (Reproduced from **ref. 30** with permission from Elsevier.).

glycosylated proteins such as G protein-coupled receptors. Secondly, the FAAH SPA exploits the nonspecific interaction that [³H]-MAEA has for the SPA bead (*see* **Fig. 4**). In contrast, the classical assays require that the radiolabels should have an extremely low binding affinity for the beads so as to limit the background signal and therefore, displacement of radiolabels by agonists or antagonists results in a decreased SPA signal. In contrast, in the assay reported here, FAAH inhibitors are expected to increase the SPA signals. This homogenous assay has been shown to be sensitive, simple, and compatible with HTS for the identification of FAAH inhibitors.

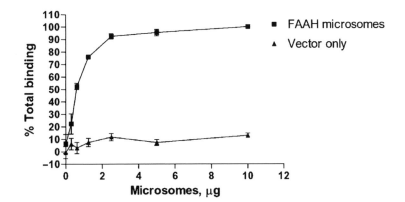

Fig. 3. The binding of [³H]-MAEA is specific for FAAH. The relative activity of FAAH microsomes or control "Vector only" microsomes was measured by increasing the microsomal protein concentration ranging from 0 to 10 μg per assay. 100 nM of [³H]-MAEA and 0.5 mg of WGA SPA beads were incubated with indicated concentrations of microsomes at 25°C for 3 h. The radioactivity was measured in a TopCount. The percentage of total binding was calculated and was plotted against amount of microsomal protein in the assay. Data shown represents the mean ±SD of triplicate data from at least two independent experiments. The result showed FAAH microsomes, but not vector microsomes, captured the [³H]-MAEA. The binding reached saturation at 4–5 μg of FAAH microsomes per assay. By comparison, vector microsomes at concentrations up to 20 μg per well did not lead to any significant binding. These results indicate that the binding is specific for the FAAH microsomes (Reproduced from **ref. 30** with permission from Elsevier.).

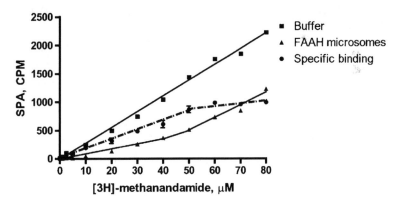

Fig. 4. Titration of the binding of [³H]-MAEA to FAAH microsomes and SPA beads. Assays performed by incubating increasing amounts of [³H]-MAEA with 0.5 mg of PVT-WGA beads in the presence or absence of 2 μg of FAAH microsomes. The result showed that the increase in SPA counts was linear with the concentration of [³H]-MAEA and no obvious saturation was observed up to 80 μM. This increase represented the total binding of [³H]-MAEA to the WGA beads. Inclusion of 2 μg of FAAH microsomes in the assay reduced the SPA counts accordingly. The specific binding of [³H]-MAEA to the FAAH microsomes was calculated as the difference between the SPA counts with and without the FAAH microsomes. The specific binding reached saturation levels at the concentration of 50 μM of [³H]-MAEA. An estimated K_D of 25 μM is comparable to the IC_{50} of 7.4 μM seen in the hydrolysis assay (Reproduced from **ref. 30** with permission from Elsevier.).

2. Materials

2.1. Molecular Cloning of Human FAAH cDNA and Generation of Mammalian Expression Constructs

1. Polymerase Chain Reaction (PCR) oligonucleotide primers are customarily synthesized by Invitrogen (Carlsbad, CA) and dissolve in ultrapure distilled water at final concentration of 10 µM. The sequence of the forward primer, which contains an *Xba*I restriction enzyme site and an optimized Kozak sequence (CCACC) immediately upstream of the initiating ATG, is: 5′ATAATCTAGAATTCGCCACCATGGTGCAG-TACGAGCTGTGGGCCGCGCTGCCT3′.

   The reverse primer, which contains an *Eco*RI restriction enzyme site, is: 5′ATATACATATAAGAATTCTCAATGA TGATGATGATGGGATGACTGCTTTTCAGGGGT CATCAGTCGCTCCACCTC3′.

2. PCR SuperMix High Fidelity kit is obtained from Invitrogen (Carlsbad, CA).

3. Restriction endonucleases such as *Xba*I and *Eco*RI and other molecular reagents can be obtained from Invitrogen.

4. Thin-walled PCR tubes DNase and RNase free are purchased from Roche (Indianapolis, IN).

5. A Thermal Cycler e.g., MJR PTC-200 from Bio-Rad (Hercules, CA).

2.2. Cell Culture and Generation of FAAH Stable CHO Cell Line

1. Chinese hamster ovary (CHO)-K1 cells are obtained from ATCC (Manassas, VA).

2. MEM-α growth medium (MEM-α medium containing 10% dialyzed fetal bovine serum, 100 IU penicillin, 100 µg/mL streptomycin, 1 mg/mL G418, 2 mM L-glutamine, and 15 mM HEPES (pH 7.4). All cell culture reagents are purchased from Invitrogen.

3. Trypsin solution (0.25%)-Invitrogen.

4. Methotrexate is purchased from Sigma (St. Louis, MO). Dissolve Methotrexate in tissue-culture grade water at 10 mM, stored in aliquots at –20°C, and then add to cell culture medium before use.

5. Lipofectamine 2000 transfection reagents and restriction endonucleases such as *Cla*I are purchased from Invitrogen.

6. Sterile cloning discs (3 mm) were from RPI (Mount Prospect, IL).

7. Cell culture dishes and cell scrapers are obtained from Corning (Corning, NY).

2.3. Preparation of FAAH Microsomes

1. Cell lysis buffer: 50 mM HEPES, pH 7.4, including Complete EDTA-free Protease inhibitor cocktail (1 table/50 mL) (Roche, Indianapolis, IN).

2. Fatty acid-free bovine serum albumin (BSA) is purchased from Sigma-Aldrich (St. Louis, MO).

3. Ultrasonic cell disrupter (VirTis, Gardiner, NY).

4. L8-M Ultracentrifuge (Beckman, Fullerton, CA).

5. Protein assay kit (Bio-Rad, Hercules, CA).

6. Microsome buffer: 50 mM HEPES, pH 7.4, 1 mM EDTA.

2.4. Scintillation Proximity Assay

1. Assay buffer: 50 mM HEPES, pH 7.4, 1 mM EDTA, and 0.1% Fatty acid-free BSA (*see* **Note 1**).

2. [^{3}H]-R(+)-methanandamide (60 Ci/mmol) is purchased from American Radiolabeled Chemicals Inc. (St. Louis, MO). [^{3}H]-MAEA (60 Ci/mmol) is supplied as a 20-μM solution in ethanol and should be stored at 4°C. All radioactive materials should be securely stored in specially designated areas and suitable Personal Protective Equipment, such as laboratory coat, safety glasses, and gloves should be worn whenever radioactive materials are handled (*see* **Note 2**).

3. R(+)-methanandamide (MAEA) is purchased from Tocris (Ellisville, MO). Dissolve MAEA (3.61 mg) of MAEA in 1 mL 100% DMSO to make a 10-mM stock solution which is stored at –20°C until used.

4. Polyvinyl toluene (PVT)-WGA SPA beads are obtained from Amersham Biosciences (Piscataway, NJ). The beads are supplied in a lyophilized form and can be stored at 2–8°C for up to 6 months. Be careful not to freeze PVT-WGA beads, as this may alter their binding capacity. Refer to **Note 3** for the potential use of other types of SPA beads.

5. White OptiPlate™-96 polystyrene microplates and TopSeal-A film are obtained from PerkinElmer Life Science (Boston, MA).

6. The FAAH inhibitor URB-597, CAY-10400, CAY-10402, and CAY-10435 are purchased from Cayman Chemicals (Ann Arbor, MI). Arachidonoyl trifluoromethyl ketone (ATFK) and other chemicals are purchased from Sigma-Aldrich (St Louis, MO). Dissolve compounds in DMSO to make a stock solution of 10 mM.

7. TopCount NXT counter (PerkinElmer, Waltham, MA).

3. Methods

FAAH is an integral membrane protein predominantly localized to the intracellular microsomal fractions of cells *(26)*. Since it is preferable to study the enzyme activity in its natural membrane-bound form, we used microsomes expressing human FAAH to develop our FAAH SPA binding assay. In this section, we first describe procedures for the generation and preparation of FAAH microsomes, including cloning of the FAAH gene, establishing a cell line stably expressing the FAAH protein and purification of microsomes. **Subheading 3.4** describes the development of the SPA assay using FAAH microsomes.

3.1. Cloning of FAAH cDNA and Generation of Mammalian Expression Constructs

1. Amplify the Human FAAH cDNA from human brain cDNA library by PCR using the FAAH gene-specific oligonucleotide primers. The reaction is performed using the SuperMix High Fidelity kit from Invitrogen to obtain the FAAH cDNA. The PCR is conducted for 30 cycles of 94°C (30 s), 60°C (1 min), and 72°C (1 min), followed by an extension for 7 min at 72°C after the last cycle.

2. Purify the PCR products using a QIAquick Gel Extraction kit following the manufacturer's protocol (Qiagen, Valencia, CA).

3. Digest the purified PCR fragments and pHTOP vector individually by incubating 8 μL of PCR product or vector (2 μg) alone with two units of *Xba*I and *EcoR*1 enzymes at 37°C for 2 h. The digests are purified using QIAquick PCR Purification Kit from Qiagen. The PCR DNA fragment is then ligated into the pHTOP vector using a Fast-Link DNA ligation kit from Epicentre (Madison, WI).

4. Confirm the DNA constructs by DNA sequencing.

3.2. Developing Stable CHO Cell Lines

1. Linearize the pHTOP-FAAH plasmids with the endonuclease *Cla*1. 100 μg of plasmid is incubated with ten units of *Cla*1 in React buffer I for 2 h at 37°C. DNA samples are purified with the GIAquick PCR purification kit following the manufacturer's protocol (QIAGEN).

2. Culture CHO-K1 cells to 70–80% confluency in MEM-α growth medium (*see* **Subheading 2.2**, **items 2**). Transfect the cells with linearized DNA using Lipofectamine 2000 reagent as per manufacturer's protocol (Invitrogen).

3. 48-h post transfection split the transfected cells 50-fold and culture at various concentrations of methotrexate (20, 50, and 100 nM), then select for 2–3 weeks allowing colonies to form.

4. Isolate colonies from each concentration of methotrexate with sterile cloning discs (3 mm) dipped in trypsin solution and propagate for screening in MEM-α growth medium, with the selected concentration of methotrexate.

5. Lyze equivalent amount of cells and analyze the cell lysates by Western blot with anti-His antibody (Penta-His, Qiagen). Clones with highest FAAH expression are chosen for large-scale culture.

3.3. Preparation of Microsomal Fractions

1. Grow the CHO cells expressing FAAH or vector alone in cell culture plate (245 mm × 245 mm × 25 mm) in MEM-α growth medium (100 mL per plate) and the chosen concentration of methotrexate (we use 100 nM for cells expressing FAAH and 5 nM for cells containing vector alone).

2. Harvest cells at 90–100% confluence (usually 2–3 days). Remove growth medium and transfer the plates to ice. Wash cells with 5–10 mL PBS three times. Add 10 mL of cell lysis buffer (*see* **Subheading 2.3**, **item 1**) to each plate. Scrap cells using cell scraper and collect cells.

3. Sonicate the cells five times for 10 s each at scale 5 with 15-s intervals on ice. Centrifuge the cell lysate at 12,000 × g for 20 min at 4°C and remove the cell debris, nucleus, and crude membrane fractions. Centrifuge the supernatant again at 100,000 × g for 30 min in a SW28 rotor in a Beckman Optima LE-80K ultracentrifuge. Resuspend the pellet, which is the microsome fraction, in ice-cold microsome buffer (*see* **Subheading 2.3**, **item 6**) (nb. without protease inhibitors).

4. Resuspend the microsomes by squeezing through a 23.5-gauge needle five times. Measure the protein concentration by Protein Assay Kit from Bio-RAD. Dilute the microsome samples to 1 mg/mL protein concentration with microsome buffer, aliquot, and store at –80°C until use.

3.4. FAAH SPA Assay

3.4.1. Preparation of Working Solutions

1. SPA bead working solution: Reconstitute the lyophilized WGA SPA beads in SPA assay buffer at a concentration of 10 mg/mL. For an assay in one 96-well plate, reconstitute 50-mg beads with 5 mL of assay buffer in a 15-mL polypropylene conical tube. The beads should be mixed to ensure a homogenous suspension while pipetting. This may be done by gentle stirring with a magnetic stirrer. Once reconstituted, the beads can be stored at 2–8°C and should be used within 1 week.

2. [³H]-MAEA working solution: for one 96-well plate assay, add 1 μL of the 20 μM [³H]-MAEA stock to 20 μL of the 100 μM "cold" MAEA stock in 2 mL SPA assay buffer, to give final concentration of MAEA of 1 μM.

3. Microsome solution: Thaw a vial of frozen microsome stock (1 mg/mL) in a 37°C water bath. Make a working solution by reconstituting the microsomes in SPA assay buffer to a final concentration of 25 μg/mL. For one 96-well plate assay, add 250 μL of the 1 mg/mL microsome stock to 13-mL assay buffer.

3.4.2. FAAH SPA Assay Protocol

1. The final SPA assay consists of 100 nM [³H]-MAEA (60 Ci/mmol), 2.5 μg FAAH microsomes, 0.5 mg WGA beads, and 2.5% DMSO (*see* **Note 4**). The final volume of solution in one well of 96-well plates is 200 μL.

2. Add 5 μL of DMSO into four control nonspecific wells (NS).

3. Add 5 μL of 5 μM URB597 in DMSO into four control wells (CTRL).

4. Add 5 μL of test compound(s) in DMSO into sample wells (SPL).

5. Add 120 μL of the FAAH microsome working solution into all wells.

6. Pipette 20 μL of the [³H]-MAEA work solution into all wells.

7. Pipette 50 μL of well-suspended SPA beads to all wells.

8. Seal the microplate using TopSeal-A film from PerkinElmer (Boston, MA).

9. Agitate assay plate on a plate shaker for 2–3 h at room temperature (*see* **Note 5**).

10. Count the plate on a TopCount NXT (PerkinElmer) for 1 min per well (*see* **Note 6**).

3.4.3. Data Analysis

1. Inhibition (%) = (SPL-NS)/(CTRL-NS) × 100.

2. Calculate the IC_{50} values using the GraphPad Prism 3.02 program based on a sigmoidal dose–response equation. An example of the result from a SPA assay is shown in **Fig. 5** (*see* **Note 7**).

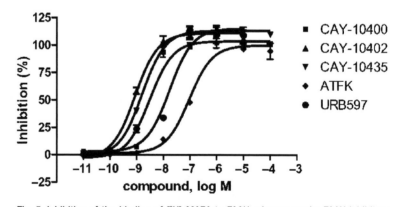

Fig. 5. Inhibition of the binding of [³H]-MAEA to FAAH microsomes by FAAH inhibitors in SPA. The assay was carried out by incubating 5 μg of FAAH microsomes with 100 nM [³H]-MAEA and 0.5 mg of WGA beads in 50 mM HEPES, 1 mM EDTA, pH 7.4, 0.1% BSA, with inhibitors in a 200 μL total reaction volume, at room temperature for 3 h. The final concentration of DMSO in the assay was 5%. IC_{50} values were calculated by fitting the curve to a nonlinear regression equation. Data shown represent the mean ±SD of triplicate data from at least two independent experiments. As shown in the figure, all five well-known FAAH selective inhibitors inhibited the binding of [³H]-MAEA to FAAH microsomes in a concentration-dependent manner, with IC_{50} values of 3.1, 0.9, 1.6, 21.5, and 87 nM, respectively. These values are comparable to IC_{50} values obtained in the anandamide hydrolysis assay (data not shown) (Reproduced from **ref. 30** with permission from Elsevier).

4. Notes

1. A final BSA concentration of 0.1% in the assay buffer is found to increase the reproducibility of the assay. MAEA is a highly lipophilic molecule and the presence of BSA might increase its solubility, prevent nonspecific binding to the walls of the plates, and may also improve the stability of the microsomes. Pipette the assay buffer gently to avoid creating unwanted air bubbles caused by the BSA.

2. Radiation safety: SPA is a radioisotopic technique and requires the use of radioactive material. Therefore, please follow your institution's instructions relating to the safe handling and use of the radioisotopic materials and local legislation governing the handling, use, storage, disposal, and transportation of radioactive materials. Good laboratory practices and Personal Protective Equipment should always be used.

3. A number of SPA beads have been developed by Amersham Bioscience. These beads differ in their manufacturing materials and their surface molecules, which are specifically for the coupling of target proteins, membrane fractions, or cells. We compared the five types of SPA beads provided by Amersham Bioscience, PVT-WGA, PVT-PEI-WGA type A, PVT-PEI-WGA type B, Ysi-WGA, and Ysi-Poly-1-lysine beads in the FAAH SPA assay. The results suggest that PVT-WGA and PVT-PEI-WGA type B are the better beads for use in the FAAH SPA.

4. The effect of DMSO on the FAAH SPA was analyzed by incubating 0.5 μg of FAAH microsomes with 100 nM [^{3}H]-MAEA and 0.5 mg of PVT-WGA beads, with increasing DMSO concentrations (0–10%v/v). The results (not shown) indicate that DMSO concentrations up to 10% of the final assay volume had no significant effect. We normally use 2–5% DMSO in our assay since compounds are more soluble at this range of DMSO.

5. The SPA assay required at least 2 h to reach steady state equilibrium (data not shown) and we therefore counted the plate after 3-h incubation.

6. Instrument setting for counting microplate in TopCount - Scintillator: Class; Energy range: Low; Efficiency: High; Isotope: ^{3}H; Window region A: 0.00–50.00; Window region B: 0.00–256.00.

7. The uniformity and signal to noise ratio of the microsome-based FAAH SPA assay was determined over multiple experiments in a 96-well format. The Z' factor, which is a measure of the reproducibility of the assay and its compatibility for HTS (31), ranged from 0.73 to 0.8 suggesting that the assay is highly suitable for screening compounds.

References

1. Deutsch, D. G., Ueda, N., and Yamamoto, S. (2002) The fatty acid amide hydrolase (FAAH). *Prostaglandins Leukot. Essent. Fatty Acids* **66**, 201–210

2. McKinney, M. K. and Cravatt, B. F. (2005) Structure and function of fatty acid amide hydrolase. *Annu. Rev. Biochem.* **74**, 411–432

3. Bisogno, T., De Petrocellis, L., and Di Marzo, V. (2002) Fatty acid amide hydrolase, an enzyme with many bioactive substrates: possible therapeutic implications. *Curr. Pharm. Des.* **8**, 533–547

4. Devane, W., Hanus, L., Breuer, A., Pertwee, R. G., Stevenson, L. A., Griffin, G., Gibson, D., Mandelbaum, A., Etinger, A., and Mechoulam, R. (1992) Isolation and structure of a brain constitute that binds to the cannabinoid receptor. *Science* **258**, 1946–1949

5. Cravatt, B. F., Prospero-Garcia, O., Siuzdak, G., Gilula, N. B., Henriksen, S. J., Boger, D. L., and Lerner, R. A. (1995) Chemical characterization of a family of brain lipids that induce sleep. *Science* **268**, 1506–1509

6. Kuehl, F. A., Jacob, T. A., Ganley, O. H., Ormond, R. E., and Meisinger, M. A. P. (1957) The identification of *N*-(2-hydroxyethyl)-palmitamide as a naturally occurring anti-inflammatory agent. *J. Am. Chem. Soc.* **79**, 5577–5578

7. Rodriguez de Fonseca, F., Navarro, F. M., Gomez, R., Escuredo, L., Nava, F., Fu, J., Murillo-Rodriguez, E., Giuffrida, A., LoVerme, J., Gaetani, S., Kathuria, S., Gall, C., and Piomelli, D. (2001) An anorexic lipid mediator regulated by feeding. *Nature* **414**, 209–212

8. Clement, A. B., Hawkins, E. G., Lichtman, A. H., and Cravatt, B. F. (2003) Increased seizure susceptibility and proconvulsant activity of anandamide in mice lacking fatty acid amide hydrolase. *J. Neurosci.* **23**, 3916–3923

9. Cravatt, B. F., Saghatelian, A., Hawkins, E. G., Clement, A. B., Bracey, M. H., and Lichtman, A. H. (2004) Functional disassociation of the central and peripheral fatty acid amide signaling systems. *Proc. Natl. Acad. Sci. U.S.A.* **101**, 10821–10826

10. Cravatt, B. F., Demarest, K., Patricelli, M. P., Bracey, M. H., Giang, D. K., Martin, B. R., and Lichtman, A. H. (2001) Supersensitivity to anandamide and enhanced endogenous cannabinoid signaling in mice lacking fatty acid amide hydrolase. *Proc. Natl. Acad. Sci. U.S.A.* **98**, 9371–9376

11. Lichtman, A. H., Hawkins, E. G., Griffin, G., and Cravatt, B. F. (2002) Pharmacological activity of fatty acid amides is regulated, but not mediated, by fatty acid amide hydrolase in vivo. *J. Pharmacol. Exp. Ther.* **302**, 73–79

12. Huitron-Resendiz, S., Sanchez-Alavez, M., Wills, D. N., Cravatt, B.F., and Henriksen, S. J. (2004) Characterization of the sleep-wake patterns in mice lacking fatty acid amide hydrolase. *Sleep* **27**, 857–865

13. Kathuria, S., Gaetani, S., Fegley, D., Valino, F., Duranti, A., Tontini, A., Mor, M., Tarzia, G., Rana, G. L., Calignano, A., Giustino, A., Tattoli, M., Palmery, M., Cuomo, V., and Piomelli, D. (2003) Modulation of anxiety through blockade of anandamide hydrolysis. *Nat Med* **9**, 76–81

14. Bracey, M. H., Hanson, M. A., Masuda, K. R., Stevens, R. C., and Cravatt, B. F. (2002) Structural adaptations in a membrane enzyme that terminates endocannabinoid signaling. *Science* **298**, 1793–1796

15. Patricelli, M. P., Lovato, M. A., and Cravatt, B. F. (1999) Chemical and mutagenic investigations of fatty acid amide hydrolase: evidence for a family of serine hydrolases with distinct catalytic properties. *Biochemistry* **38**, 9804–9812

16. Ueda, N., Kurahashi, Y., Yamamoto, S., and Tokunaga, T. (1995) Partial purification and characterization of the porcine brain enzyme hydrolyzing and synthesizing anandamide. *J. Biol. Chem.* **270**, 23823–23827

17. Boger, D. L., Sato, H., Lerner, A. E., Austin, B. J., Patterson, J. E., Patricelli, M. P., and Cravatt, B. F. (1999) Trifluoromethyl ketone inhibitors of fatty acid amide hydrolase: a probe of structural and conformational features contributing to inhibition. *Bioorg. Med. Chem. Lett.* **9**, 265–270

18. Boger, D. L., Fecik, R.A., Patterson, J. E., Miyauchi, H., Patricelli, M. P., and Cravatt, B. F. (2000) Exceptionally potent inhibitors of fatty acid amide hydrolase: the enzyme responsible for degradation of endogenous oleamide and anandamide. *Proc. Natl. Acad. Sci. U.S.A.* **97**, 5044–5049

19. Boger, D. L., Miyauchi, H., Du, W., Hardouin, C., Fecik, R. A., Cheng, H., Hwang, I., Hedrick, M. P., Leung, D., Acevedo, O., Guimaraes, C. R. W., Jorgensen, W. L., and Cravatt, B. F. (2005) Discovery of a potent, selective, and efficacious class of reversible – ketoheterocycle inhibitors of fatty acid amide hydrolase effective as analgesics. *J. Med. Chem.* **48**, 1849–1856

20. Deutsch, D. G. and Chin, S. A. (1993) Enzymatic synthesis and degradation of anandamide, a cannabinoid receptor agonist. *Biochem. Pharmacol.* **46**, 791–796

21. Koutek, B., Prestwich, G., Howlett, A., Chin, S., Salehani, D., Akhavan, N., and Deutsch, D. (1994) Inhibitors of arachidonoyl ethanolamide hydrolysis. *J. Biol. Chem.* **269**, 22937–22940

22. Lang, W., Qin, C., Hill, W. A. G., Lin, S., Khanolkar, A. D., and Makriyannis, A. (1996) High-performance liquid chromatographic determination of anandamide amidase activity in rat brain microsomes. *Anal. Biochem.* **238**, 40–45

23. Omeir, R. L., Chin, S., Hong, Y., Ahern, D. G., and Deutsch, D. G. (1995) Arachidonoyl ethanolamide-[1,2–14C] as a substrate for anandamide amidase. *Life Sci.* **56**, 1999–2005

24. Maccarrone, M., Bari, M., and Agro, A. F. (1999) A sensitive and specific radiochromatographic assay of fatty acid amide hydrolase activity. *Anal. Biochem.* **267**, 314–318

25. Boldrup, L., Wilson, S. J., Barbier, A. J., and Fowler, C. J. (2004) A simple stopped assay for fatty acid amide hydrolase avoiding the use of a chloroform extraction phase. *J. Biochem. Biophys. Methods* **60**, 171–177

26. Ramarao, M. K., Murthy, E. A., Shen, M. W. H., Wang, Y., Bushell, K. N., Huang, N., Pan, N., Williams, C., and Clark, J. D. (2005) A fluorescence-based assay for fatty acid amide hydrolase compatible with high-throughput screening. *Anal. Biochem.* **343**, 143–151

27. Nelson, N. (1987) A novel method for the detection of receptors and membrane proteins by scintillation proximity radioassay. *Anal. Biochem.* **165**, 287–293

28. Cook, N. D., Jessop, R. A., Robinson, P. S., et al. (1991) Scintillation proximity enzyme assay. A rapid and novel assay technique applied to HIV proteinase. *Adv. Exp. Med. Biol.* **306**, 525–528

29. Frolik, C. A., Black, E. C., Chandraselhar, S., and Adrian, M. D. (1998) Development of a scintillation proximity assay for high-throughput measurement of intact parathyroid hormone. *Anal. Biochem.* **265**, 216–224

30. Wang, Y., Xu, J., Uveges, A., Ramarao, M. K., Rogers, K. E., and Jones, P. G. (2006) A novel scintillation proximity assay for fatty acid amide hydrolase compatible with inhibitor screening. *Anal. Biochem.* **354**, 35–42

31. Zhang, J., Chung, T. D., Oldenburg, K. R., (1999) A simple statistical parameter for use in evaluation and validation of high throughout screening assays. *J. Biomol. Screen.* **4**, 67–73

Chapter 17

A Natural Products Approach to Drug Discovery: Probing Modes of Action of Antitumor Agents by Genome-Scale cDNA Library Screening

Hendrik Luesch and Pedro Abreu

Summary

In the last few years, genomic tools have been incorporated in natural product approaches to drug discovery, including understanding mechanisms of action which cannot be elucidated from phenotypic screens such as cell viability assays. The characterization of perturbed biological pathways and target identification are important for the evaluation of the compounds' potential as drug leads and for subsequent medicinal chemistry efforts; however, general procedures to tackle this task are lacking. The combination of high-throughput screening and genomic-scale assays has proven to be a powerful tool to aid in the identification of mechanisms and potentially of protein targets, not only in yeast but also mammalian cells. Arrayed libraries of cDNAs can be transfected into cancer cell lines in a high-throughput fashion to generate variants of spatially separated cancer cells with increased gene dosages for one particular cDNA. Cells overexpressing gene products that are directly targeted by a small molecule or that lie in the perturbed pathway may be less susceptible to the effects of the compound. This fact provides the basis for drug susceptibility screens employing cDNA libraries. The general procedures to optimize and execute those screens and subsequently validate putative screening hits are discussed in detail.

Key words: Natural products, High-throughput screening, Genome-wide cDNA overexpression, Functional genomics, Gene dosage screen, Drug susceptibility

1. Introduction

In the course of the evolutionary and selection process, organisms have evolved the ability to biosynthesize secondary metabolites (natural products) because of the selectional advantages they obtain as a result of the interactions of the compounds with specific receptors in other organisms (1). The specificity to bind

Ana Cecília A. Roque (ed.), *Ligand-Macromolecular Interactions in Drug Discovery: Methods and Protocols*,
Methods in Molecular Biology, vol. 572,
DOI 10.1007/978-1-60761-244-5_17, © Humana Press, a part of Springer Science+Business Media, LLC 2010

with and modulate the function of biological macromolecules, including targets within the human cell, confers to natural products a dominant role in drug design approaches (2–4). Moreover, the discovery of new molecular targets associated with various disease types, as a result of the sequencing of the human genome, led to a reviving interest in "well-known" natural products, some of which showed promising selective activity against those targets (5, 6). From 01/1981 to 06/2006, natural products and natural product-derived compounds accounted for approximately 28% of the 1,184 new chemical entities reported for this time period, whereas another 42% included related categories as natural product mimics, synthetic compounds with pharmacophores of natural origin, and biomacromolecules. More than 60% of the anticancer and 70% of the antiinfective antibiotics currently in clinical use are natural products or natural product-based. Current marketed drugs obtained from plant, microbial, and marine sources, as well as drug candidates from natural product leads, cover all areas of human diseases, namely, infectious (antibacterial, antifungal, antiparasitic, antiviral), neurological (central nervous system, neurodegenerative, Parkinson's disease, Alzheimer's disease), cardiovascular and metabolic, oncologic, and immunological, inflammatory and related disease areas (7–14).

To continue to be competitive with other drug discovery methods, natural product research needs to continually improve the speed of the screening, dereplication of known compounds at an early stage, bioassay-guided isolation and purification of the targeted molecules, and structure elucidation of new chemical entities as potential candidates for drug development. Perhaps the strongest impetus for development of natural product approaches to drug discovery is the advancement in bioassay technology over the last several years, in particular, those using mechanism-based high-throughput screening (HTS), genomics, and newer molecular biology tools (9, 11–18). Historically, hit-to-lead process in most academic laboratories followed a methodology of bioguided isolation using thin-layer chromatography bioautographic methods for detection of antimicrobial and antioxidant compounds, acetylcholinesterase inhibitors, and detection of binding properties of secondary metabolites to biomacromolecules, as well as traditional cell-based in vitro assays before real molecular biological targets were identified (19, 20). Cell-based assays are extensively used in HTS, particularly for the screening of antibacterial, antifungal, anticancer, and antiviral natural products. An early indication of cytotoxicity, and an approximation of the degree of cell penetration of the compound are here obtained using spectrophotometric or turbidimetric methods for detection of activity (11, 21). Recent developments include the use of reporter genes for measuring the activation of genes upstream cellular functions like proliferation and differentiation. Biochemical assays have the advantage of providing target-specific information, by the use of human receptors or enzymes as assay reagents. One of the newer biochemical assays is a capillary electrophoresis (CE) technique that allows the identification of active compounds in natural product extracts, as well as their binding affinity to targeted proteins, even in the presence of interferences *(22)*. This technology was recently used to evaluate the inhibition effects of the alkaloid geldanamycin, and macrolide radicicol on the heat-shock protein 90 (HSP90), a chaperone that regulates the function and stability of many key signaling proteins in cancer cells *(22, 23)*. Besides CE, other powerful analytical tools have been coupled on-line with bioassay platforms for detection and identification of active compounds. These include HPLC-based activity profiling of extracts, HPLC-based on-flow bioassays using several detection systems, MS-based methods, and NMR-based approaches for detecting protein–ligand interactions. In addition, microfluidic systems have been applied to bioassays for the development of binding assays on microchip format, whereas enzyme-based biosensors have been designed for screening substances with particular chemical functionalities *(18, 24–26)*. The application of molecular imprinted polymers for screening natural products has also been reported. These materials possess high selectivity and affinity for the target molecule, and so they can be used as mimics of enzymes, receptors, and antibodies, for trapping pharmacophoric compounds in extracts *(18, 27)*.

To reduce time and costs associated with the discovery and development of a new drug, computational methods for the discovery of lead structures appear as a complementary tool of the classical pharmacognostic approaches. The identification of a promising candidate with a desirable biological activity is carried out by mining compound databases in silico, employing virtual screening filtering experiments, structure-based 3D and ligand-based pharmacophore models, docking studies, and neural networks *(17)*.

Molecular diversity and biological relevance of natural products led to their integration in combinatorial strategies to discover lead compounds from natural product and natural product-like libraries that match the elements of conservation and diversity simultaneously expressed by biological targets. Library design approaches include: the use of natural products as core scaffolds for derivatization in either solid- or solution-phase; the generation of hybrid libraries by diversity-oriented synthesis; the modification of natural scaffolds of microbial origin (diversity-modified natural scaffolds); the construction of libraries from chiral building blocks obtained from selective chemical fragmentation of natural products; the production of combinatorial libraries around privileged structures; target-oriented total synthesis of natural product analogues; and the use of protein structure similarity clusters (PSSCs) as guiding structures for compound library synthesis *(28, 29)*. In the context of combinatorial chemistry strategies, the concepts of chemical genetics, chemical genomics, and chemical proteomics have been introduced. In chemical genetics approaches, libraries of small molecules are screened to identify ligands that can modulate a specific function of a gene product in vivo, which need not necessarily be a target for therapeutic intervention. Once a suitable molecule has been identified, the gene product that the molecule is modulating must be identified (forward chemical genetics). Reverse chemical genetics involves the use of small molecules against a protein (gene product) of interest. Once a binding partner has been chosen it is studied to identify the phenotypic effect of adding the small molecule *(30)*. Chemical genomics incorporates the PSSCs concept and can be defined as the systematic search for a selective small molecule modulator for each function of all gene family products, which are predominantly classified on the basis of sequence similarities and function (e.g., kinases, phosphatases, and proteases) *(29, 30)*. Other tools for identifying a natural product's receptor and its corresponding gene are chemical proteomics and reverse chemical proteomics *(31)*. The former uses a tagged small molecule to isolate a single protein or family of proteins from an entire proteome, whereas in the latter, the starting point is the transcriptome of a phenotype of interest. The transcriptome is cloned into a bacteriophage vector such that the phage particles display the proteins encoded by the DNA on their surface. A tagged biologically active small molecule can then be used to isolate those phages displaying proteins capable of binding to the small molecule. Combinatorial biosynthesis is another approach for expanding molecular diversity of natural products, through modification or exchange of genes between organisms to create hybrid molecules. This approach is rendered feasible by the fact that all of the genes encoding the large number of enzymes required for the synthesis of a typical secondary metabolite are clustered in a tight locus. Recombinant DNA (rDNA) methods are used to introduce genes coding for natural products synthetases and synthases into producers of other natural products or into nonproducing strains to obtain modified or hybrid antibiotics *(32, 33)*.

Most recent natural product approaches to drug discovery include the use of advances of genomics. For example, the application of tools for functional analysis of genomic responses by simultaneous and quantitative analysis of messenger RNAs provided an effective mean to define the bioactivity of an extract of *Ginkgo biloba*, showing that the extract affects transcription of functionally diverse groups of genes in vitro and in vivo *(34)*. The molecular pathways underlying the diverse biological activity of phytochemicals are currently evaluated by expression-based bioassays, using cDNA microarrays and real-time reverse transcription polymerase chain reaction to quantify gene expression *(35)*. Another genomic screening methodology based on DNA microarray was established to identify natural products from medicinal plant extracts that can regulate cellular gene expression. The identification of active compounds is guided by real-time polymerase chain reaction and LC-MS *(36)*. An enormous impact on genomics-guided natural product discovery was caused by recent advances in microbial genomics, and in particular, genome sequencing of actinomycetes, fungi, and myxobacteria *(33, 37–39)*. It is now possible to estimate the biosynthetic potential for a given organism by mining the whole-genome sequence, because natural-product biosynthetic genes are present in clusters in microbial genomes. As an alternative approach to whole-genome sequence mining, genome scanning provides an efficient way to discover natural-product biosynthetic gene clusters without having the complete genome sequence. This approach takes advantage of the fact that the genes for natural-product biosynthesis form clusters in a microbial genome, the size of which range from 20 to 200 kilobases. By shotgun-sequencing a small number of random genome sequence-tags (GSTs) from a library of genomic DNA, it is expected that, when analyzed, any given gene cluster will be represented by multiple GSTs. GSTs derived from genes that are likely to be involved in the biosynthesis of natural products are identified and used as probes in order to localize entire biosynthetic gene clusters. Following a guided-culturing approach, novel natural products have also been discovered from rare microorganisms using specific enrichment techniques of media, primarily by varying the media. A valuable alternative to cultivating rare or slow-growing organisms is to extract community DNA and produce clone libraries in a cultivation-independent approach termed metagenomics *(37, 40)*. Furthermore, improvements in heterologous expression have not only helped to identify gene clusters but have also made it easier to manipulate these genes in order to generate new compounds. It enables the (over)production of structurally complex substances through transfer of the biosynthetic genes from the original producer to more amenable heterologous hosts, and provides the basis to generate novel analogs through biosynthetic engineering *(37, 41)*.

When combined with HTS, genomic-scale assays constitute a powerful tool to aid in the identification of mechanisms and potentially of protein targets. Drug susceptibility screens with a variety of genetically modified cells can aid in this endeavor. Changing the gene dosage of the target (protein) of a growth-inhibitory small molecule is known to affect the susceptibility of the cell to the compound; thus, gene products identified in this manner can at least hint at the drug's mechanism of action *(42)*. Gene dosages can be increased by overexpression of genomic DNA (yeast) or ectopic cDNA overexpression (mammalian cells) *(42)*. As of recently, this can be achieved on a genomic scale in yeast and cancer cells where individual cell populations overexpressing the same protein are spatially separated from other unique cell populations *(43, 44)*. Conversely, genome-wide haploid and diploid yeast deletion strains have been generated, allowing for the investigation of phenotypes or comparison of gene expression profiles upon gene dosage reduction *(45–48)*. Reduction of gene dosage in mammalian cells is achievable by exploiting the RNA interference mechanism *(49)*. While yeast has primarily been the model organism of choice in the past, drug susceptibility screens are increasingly executed in cancer cells where more complex pathways can be interrogated to draw more relevant conclusions with respect to the mechanism of action of antitumor agents.

This chapter focuses exclusively on overexpression methods in cancer cells, developed by Luesch et al. *(42–44)*. Briefly, libraries of cDNAs can be transiently transfected using a variety of lipid-based reagents. After treatment of the cells with an antiproliferative agent followed by an appropriate incubation time, cell viability can be assessed using one of many detection reagents available; however, not every reagent may be suitable (see below). The cDNAs that attenuate the antitumor effect of the compound will need to be reliably detected over assay noise and subsequently validated in secondary assays. Based on the identity of all putative resistance-conferring hits, a unifying picture may emerge that can lead to a testable hypothesis regarding the mechanism and/or target of the small molecule.

2. Materials

2.1. Cell Culture

1. Dulbecco's Modified Eagle Medium (DMEM) (1×) liquid, high glucose (Invitrogen, Carlsbad, CA).
2. DMEM supplemented with 10% fetal bovine serum (FBS, HyClone, Logan, UT) and 1% Antibiotic-Antimycotic (Invitrogen).
3. Phosphate buffered saline (PBS, Invitrogen).
4. Cancer cell line (e.g., U2OS, HeLa; ATCC, Manassas, VA).

2.2. Assay Optimization

1. Antiproliferative natural product or synthetic small molecule (e.g., identified in drug screens).
2. Clear-bottom white tissue culture (TC) treated 96-well plates (Corning Incorporated, Corning, NY).
3. Clear and solid-bottom white TC treated 384-well plates (Corning Incorporated).

4. FuGENE 6 (Roche, Indianapolis, IN) or other transfection reagent.

5. Reporter plasmids: CMV-*GFP* (green fluorescent protein), CMV-luciferase (Clontech, Palo Alto, CA).

6. Detection reagent:

   (a) Whole-well assays: CellTiter-Glo® Luminescent Cell Viability Assay (Promega, Madison, WI) or AlamarBlue™ (Invitrogen).

   (b) To detect viability using reporter plasmid (e.g., CMV-luciferase): Bright-Glo™ luciferase assay system (Promega).

7. Luminescence plate reader (SpectraMax M5 or CLIPR from Molecular Devices, Sunnyvale, CA).

8. Fluorescence microscope (Nikon TE2000-U, Nikon Instruments, Melville, NY).

2.3. High-Throughput cDNA Library Screen

1. Expression cDNAs can be obtained from Origene Technologies (Rockville, MD), Mammalian Gene Collection (MGC; e.g., through ATCC) or purchased arrayed and assay-ready (Open Biosystems, Huntsville, AL).

2. Dispense instrument (WellMate from Matrix Technologies Corp., Hudson, NH).

3. All items required for assay optimization **Subheading 2.2** except **items 2** and **8**.

2.4. Validation

1. Fluorescence microscope (Nikon TE2000-U, Nikon Instruments).

2. Fluorescent reporter (CMV-*GFP*, Clontech).

3. Clear-bottom black tissue culture (TC) treated 96- or 384-well plates (Corning Incorporated, Corning, NY).

4. Becton Dickinson LSRII Flow Cytometer (Becton Dickinson, Franklin Lakes, NJ) or similar instrument.

5. FlowJo Flow Cytometry Analysis Software (Tree Star, Inc., Ashland, OR).

6. [Optional:] Pathway analysis database (e.g., Ingenuity Systems, Redwood City, CA).

3. Methods

Various requirements should be fulfilled in order to execute a successful cDNA screening assay. Thorough optimization is a prerequisite for obtaining useful data from the genomic screen.

Automated or partially automated screens are likely to decrease variability. False positives from the high-throughput screen need to be filtered out in secondary assays and biologically meaningful hits rigorously validated.

3.1. Choice of Cell Line, Drug Concentration, and Detection Reagent

1. The cell line has to be susceptible to the compound; the more potent the compound is against the cell line, the less material is needed to execute the cDNA library screen. This is especially important for natural products, since oftentimes only small quantities can be isolated from the biological source and compounds are not readily synthesizable.

2. To reliably detect an increase in viability upon overexpression of certain cDNAs, one should take advantage of the entire dynamic range of the dose-response curve and choose an inhibitory concentration at the foot of the steepest part of the curve. Thus, a drug concentration near the IC_{90} is recommended (**Fig. 1**).

3. A suitable compound for this assay would ideally be cytotoxic rather than cytostatic. The greater differential between the number of cells in control wells versus number of cells in drug-treated wells leads to an increase in the dynamic range during detection. This increases the chances of reliably detecting outliers from the cDNA screen and thus reduces the potential for false positives (**Fig. 1a**).

4. The dynamic range is also expected to be greater when the doubling time of the cancer cell line used is shorter, since it is proportional to cell number differential (control vs.

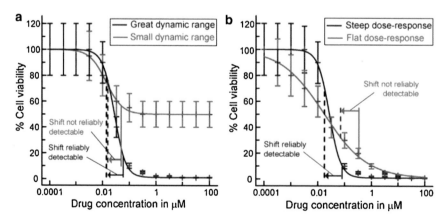

Fig. 1. Combination of small molecule, cell line, and detection method determine assay feasibility. Great dynamic range and a steep dose–response curve are prerequisites to reliably detect anticipated small changes in dose-response upon cDNA overexpression. (**a**) Fictitious dose–response curves with great dynamic range (*red*) vs. small dynamic range (*green*). At a screening concentration of the IC_{90}, an increase in viability due to shift of the dose–response curve (upon cDNA overexpression) by ca. threefold (log/2) cannot be reliably detected if the dynamic range is too small. Variability (see error bars) under high-throughput conditions may be too large. (**b**) Fictitious steep (*red*) vs. flat (*green*) dose–response curves. A shift of the curve by ca. threefold can only be reliably detected for the red curve (see error bars).

treated). Therefore, it is recommended to use a cell line with high doubling rate.

5. The dose–response curve should be steep since the expected IC_{50} shift upon overexpression of the target or target pathway member may be subtle, possibly only twofold to threefold (x-axis). If the dose–response curve is fairly flat, and considering that inherent assay variability under high-throughput conditions may correspond to a twofold difference in signal intensity, it is unlikely that an increase in viability (at the original IC_{90}) upon cDNA overexpression can be reliably detected (**Fig. 1b**).

6. The dynamic range for the detection window will also depend on the detection reagent. Under certain circumstances it may be better to measure cell viability based on ATP content (e.g., using CellTiter-Glo), while for other compounds reduction of MTT [3-(4,5-dimethylthiazol-2-yl)-2,5-diphenyltetrazolium bromide] (Promega) or other redox indicators such as AlamarBlue may be preferable to measure metabolically active cells (*see* **Note 1**). These whole-well assays measure viability across the entire well, i.e., transfected and nontransfected population. Since the nontransfected cells (without overexpressed cDNA) may increase the assay background, it may be preferable to assay only the transfected population. By exploiting the method of cotransfection *(50)*, viability can be determined by introducing a reporter plasmid (e.g., CMV-luciferase) into cDNA-transfected cells and then measuring luciferase activity as an indicator of cell viability. Choice of detection reagent will also largely influence the overall screening cost.

3.2. Assay Optimization

1. Antiproliferative agents (growth-inhibitory and/or cytotoxic) are commonly obtained from phenotypic screens of natural product or synthetic small molecule libraries for compounds that significantly decrease viability within 48 h.

2. Candidate cancer cell lines will be evaluated for transfection efficiency using CMV-*GFP* as a reporter plasmid.

3. Dose–response curves are obtained in various transfectable cancer cell lines by using several different detection reagents. For example, in 96-well plates, cells are seeded in 100 μL of propagation medium, placed in a cell culture incubator (37°C, 5% CO_2), and after 24 h (30–40% confluency), treated with 1 μL of a dilution series of the compound (100× stock) in DMSO or ethanol (final solvent concentration 1%, v/v) and returned to the incubator. Viability is assessed 48 h later using a commercially available whole-well assay detection kit (AlamarBlue or CellTiter-Glo) and plate reader in fluorescence or luminescence mode.

4. Next, the assay is miniaturized to accommodate 384-well format; this requires reduction in cell number and reagents. The protocol depends on whether a whole-well assay detection method (a) is used or cotransfection with a reporter for cell viability (b) is followed.

   (a) Cells are seeded in 40 μL of growth medium at the optimal density (*see* **step 3**) and, after 24 h of incubation (37°C, 5% CO_2), treated with the antiproliferative agent (final solvent concentration ≤1%, v/v; *see* **Note 2**). After 48 h of treatment time (37°C, 5% CO_2), cells are lysed and either luminescence is detected upon addition of CellTiter-Glo or fluorescence is recorded when using AlamarBlue.

   (b) If a reporter plasmid is used as an indicator of cell viability (e.g., CMV-luciferase), cancer cells are batch-transfected: a transfection mix (per well of a 384-well plate) consisting of 20 μL serum-free DMEM, 20 ng reporter plasmid, and transfection reagent in a ratio optimized for this cell line (e.g., 30 or 60 nL FuGENE 6; *see* **Note 3**) are incubated for 30–45 min at room temperature and then 20 μL cells in DMEM-10% FBS are added. After 24 h of incubation in a cell culture incubator (37°C, 5% CO_2), antiproliferative compound is added (final solvent concentration ≤1%, v/v; *see* **Note 2**). After an additional 48 h of incubation (37°C, 5% CO_2), cells are lysed and luminescence detected upon addition of Bright-Glo.

5. The condition(s) that emerge as most favorable (great dynamic range, steep dose-response curve, **Fig. 1**) will be used for a test screen consisting of approximately 1,000 cDNAs prespotted in 384-well plates ("1K set," *see* **Note 4**).

3.3. High-Throughput cDNA Library Screen

The general procedure is depicted in **Fig. 2a**.

1. An expression cDNA library can be obtained from commercial sources or needs to be prepared (*see* **Note 5**).

2. Screening plates are prepared by spotting normalized DNA (e.g., 62.5 or 40 ng per well) into 384-well plates (*see* **Note 6**). Each plate is sealed with adhesive aluminum foil. To minimize the number of freeze–thaw cycles, several screening sets are prepared simultaneously and stored at −80°C. Alternatively, assay-ready screening sets can be obtained commercially (Open Biosystems, *see* **Note 7**).

3. One screening set is thawed at room temperature and then plates are centrifuged ($1,000 \times g$) to spin down any DNA to the bottom of the plate. Plates are wiped off with ethanol and transferred into a sterile hood where the aluminum foil is removed. Control DNA, i.e., a negative control (empty vector) and a positive control (if available or known) should be spotted on each plate.

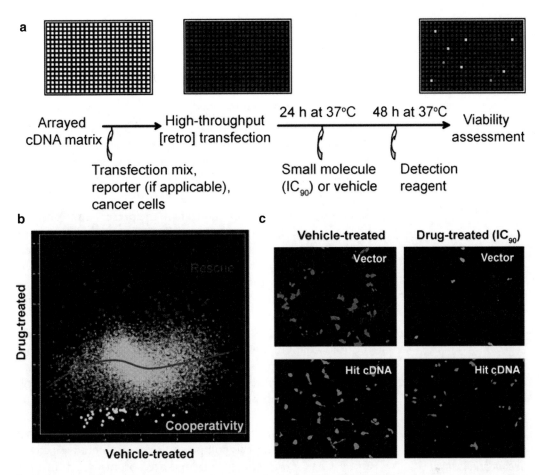

Fig. 2. Genome-wide drug resistance screen in cancer cells. (a) Overview of the high-throughput screening procedure. (b) Statistical analysis of vehicle-treated and drug-treated screens using LOESS as the 2D fitting model identifies outlier cDNAs that potentially attenuate or cooperate with drug activity. LOESS is a nonparametric, flexible tool for fitting almost any curve shape, and the *blue line* represents the fitted curve from such analysis. *Red dots* indicate cDNAs that appear to confer resistance, whereas *yellow dots* represent cDNAs that exhibit cooperative properties. (c) Cotransfected CMV-GFP as an indicator of cell viability. The putatively resistance-conferring cDNAs identified in the primary screen are drug-treated with the IC_{90} (or vehicle control) and viability assessed microscopically. *Green* cells correspond to live cells (Reproduced in part from **ref. 44** with permission).

4. Transfection mix is prepared based on the number of plates to be screened (*see* **Note 8**). Then, 20 μL of serum-free DMEM containing transfection reagent (and reporter plasmid, if applicable) are dispensed into the screening plates with prespotted cDNAs (e.g., using the WellMate).

5. After an appropriate incubation time (30–45 min for FuGENE 6) at room temperature, 20 μL of DMEM supplemented with 10% FBS and containing cancer cells at the proper density (e.g., 3,000 U2OS cells per well) are added using a reagent dispenser (same as above). Lids are placed on the plates which are then stored in a humidified tissue culture incubator at 37°C (5% CO_2) for 24 h (*see* **Note 9**).

6. The growth-inhibitory compound dissolved in DMSO or ethanol (200× stock) is further diluted (40 times) into DMEM (5× stock). Screening plates are removed from the incubator and 10 μL-aliquots of this stock solution are dispensed using the reagent dispenser above (*see* **Note 10**). Plates are returned to the incubator for another 48 h.

7. With the aid of a reagent dispenser, detection reagent is added and then viability determined using a multiwell plate reader in fluorescence mode (e.g., AlamarBlue for whole-well assay) or by luminescence (e.g., CellTiter-Glo for whole-well assay or Bright-Glo when using luciferase reporter).

8. Screens are carried out in duplicate in the presence and absence of compound. Outliers are determined after statistical analysis of choice. It has been found useful to determine signal averages and standard deviations for each plate. Fold activation (or repression) is calculated for each cDNA with respect to the corresponding plate average. To take into account compound-independent effects of the cDNAs in the simplest yet effective manner (*see* **Note 11**), the ratio can be calculated for the fold changes with and without drug treatment. Alternatively, a local regression (LOESS) can be used to normalize the data and identify cDNAs that attenuate the small molecule's growth-inhibitory activity (**Fig. 2b**) *(51)*.

3.4. Validation of Screening Hits in Secondary Assays

1. The (now presumably manageable number of) cDNAs that putatively attenuate the compound's antiproliferative activity and a CMV-*GFP* reporter plasmid are cotransfected into the cancer cell line, plated in 96-well plates, and 24 h later treated with compound as above for 48 h. Using an inverted fluorescence microscope, the number of GFP positive (and thus live) cells can be assessed. Due to the principle of cotransfection *(50)*, these cells presumably also contain the cDNA so that fluorescence is a direct measure of the rescue effect exerted by the cDNA of interest (**Fig. 2c**, *see* **Note 12**).

2. If the antiproliferative compound induces a cell cycle stage specific growth arrest (**Fig. 3a** vs. **Fig. 3d**), it can be determined whether the validated resistance-conferring cDNAs prevent or rescue from compound-induced growth arrest or simply prevent cell death (e.g., apoptosis) through other mechanisms. The cDNAs and *GFP* (or variants thereof, e.g., enhanced *GFP*, *EGFP*) are transfected into cancer cells and treated with compound or solvent control as above. After 48 h of treatment, cells are fixed, stained with propidium iodide (PI), and subjected to DNA content analysis by fluorescence-activated cell sorting (FACS) by standard procedures (**Fig. 3**) *(44)*. Certain cDNAs that specifically antagonize the small molecule's activity are expected to prevent compound-induced

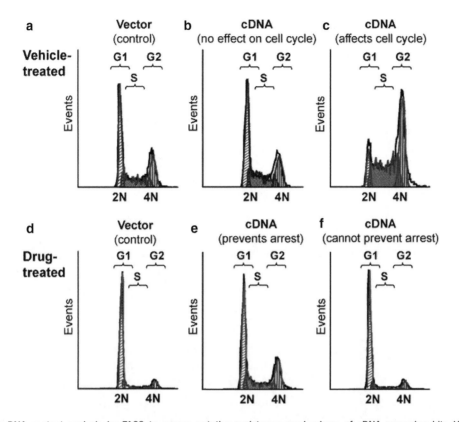

Fig. 3. DNA content analysis by FACS to assess putative resistance mechanisms of cDNA screening hits. Various scenarios are described. (**a–c**) Cell cycle profiles of vehicle-treated cells transfected with (**a**) control (vector), (**b**) a cDNA that does not affect the profile upon overexpression, and (**c**) a cDNA that changes the profile upon overexpression. (**d–f**) Cell cycle profiles of cells treated with a small molecule that causes G1-specific cell cycle arrest. Approximately 24 h before treatment, cells were transfected with (**d**) vector control, (**e**) a cDNA that prevents the drug-induced G1 arrest (populations in S and G2 phase are increased compared with (**d**) and are similar as in (**a**)), and (**f**) a cDNA that suppresses cytotoxicity yet does not change the population distribution of the cell cycle. For all experiments described in **a–f**, *GFP* or *EGFP* are cotransfected to mark the transfected populations. Populations in G1, S, or G2 phase are indicated.

cell cycle arrest and would restore the original cell cycle profile **(Fig. 3e)**, usually without affecting the cell cycle in the absence of compound **(Fig. 3b)**. Other cDNAs may be purely antiapoptotic and not affect the compound-induced cell cycle profile **(Fig. 3f)**. A third class of cDNAs will affect the cell cycle even in the absence of compound **(Fig. 3c)**; this is a reason to test the effects of each cDNA without compound as this could otherwise result in misinterpretation of data from FACS analysis.

3. Subsequent experiments are dependent upon the nature of the cDNAs. Ideally, a logical connection can be made, e.g., several gene products may lie in the same signaling pathway, or gene expression may be controlled by the same transcription factor, etc. Commercial or public databases may assist in this endeavor.

4. Notes

1. When using reagents that require absorbance measurements such as MTT, obviously clear bottom plates must be utilized.

2. In 384-well plates, this can be achieved by preparing a 5× concentrated stock in DMEM from a 200× of the small molecule in ethanol (or DMSO) and adding 10 μL to each assay well.

3. The optimal ratio of FuGENE 6 (μL) to DNA (μg) amount for U2OS cells was found to be 3:2 and for HeLa cells 3:1. The ratio should be optimized before starting the experiment by cotransfecting CMV-*GFP* to monitor transfection efficiency. Transfection efficiency varies depending on cell type and transfection reagent.

4. The amount of transfection reagent needs to be recalculated based on amount of cDNA per well (e.g., 62.5 or 40 ng) and reporter plasmid (e.g., 20 ng per well, if applicable).

5. The cDNAs are cloned into mammalian expression vectors, and after bacterial transformation, plasmid DNA is obtained, DNA concentration measured and normalized.

6. To simplify spotting, a stock concentration of 31.25 or 20 ng/μL is suitable; this would require 2 μL per well which can easily be achieved with robotics (when screening large scale/genomic libraries) or manually (when evaluating a small subset of cDNAs).

7. The cDNA amount can be varied. Assay-ready plates from Open Biosystems contain 35 ng of DNA per well.

8. It is recommended to calculate with one extra plate to account for dead volume in reagent dispenser, foam formation during mixing, etc.

9. Transfection by addition of cells to plates containing oligonucleotides/DNA and transfection reagent is commonly referred to as "reverse" or "retro" transfection necessitated by the arrayed screening format; the order is opposite to standard procedures.

10. The additional dilution step in DMEM allows dispensing with fairly high accuracy using the reagent dispenser from above without any additional instrument requirements.

11. For example, some cDNAs enhance or retard proliferation in the absence of compound; when using reporter construct as indicator of viability, some cDNAs may also nonspecifically activate the promoter.

12. This assay can be scaled up to a larger format. The cDNA and reporter concentrations require adjustment accordingly.

Acknowledgments

H. L. received a Junior Investigator Award (pilot grant) from the University of Florida Shands Cancer Center, American Cancer Society, Institutional Research Grant, ACS-IRG-01–188–01. Further funding to H. L. to study mechanisms of anticancer drug action is provided by the James & Esther King Biomedical Research Program, Grant No. 06-NIR07.

References

1. Stone, M. J. and Williams, D. H. (1992) The evolution of functional secondary metabolites (natural products). *Mol. Microbiol.* **6**, 29–34

2. Dixon, N., Wong, L. S., Geerlings, T. H., and Micklefield, J. (2007) Cellular targets of natural products. *Nat. Prod. Rep.* **24**, 1288–1310

3. Paterson, I. and Anderson, E. A. (2005) The renaissance of natural products as drug candidates. *Science* **310**, 451–453

4. Breinbauer, R., Vetter, I. R. and Waldmann, H. (2002) From protein domain to drug candidates. *Angew. Chem. Int. Ed.* **41**, 2878–2890

5. Kramer, R. and Cohen, D. (2004) Functional genomics to new drug targets. *Nat. Rev. Drug Discov.* **3**, 965–972

6. Balunas, M. J. and Kinghorn, A. D. (2005) Drug discovery from medicinal plants. *Life Sci.* **78**, 431–441

7. Newman, D. J. and Cragg, G. M. (2007) Natural products as sources of new drugs over the last 25 years. *J. Nat. Prod.* **70**, 461–477

8. Butler, M. S. (2005) Natural products to drugs: natural product derived compounds in clinical trials. *Nat. Prod. Rep.* **22**, 162–195

9. Shu, Y.-Z. (1998) Recent natural products based drug development: a pharmaceutical industry perspective. *J. Nat. Prod.* **61**, 1053–1071

10. Newman, D. J. and Cragg, G. M. (2004) Marine natural products and related compounds in clinical and advanced preclinical trials. *J. Nat. Prod.* **67**, 1216–1238

11. Baker, D. D., Chu, M., Oza, U., and Rajgarhia, V. (2007) The value of natural products to future pharmaceutical discovery. *Nat. Prod. Rep.* **24**, 1225–1244

12. Lee, K.-H. (2004) Current developments in the discovery and design of new drug candidates from plant natural product leads. *J. Nat. Prod.* **67**, 273–283

13. McChesney, J. D., Venkataraman, S. K., and Henri, J. T. (2007) Plant natural products: back to the future or into extinction? *Phytochemistry* **68**, 2015–2022

14. Lam, K. S. (2007) New aspects of natural products in drug discovery. *Trends Microbiol.* **15**, 279–289

15. Butler, M. S. (2004) The role of natural product chemistry in drug discovery. *J. Nat. Prod.* **67**, 2141–2153

16. Atta-ur-Rahman, Choudhary, M. I., and, Thomsen, W. J. (eds.) (2001) *Bioassay Techniques for Drug Development.* Harwood Academic Publishers, Amsterdam

17. Rollinger, J. M., Langer, T., and Stuppner, H. (2006) Strategies for efficient lead structure discovery from natural products. *Curr. Med. Chem.* **13**, 1491–1507

18. Potterat, O. and Hamburger, M. (2006) Natural products in drug discovery – concepts and approaches for tracking bioactivity. *Curr. Org. Chem.* **10**, 899–920

19. Choma, I. (2005) The use of thin-layer chromatography with direct bioautography for antimicrobial analysis. *LC-GC Eur.* **18**, 482–488

20. Bohlin, L. and Bruhn, J. G. (eds.) (1999) *Bioassay Methods in Natural Product Research and Drug Development.* Kluwer, Dordrecht, pp.119–142

21. Garipa, R. J. (2009) The emerging role of cell-based assays in drug discovery. In: Minor, L. K. (ed.) *Handbook of Assay Development in Drug Discovery.* CRC Press, Taylor & Francis Group, Boca Raton, FL, pp. 221–226

22. Gullo, V. P., McAlpine, J., Lam, K. S., Baker, D., and Petersen, F. (2006) Drug discovery from natural products. *J. Ind. Microbiol. Biotechnol.* **33**, 523–531

23. Jarvis, L. M. (2007) Living on the edge. Drugs targeting the protein Hsp90 push already unstable cancer cells to the brink. *Chem. Eng. News* **26**, 15–23

24. van Elswijk, D. A. and Irth, H. (2003) Analytical tools for the detection and characterization of biologically active compounds from nature. *Phytochem. Rev.* **1**, 427–439

25. Vuorela, P., Leinonen, M., Saikku, P., Tammela, P., Rauha, J.-P., Wennberg, T., and Vuorela, H. (2004) Natural products in the process of finding new drug candidates. *Curr. Med. Chem.* **11**, 1375–1389

26. Potterat, O. (2006) Targeted approaches in natural product lead discovery. *Chimia* **60**, 19–22

27. Xu, X., Zhu, L., and Chen, L. (2004) Separation and screening of compounds of biological origin using molecularly imprinted polymers. *J. Chromatogr. B* **804**, 61–69

28. Boldi, A. M. (ed.) (2006) *Combinatorial Synthesis of Natural Product-Based Libraries.* CRC Press, Taylor & Francis, Boca Raton, FL

29. Dekker, F. J., Koch, M. A., and Waldmann, H. (2005) Protein structure similarity clustering (PSSC) and natural product structure as inspiration sources for drug development and chemical genomics. *Curr. Opin. Chem. Biol.* **9**, 232–239

30. Spring, D. R. (2005) Chemical genetics to chemical genomics: small molecules offer big insights. *Chem. Soc. Rev.* **34**, 472–482

31. Piggott, A. M. and Karuso, P. (2004) Quality, not quantity: the role of natural products and chemical proteomics in modern drug discovery. *Comb. Chem. High Throughput Screen.* **7**, 607–630

32. Hutchinson, C. R. (2005) Manipulating microbial metabolites for drug discovery and production. In: Zhang, L. and Demain, A. L. (eds.), *Natural Products. Drug Discovery and Therapeutic Medicine.* Humana, Totowa, NJ, pp. 77–93

33. Farnet, C. M. and Zazopoulos, E. (2005) Improving drug discovery from microorganisms. In: Zhang, L. and Demain, A. L. (eds.) *Natural Products. Drug Discovery and Therapeutic Medicine.* Humana, Totowa, NJ, pp. 95–106

34. Gohil, K. (2002) Genomic responses to herbal extracts: lessons from in vitro and in vivo studies with an extract of *Ginkgo biloba. Biochem. Pharmacol.* **64**, 913–917

35. Coldre, C. D., Hashim, P., Ali, J. M., Oh, S.-K., Sinskey, A. J., and Rha, C. (2003) Gene expression changes in the human fibroblast induced by *Centella asiatica* triterpenoids. *Planta Med.* **69**, 725–732

36. Kawamura, A., Brekman, A., Grigoryev, Y., Hasson, T. H., Takaoka, A., Wolfe, S., and Soll, C. E. (2006) Rediscovery of natural products using genomic tools. *Bioorg. Med. Chem. Lett.* **16**, 2846–2849

37. Van Lanen, S. G. and Shen, B. (2006) Microbial genomics for the improvement of natural product discovery. *Curr. Opin. Microbiol.* **9**, 252–260

38. McAlpine, J. B., Bachmann, B. O., Piraee, M., Tremblay, S., Alarco, A.-M., Zazopoulos, E., and Farnet, C. M. (2005) Microbial genomics as a guide to drug discovery and structural Elucidation: ECO-02301, a novel antifungal agent, as an example. *J. Nat. Prod.* **68**, 493–496

39. Hornung, A., Bertazzo, M., Dziarnowski, A., Schneider, K., Welzel, K., Wohlert, S.-E., Holzenkämpfer, M., Nicholson, G. J., Bechthold, A., Süssmuth, R. D., Vente, A., and Pelzer, S. (2007) A genomic screening approach to the structure-guided identification of drug candidates from natural sources. *Chembiochem.* **8**, 757–766

40. Martinez, A., Hopke, J., MacNeil, I. A., and Osburne, M. S. (2005) Accessing the genomes of uncultivated microbes for novel natural products. In: Zhang, L. and Demain, A. L. (eds.) *Natural Products. Drug Discovery and Therapeutic Medicine.* Humana, Totowa, NJ, pp. 295–312

41. Wenzel, S. C. and Müller, R. (2005) Recent developments towards the heterologous expression of complex bacterial natural product biosynthetic pathways. *Curr. Opin. Biotechnol.* **16**, 594–606

42. Luesch, H. (2006) Towards high-throughput characterization of small molecule mechanisms of action. *Mol. BioSyst.* **2**, 609–620

43. Luesch, H., Wu, T. Y., Ren, P., Gray, N. S., Schultz, P. G., and Supek, F. (2005) A genome-wide overexpression screen in yeast for small-molecule target identification. *Chem. Biol.* **12**, 55–63

44. Luesch, H., Chanda, S. K., Raya, R. M., DeJesus, P. D., Orth, A. P., Walker, J. R., Izpisúa Belmonte, J. C., and Schultz, P. G. (2006) A functional genomics approach to the mode of action of apratoxin A. *Nat. Chem. Biol.* **2**, 158–167

45. Giaever, G., Shoemaker, D. D., Jones, T. W., Liang, H., Winzeler, E. A., Astromoff, A., and Davis, R. W. (1999) Genomic profiling of drug sensitivities via induced haploinsufficiency. *Nat. Genet.* **21**, 278–283

46. Lum, P. Y., Armour, C. D., Stepaniants, S. B., Cavet, G., Wolf, M. K., Butler, J. S., Hinshaw, J. C., Garnier, P., Prestwich, G. D., Leonardson, A., Garrett-Engele, P., Rush, C. M., Bard, M., Schimmack, G., Phillips, J. W., Roberts, C. J., and Shoemaker, D. D. (2004) Discovering modes of action for therapeutic compounds using a genome-wide screen of yeast heterozygotes. *Cell* **116**, 121–137

47. Giaever, G., Flaherty, P., Kumm, J., Proctor, M., Nislow, C., Jaramillo, D. F., Chu, A. M., Jordan, M. I., Arkin, A. P., and Davis, R. W. (2004) Chemogenomic profiling: identifying the functional interactions of small molecules

in yeast. *Proc. Natl. Acad. Sci. U.S.A.* **101**, 793–798

48. Hughes, T. R., Marton, M. J., Jones, A. R., Roberts, C. J., Stoughton, R., Armour, C. D., Bennett, H. A., Coffey, E., Dai, H., He, Y. D., Kidd, M. J., King, A. M., Meyer, M. R., Slade, D., Lum, P. Y., Stepaniants, S. B., Shoemaker, D. D., Gachotte, D., Chakraburtty, K., Simon, J., Bard, M., and Friend, S. H. (2000) Functional discovery via a compendium of expression profiles. *Cell* **102**, 109–126

49. Brummelkamp, T. R., Fabius, A. M. W., Mullender, J., Madiredjo, M., Velds, A., Kerkhoven, R. M., Bernards, R., and Beijersbergen, R. L. (2006) An shRNA barcode screen provides insight into cancer cell vulnerability to MDM2 inhibitors. *Nat. Chem. Biol.* **2**, 202–206

50. Espinet, C., Gómez-Arbonés, X., Egea, J., and Comella, J. X. (2000) Combined use of the green and yellow fluorescent proteins and fluorescence-activated cell sorting to select populations of transiently transfected PC12 cells. *J. Neurosci. Methods* **100**, 63–69

51. Chambers, J. M. and Hastie, T. J. (eds.) (1992) *Statistical Models in S.* Chapman & Hall/CRC, London

Chapter 18

Ligand–Macromolecule Interactions in Live Cells by Fluorescence Correlation Spectroscopy

Aladdin Pramanik

Summary

The receptor concept is the primary theoretical basis for modern pharmacology. Drugs, hormones, neurotransmitters, toxin, and other biologically active substances are referred to as ligands. Ligands exert their actions by way of interaction with receptors/macromolecules. The resulting receptor/macromolecule–ligand complexes produce alterations in physiological processes. Receptor/macromolecule-binding studies most often require the use of radioactively labeled ligands. When the numbers of receptors/macromolecules are few per cell, it is impossible to detect the specific binding because of a high background. Specific interactions between certain ligands and their receptors/macromolecules are, therefore, often overlooked by the conventional binding technique. Fluorescence correlation spectroscopy (FCS) allows detection a ligand–macromolecule interaction in live cells in a tiny confocal volume element (0.2 femtoliter (fL)) at single-molecule detection sensitivity. FCS permits the identification of macromolecules that were not possible to detect before by isotope labeling. The beauty of the FCS technique is that there is no need for separating an unbound ligand from a bound one to calculate the macromolecule bound and free ligand fractions. This study will demonstrate FCS as a sensitive and a rapid technique to study ligand–macromolecule interaction in live cells using fluorescently labeled ligands (Fl-L). This study is of pharmaceutical significance since FCS assay of ligand–macromolecule interactions in live cells is one step forward toward a high throuput drug screening in cell cultures.

Key words: Fluorescence, Ligand, Receptor, Macromolecule, Cell culture, Spectroscopy, Ultra sensitivity, Drug discovery

1. Introduction

For certain ligands (e.g., peptides, natural products) it is not possible to detect macromolecules with the conventional "binding method radiography." The background of the radiography method is high and it includes several washing steps to remove unbound ligand. The half-life of the macromolecule–ligand complex is

Ana Cecília A. Roque (ed.), *Ligand-Macromolecular Interactions in Drug Discovery: Methods and Protocols*,
Methods in Molecular Biology, vol. 572,
DOI 10.1007/978-1-60761-244-5_18, © Humana Press, a part of Springer Science + Business Media, LLC 2010

often shorter or similar to the time required for separations of free and bound ligands in this method. Therefore, interactions between certain ligands and their different macromolecules/receptors are often overlooked by the conventional binding technique when numbers of macromolecules/receptors are few per cell. The single-molecule detection sensitivity of Fluorescence correlation spectroscopy (FCS) and its short measurement time *(1–4)* allows ligand–macromolecule interactions in live cells, even if the macromolecules are present only in very sparse numbers. This study will illustrate how FCS can be used for monitoring the interaction of ligands with their cognate macromolecules.

2. Materials

2.1. Labeling of Ligands with Fluorescent Dyes

1. Fluorescently labeled ligands (Fl-L) are used for macromolecule binding by FCS (*see* **Note 1**).
2. Dye-labeled ligands should not contain unbound dye molecules (*see* **Note 2**).
3. Dye-ligand complexes are stable for 2–3 months (*see* **Note 3**).
4. Ratio between molecular weight of a dye molecule and a ligand must be at least 2 or bigger (*see* **Note 4**).

2.2. Cell Culture

1. The cells are cultured in Dulbecco's Modified Eagle's Medium (DMEM) containing 10% new-born calf serum and antibiotics (penicillin 50 U/mL and streptomycin 50 µg/mL) at 37°C in 5% CO_2 humidified atmosphere.
2. The cells are passaged once every week, and the medium is changed every second day. Cells of the third to fifth passages are used for experiments.
3. The cells are plated on chamber slides (Nunc chambers) 48 h before the studies, allowed to adhere for 36 h, and are then serum starved until the start of the experiments.
4. Washing of the cells (*see* **Note 5**).

3. Methods

3.1. Binding Procedure

1. Binding studies by FCS are carried out on cells cultured in eight-well Nunc chambers (Nalge Nunc Inc., IL, USA) at 20°C.

2. Prior to the experiments cells are washed 5 times with PBS (phosphate buffer saline) and incubated with binding buffer containing Fl-L at room temperature (*see* **Note 6**).

3. Binding of ligands is measured after 60 min incubation of cells in the presence of 5 nM Fl-L.

4. The specificity of the ligand binding is performed (*see* **Note 7**).

5. The focus of the laser beam is properly positioned on the membrane surface if ligand–macromolecule interactions are followed on membranes of cultured cells (*see* **Note 8**).

3.2. FCS Setup

1. To follow ligand–macromolecule interactions, FCS measurements are performed with confocal illumination of a laser volume element of 0.2 fL in a ConfoCor instrument of Carl Zeiss (*see* **Fig. 1** and **Note 9**).

2. The FCS measurements are performed on cells cultured in Nunc chambers (*see* **Fig. 1** and **Note 10**).

3.3. FCS Data Evaluation

1. Diffusion times of ligand-bound macromolecules and their numbers are calculated using the autocorrelation function (*see* **Note 11**).

2. The binding curve is made (*see* **Note 12**).

3.4. Binding of Fluorescently Labeled Ligand to Macromolecules

Ligand–macromolecule interactions in live cells using FCS have been performed in several laboratories *(5–20)*. As examples data on the C-peptide (CP) and the GAL binding will be presented to illustrate how FCS can be used to study the interaction of ligands with their cognate receptors/macromolecules.

1. In FCS measurements, fluorescence intensity fluctuations are recorded from only those molecules that diffuse through the confocal laser volume element (*see* **Figs. 1** and **2a, d** and **Note 13**).

2. Diffusion time of binding complexes (*see* **Fig. 2b, e** and **Note 14**).

3. Distribution of diffusion times of binding complexes (*see* **Fig. 3** and **Note 15**).

3.5. Specificity of the Binding

1. Displacement of the binding (*see* **Fig. 2c, f** and **Notes 7** and **16**).

2. Binding curve (*see* **Fig. 4** and **Notes 12** and **17**).

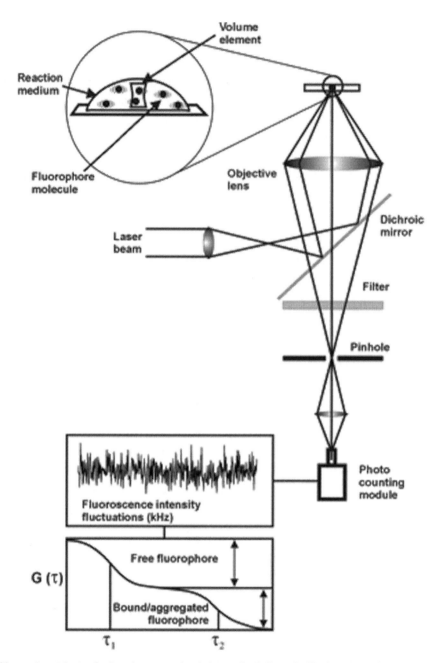

Fig. 1. FCS experimental setup for ligand–macromolecule interaction in live cells. The laser beam from an argon ion laser is sharply focused via a dichroic mirror and a lens to form a tiny confocal volume element of 0.2 fL. The laser beam is projected from below into an eight-well Nunc chamber containing a monolayer of cultured cells and fluorescently labeled ligands (Fl-L) (e.g., fluorescently labeled CP (the proinsulin C-peptide) or GAL (the neuropeptide galanin)). After excitation of the Fl-L, emitted light is transmitted via the dichroic mirror, a bandpass filter, and a pinhole to a photodetector. The volume element is positioned onto the cell surface using a microscope for detection of the ligand binding. The dimensions of the laser beam focus and the pinhole together define the confocal volume element (see magnification) from which fluorescent light is collected. The photodetector operates in photon-counting mode, responding with an electrical pulse to each detected photon. The electrical pulses are fed into a digital signal correlator which computes online in real time the autocorrelation function of the detected fluorescence intensity fluctuations. G(τ), autocorrelation function. τ_1 and τ_2, diffusion times for free and bound dye-labeled ligands, respectively; $\tau_2 \gg \tau_1$.

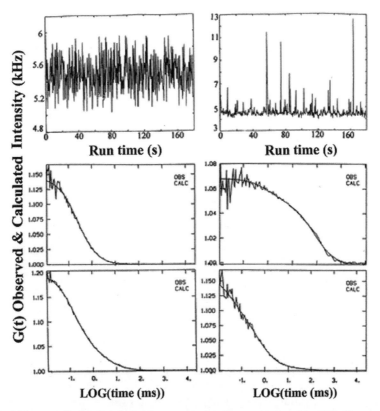

Fig. 2. Binding of fluorescently labeled ligand to macromolecules as monitored by FCS. Fluorescence intensity fluctuations (**a**) and autocorrelation function (**b**) for rhodamine-labeled CP (Rh-CP) (5 nM) free in solution, $\tau_D = 0.15$ ms. Fluorescence intensity fluctuations (**d**) and autocorrelation function (**e**) for Rh-CP bound to membranes on the cell surfaces. Diffusion times (τ_D) and corresponding fractions (y): $\tau_{D1} = 0.15$ ms, $y = 0.1$; $\tau_{D2} = 1$ ms, $y_2 = 0.15$; $\tau_{D3} = 80$ ms, $y_1 = 0.75$. Autocorrelation functions of displacement of membrane-bound Rh-CP by postincubation (**c**) and preincubation (**f**) of a 1,000-fold molar excess of nonlabeled CP.

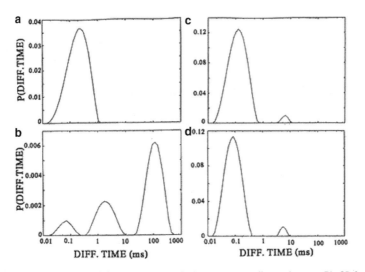

Fig. 3. Distributions of diffusion times of CP binding and displacement to cell membranes. Rh-CP free in solution (**a**), binding of Rh-CP to the cell membranes: three components – free ligand, medium complex, big complex (**b**), displacement of membrane-bound Rh-CP by postincubation of a 1,000-fold molar excess of nonlabeled CP (**c**), and inhibition of membrane-bound Rh-CP by pertussis toxin (**d**).

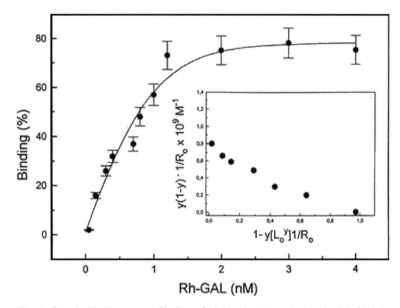

Fig. 4. Galanin-binding curve. Binding of rhodamine-labeled galanin (Rh-GAL) to cell membranes of insulinoma cells. Fractional saturation of the membrane-bound Rh-GAL (y) as a function of the ligand concentration (L_0) in the binding medium. Cells were incubated with binding buffer containing different concentrations of Rh-GAL for 60 min. Scatchard plot is shown as an insert. Each data point represents the mean of at least six separate measurements. y bound ligand; $1-y$ free ligand; R_0 total receptor concentration; L_0 total ligand concentration.

4. Notes

1. In order to follow ligand–macromolecule interactions using FCS, it is necessary to covalently label ligands with at least one fluorophore with an account on steric effects of the fluorophore molecules. The proper incorporation of a fluorophore into the ligand is an important factor in receptor/macromolecule-binding studies withFl-L.

2. Ideally, dye–labeled ligands should not contain unbound dye molecules since they may contribute to nonspecific interactions.

3. Since dye-ligand complexes usually have a certain half-lifetime, the purity of dye-labeled ligand complexes should be checked from time to time. It is observed that even 100% dye-labeled ligand loses dye molecules (5–10%) after 2–3 months.

4. A problem concerning ligand tagging is the labeling of low molecular weight ligands, which is related to the limitation of FCS to distinguish molecular sizes by translational diffusion. The increase in molecular weight of a labeled ligand by a factor of 8 only doubles its diffusion time. It is, thus, sometimes difficult to distinguish the diffusion time of a dye-labeled ligand

from that of the dye, particularly if the molecular weight of the ligand is less than that of the dye molecule. FCS differentiates species based on their diffusive speed, which scales with the cubic root of the molecular size.

5. It is well known that almost all types of cells are found to produce background fluorescence which naturally can lead to an overestimation of fluorescence signals. Washing of the cells is a way to reduce background fluorescence, in this case usually done 4–5 times with PBS (NaCl 137 mM, KCl 2.7 mM, KH_2PO_4 1.5 mM, $Na_2HPO_4 \cdot 2H_2O$ 8.1 mM, pH = 7.4). Washing of the cells has to be done with care since this washing can also induce damages to cells.

6. It is extremely important to pay attention to the buffer composition and its pH value used for the binding experiment in live cells.

7. The specificity of the Fl-L binding is demonstrated by the competitive displacement of bound Fl-L from macromolecules after 1–3 h incubation of 5 µM nonlabeled ligand added to the cell incubations (postincubation). The specific binding is also demonstrated by the inability of the Fl-L to bind to macromolecules when the cells were preincubated with nonlabeled ligand for 1–3 h.

8. It is obviously a difficult task to be sure that the focus is exactly on the membrane. However, here is a hint for this. If receptor binding is observed on a membrane, a change of the focus 1 µm upward should lead to a reduction of the binding. One should include that if the receptor is membrane bound, the binding observed in the cytoplasm mostly is nonspecific. In certain cases, if the focus is in the cytoplasm, one detects the internalization of receptors/macromolecules. If cells are cultured in eight-well Nunc chambers, it is useful to have them attached to the bottom of the chamber. Nunc chamber is usually used for the cell culture. During the FCS measurement cells should not be allowed to move, since a change of cell position during the measurement is likely to introduce errors.

9. In most cases, as focusing optics, a Zeiss Neofluar 40× numerical aperture 1.2 objective for water immersion is used in an epi-illumination setup. Separation of exciting from emitted radiation is achieved by dichroic (Omega 540 DRL PO_2, Omega Optical, Brattleboro, VT) and bandpass (Omega 565 DR 50) filters. The Fl-L are excited, e.g., with the 514.5-nm line of an Argon laser or with other lines depending on what dye molecules are used. Intensity fluctuations are detected by an avalanche photodiode (SPCM 200, EG & G, Quebec, Canada) and processed with a digital correlator (ALV 5000, ALV, Langen, Germany).

10. Nunc chamber is usually used for the cell culture. In FCS technique it is used as a sample chamber. It has a cover glass in bottom, suitable for an inverted microscope. With the objective used, 0.2-fL volume element is illuminated with dimensions of $w = 0.25$ μm and $z = 1.25$ μm, respectively. To avoid photobleaching the exciting intensity is adjusted such that the detected photon count rate does not exceed 3,000–4,000 counts per molecule and s.

11. For the analysis of the FCS data on ligand-receptor inter- actions, the autocorrelation function for 3D diffusion of the unbound Fl-L in solution and 2D diffusion of the bound Fl-L to membranes on the cell surface is given by:

$$G(\tau) = 1 + \frac{1}{N}\left[\frac{\left(1 - \sum y_i\right)Q_f^2\left(\dfrac{1}{1 + \dfrac{\tau}{\tau_D^f}}\right)}{\left(\dfrac{1}{1 + \left(\dfrac{\omega}{z}\right)^2 \dfrac{\tau}{\tau_D^f}}\right)^{\frac{1}{2}} + \sum y_i Q_i^2\left(\dfrac{1}{1 + \dfrac{\tau}{\tau_D^{bi}}}\right)}\right], \qquad (1)$$

where diffusion time τ_D and diffusion coefficient D are related as $\tau_D = w^2/4D$, where $\sum y_i$ is the fraction of the membrane-bound Fl-L diffusing with diffusion time τ_D^{bi}, and $(1 - \sum y_i)$ is the fraction of the unbound Fl-L diffusion with diffusion time τ_D^f. Q_i is the quantum yield of the bound ligand represented by y_i. Q_f is the quantum yield of the unbound ligand.

12. One of the important criteria of the ligand binding is to make a binding curve where the fraction of receptor-bound ligand $\sum y_i$ is obtained at increasing concentration of Fl-L. It is always useful to further present these data by Scatchard plot. Thus, the Scatchard representation of the mass action law for the FCS data on the ligand binding can be obtained from:

$$\frac{\sum y_i}{\left(1 - \sum y_i\right)R_0} = \sum K_i\left(n_i - \frac{y_i L_0}{R_0}\right), \qquad (2)$$

where K_i and n_i are the association constant and the number of ligand-binding sites per receptor molecule, respectively. L_0 = total ligand and R_0 = total receptor. For details of this kind and other analyses see **refs.** *(4, 8, 9)*.

13. In FCS measurements fluorescence intensity fluctuations are recorded from only those molecules that diffuse through the confocal laser volume element (**Fig. 1**). When an Fl-L is bound to a receptor/macromolecule, there is a change not only in diffusion times but also in fluorescence intensity traces. This is well pronounced in the CP binding using rhodamine-labeled CP (Rh-CP) (**Fig. 2d**). **Fig. 2** presents fluorescence intensity fluctuations of Rh-CP (5 nM) free in solution (**Fig. 2a**) and bound to human renal tubular cell membranes (**Fig. 2d**). When the volume element is positioned at the level of the cell membrane, the fluorescence intensity traces exhibited several prominent peaks (**Fig. 2d**). These peaks, with up to three times higher intensity than the baseline, are a clear representation of the Brownian motion of ligand–receptor complexes containing several Rh-CP molecules.

14. The time required for the passage of fluorescent molecules through the volume element is determined by the diffusion coefficient, which is related to the size and shape of a molecule. Thus, diffusion times obtained from the analysis of fluorescence intensity fluctuations with autocorrelation functions allow to differentiate faster diffusing and slower diffusing molecules, i.e., in this context to distinguish free ligands from ligand–receptor/macromolecule complexes. In binding studies with the FCS technique, a mixture of several ligand–macromolecule complexes (components) with different molecular weights and corresponding different diffusion times can be analyzed without any need to physically separate unbound components from bound ones (**Fig. 2b, c, e, f**). The figure presents autocorrelation functions of Rh-CP (5 nM) free in solution and bound to human renal tubular cell membranes. The analysis of correlation function shows a diffusion time of unbound Rh-CP (τ_{D1}) of 0.15 ms (**Fig. 2b**). Correlation analysis of the intensity fluctuations then reveals a diffusion process of the cell-bound CP with at least two components characterized by diffusion times of $\tau_{D2} = 1$ ms and $\tau_{D3} = 80$ ms, respectively (**Fig. 2e**).

15. Since in the FCS technique diffusion times of a ligand molecules and ligand–receptor complexes are used as characteristic parameters for binding analysis, distribution of diffusion times can be obtained via a special algorithm that can reveal the appearance of several binding complexes. For the evaluations of the distribution functions, the CONTIN program using restrained regularization *(21)* is applied. Instead of an evaluation of discrete diffusion processes with distinct species and characteristic diffusion times, the dynamic motion of receptor complexes in membranes is considered to be distributed with regard to their diffusion times. This program

can thus be used to validate the presence of discrete ligand–macromolecule complexes **(Fig. 3)** where the appearance of two binding complexes found by distribution analysis of diffusion times may suggest that these are representations of two different forms of ligand-macromolecule complexes.

16. For ligand binding, it is foremost necessary to fulfill all the criteria of ligand-receptor/macromolecule interactions such as affinity (the ligand and the receptor/macromolecule have affinity for one another), specificity (receptors/macromolecules show specificity in their interactions with ligands), concentration dependency (ligand–receptor interactions are concentration dependent, leading to a binding curve), and reversibility (ligand–receptor interactions are reversible, leading to a dissociation curve). In order to highlight specificity and binding curve for ligand–macromolecule interactions in live cells (cell cultures) using FCS, we have presented data from the CP and the GAL-binding studies. It is important to take background fluorescence into account for the judgment of the specific ligand–macromolecule interaction in live cells when a ligand is tagged with a fluorescent dye. Background fluorescence usually originates from endogenous proteins if their excitation wavelength is close to that of the dye used. The endogenous proteins such as NADH and FAD produce autofluorescence in blue-green range. Therefore, sometimes it is advantageous to make excitation in near-IR range, where autofluorescence is less prominent. Fluorescence signals will be overestimated if they are not corrected for background fluorescence. In order to check background fluorescence the autofluorescence is to be measured in cells without addition of fluorophore-labeled ligands. To check the specificity, the displacement experiment is performed in the presence of 1,000-fold molar excess of unlabeled ligand, in our case CP. This gives a hint how one should perform the specific binding. To determine the specificity of binding of Fl-L, it is necessary to examine competitive displacement with nonlabeled ligand in excess. In general, cells are incubated with Fl-L and after 30 or 60 min a 1,000-fold molar excess of nonlabeled ligand are added. This results in a reduction of the Fl-L, and after 1 h or longer about 80–85% of the total binding is found to displaced for CP (**2c, f** and **3c**) and for EGF (*8*)).

17. As a rule, saturation of the high affinity binding process occurs at nanomolar ligand concentration, e.g., the GAL binding is found to be saturated at about 2 nM Rh-GAL **(Fig. 4)**. From the binding curve it is possible to obtain K_D value. Thus, K_D for the GAL binding is found to be at 0.8 nM **(Fig. 4)**. Scatchard analysis is used to obtain an equilibrium association constant, K_{ass}. Thus, the value of K_{ass} for GAL is of $8 \times 10^9 M^{-1}$ **(Fig. 4**, inset).

Acknowledgments

I dedicate this article to the memory of my cousin Shahjahan Ali, B.Sc. in Chemistry, M.A. in English, AGM (Assistant General Manager) of Sonali Bank, Bangladesh, who expired in 2004 by "stroke" at the age of 54. I have his efforts and inspiration to thank for my becoming what I always wanted to be. This work was financially supported by grants from the Swedish Medical Research Council, and the Novo Nordisk and the Karolinska institute Foundations.

References

1. Rigler, R., Mets, U., Widengren, J., and Kask, P. (1993) Fluorescence correlation spectroscopy with high count rate and low background: analysis of translational diffusion. *Eur. Biophys. J.* **22**, 169–175

2. Eigen, M. and Rigler, R. (1994) Sorting single molecules: application to diagnostics and evolutionary biotechnology. *Proc. Natl. Acad. Sci. U.S.A.* **91**, 5740–5747

3. Rigler, R. (1995) Fluorescence correlations, single molecule detection and large number screening. Applications in biotechnology. *J. Biotech.* **41**, 177–186

4. Pramanik, A. and Widengren, J. (2004) Fluorescence correlation spectroscopy (FCS). In: Meyers, R. A. (ed.), *Encyclopedia of Molecular Cell Biology and Molecular Medicine.* Willey-VCH, Weinheim, pp. 461–500

5. Widengren, J. and Rigler, R. (1998) Fluorescence correlation spectroscopy as a tool to investigate chemical reactions in solutions and on cell surfaces. *Cell Mol. Biol.* **44**, 857–879

6. Pramanik, A., Juréus, A., Langel, Ü., Bartfai, T. and Rigler, R. (1999) Galanin receptor binding in the membranes of cultured cells measured by fluorescence correlation spectroscopy. *Biomed. Chromatogr.* **13**, 119–120

7. Boonen, G., Pramanik, A., Rigler, R. and Häberlein, H. (2000) Evidence for specific interactions between a kavain derivative and human cortical neurons measured by fluorescence correlation spectroscopy. *Planta Med.* **66**, 7–10

8. Pramanik, A. and Rigler, R. (2001) Ligand-receptor interactions in the membrane of cultured cells monitored by fluorescence correlation spectroscopy. *Biol. Chem.* **382**, 371–378

9. Pramanik, A. and Rigler, R. (2001) FCS-assay of ligand-receptor interactions in living cells. In: Rigler, R. and Elson, E. L. (eds.), *Fluorescence Correlation Spectroscopy (FCS). Theory and Applications.* Springer, Berlin Heidelberg, pp. 101–112

10. Pramanik, A., Ekberg, K., Zhong, Z, Shabqat, J., Henriksson, M., Tibell, A., Tally, M., Wahren, J., Jörnvall, H., Rigler, R. and Johansson, J. (2001). C-peptide binding to human cell membranes: importance of Glu27. *Biochem. Biophys. Res. Commun.* **284**, 94–98

11. Zhong, Z., Pramanik, A., Ekberg, K., Kratz, G., Wahren, J. and Rigler, R. (2001) Insulin binding monitored by fluorescence correlation spectroscopy. *Diabetologia* **44**, 1184–1188

12. Dittrich, P., Malvezzi-Campeggi, F., Jahnz, M. and Schwille, P. (2001) Accessing molecular dynamics in cells by fluorescence correlation spectroscopy. *Biol. Chem.* **382**, 491–494

13. Waizenegger, T., Fischer, R. and Brock, R. (2002) Intracellular concentration measurements in adherent cells: a comparison of import efficiencies of cell-permeable peptides. *Biol.Chem.* **383**, 291–299

14. Chen, Y., Muller, J.D., Ruan, Q. and Gratton, E. (2002) Molecular brightness characterization of EGFP in vivo by fluorescence fluctuation spectroscopy. *Biophys. J.* **82**, 133–144

15. Pick, H., Preuss, A.K., Mayer, M., Wohland, T., Hovius, R. and Vogel, H. (2003) Monitoring expression and clustering of the ionotropic 5HT(3) receptor in plasma membranes of live biological cells. *Biochem.* **42**, 877–884

16. Meissner, O. and Haberlein, H. (2003) Lateral mobility and specific binding to GABA(A) receptors on hippocampal neurons monitored by fluorescence correlation spectroscopy. *Biochemistry* **42**, 1667–1672

17. Halbsguth, C., Meissner, O. and Haberlein, H.H. (2003) Positive cooperation of protoberberine type 2 alkaloids from Corydalis cava on the GABA(A) binding site. *Planta Med.* **69**, 305–309

18. Haustein, E. and Schwille, P. (2003) Ultra-sensitive investigations of biological systems by fluorescence correlation spectroscopy. *Methods* (Duluth). **29**, 153–166

19. Bacia, K. and Schwille, P. (2003) A dynamic view of cellular processes by in vivo fluorescence auto- and cross-correlation spectroscopy. *Methods* (Duluth). **29**, 74–85

20. Weiss, M., Hashimoto, H. and Nilsson, T. (2003) Anomalous protein diffusion in living cells as seen by fluorescence correlation spectroscopy. *Biophys. J.* **84**, 4043–4052

21. Provencher, S.W. (1982) Contin: a general purpose constrained regulation program for inverting noisy linear algebraic and integral equations. *Comput. Phys. Commun.* **27**, 229–242

INDEX